Natural Photonics and Bioinspiration

For a listing of recent titles in the *Applied Photonics Library*,
turn to the back of this book.

Natural Photonics and Bioinspiration

Sébastien R. Mouchet
Olivier Deparis

ARTECH
HOUSE
BOSTON | LONDON
artechhouse.com

Library of Congress Cataloging-in-Publication Data
A catalog record of this book is available from the U.S. Library of Congress.

British Library Cataloguing in Publication Data
A catalog record for this book is available from the British British Library.

ISBN 978-1-63081-797-8

Cover design by Charlene Stevens

685 Canton Street
Norwood, MA 02062

10 9 8 7 6 5 4 3 2 1

In memory of Jean-Pol Vigneron

Contents

Foreword by Serge Berthier

Tapanahoni, in Suriname, in September 2010. The canoe slowly descends the river, that has now calmed, after the perilous passage of the last rapid. Sitting at the back, I contemplate the long line of students that Jean-Pol Vigneron, by my side, has led on this adventure. A modern-day Leonardo da Vinci, Jean-Pol! He had endorsed the recommendation of the brilliant artist and visionary scientist: "Go see nature, there is your future." And nature is there, majestic, all around us, still relatively unspoiled. And our future is here, too—this happy troop of young scientists from the University of Namur, who will inherit the planet that we are going to bequeath to them and that they will have to manage as best as possible, repair, and cure. And among them, I see Sébastien, one of the authors of this book, who apparently will have learned the lesson of his teacher. Nature is our future, she has lessons to teach us, to save our future, and Jean-Pol guides us all in a new bioinspired and resilient world.

Aware very early on that our current economic model was not tenable, he put his talent, knowledge, and energy at the service of this new approach: bioinspiration, the common thread of this beautiful book. That one of his students is the author is certainly the greatest reward for him, who nevertheless received so much. Through Sébastien Mouchet and his colleague and friend Olivier Deparis, this book can be considered as his posthumous work, not because he would have written it, but because he would have inspired it and he will have done everything to make it reach us.

We know the extraordinary quality of his research. As a great scientist, he will have been recognized for this. But he was also, I would add, more importantly, much more than that. There are scientists who place their specialty at the top of the tree of knowledge. Multidisciplinary work is there only undertaken to serve their ends. Jean-Pol has made it a *sine qua non*, a richness which floods all his work.

He, as a great theorist and unlike many others, never despised the possibilities of applications of his science and his work. This book, which offers a synthesis of all his bioinspired approaches, is a magnificent illustration of this exact point.

Responsible for research, he loathed secrecy, enthusiastically disseminated his information, his ideas, and delegated as much as possible, aware of the tremendous potential for training responsibilities among young researchers. Some leaders divide for better reign, Jean-Pol federated to better progress.

Open to applications, disinterested and allergic to hierarchical stiffness, understanding and friendly with his colleagues and students, he nevertheless knew how to negotiate with diplomacy and firmness. The European "Biophot" program, which he initiated and guided, carefully and closely throughout its course, has been for many of us a privileged episode in our scientific careers. For the youngest like

Sébastien—this book is the best proof of it—but also for all the others (and there were many of them), these expeditions showed that there were several ways of doing physics, science in general: go and see nature, understand its strategies, its functioning from which we have strayed so far, and get the best out of it, for us, but above all, for nature. Biomimicry must be at the service of life!

All the participants in the Biophot program, brought together by the sole will, the sole talent of Jean-Pol Vigneron, will keep from him, I am sure, the memory of his radiant good mood, of his humor, of his "Belgicisms," for his love of life, his human warmth, and his never-satisfied passion for science and the beauties of nature. This book is the most beautiful tribute that his young students, whom he loved so much, and his friend Olivier, who continues his work, could pay him!

On a subsequent expedition to Suriname, at the foot of an inselberg that we were to explore, our guide and boatman recited a prayer in his memory. These inselbergs are magical places for the Wayana Indians, sacred mountains where we communicate with souls. He had only met him twice. But Jean-Pol was like that—he was not forgotten! He now hovers in these magical places, with the great Morphos, the revered souls of the Indians, whose secrets this book beautifully reveals.

Prof. Serge Berthier
Professeur émérite Université de Paris
UNESCO-UNISA chair in Nanotechnology, Cape Town, South Africa
September 2021

Jean-Pol Vigneron (front) and Serge Berthier (back) carrying out some fieldwork along the Tapanahoni River in Suriname in September 2010.

Foreword by Pete Vukusic

The second natural photonics scientific paper I ever read had been far ahead of its time. I still recall that afternoon in a corner of my institution's quiet library basement. It was a paper copy, buried in a thick bound journal, a treatise "On the Optical Character of Some Brilliant Animal Colours" written by the third Lord Rayleigh, a polymath and distinguished scientist of the late nineteenth century. Elegantly and intuitively it discussed the nature and possible origin of structural colors across a host of animals. It was published in 1919. I read it in 1997. That was a turning point for me.

But I suppose, on reflection now, I had already been drawn irreversibly to the field. This had come about because of the first natural photonics paper I had read, "Light and Colour on the Wing: Structural Colours in Butterflies and Moths." Helen Ghiradella, its eminent, collegial, and gifted author, precisely and insightfully painted a glorious scientific picture of the way in which precise structural pattern formation in lepidopteran systems arose and gave rise to vivid color and appearance. This single publication from 1991 resonated profoundly with me and many others. Without doubt, it has been one of the triggers for the twenty-first century renaissance of scientific interest in natural photonics.

By any standard, the last two decades of collective international study have moved the horizon of understanding of natural photonics an immeasurable distance from where it was in the days of the third Lord Rayleigh. It is an exciting, scientifically rich, marvelously interdisciplinary and genuinely a living path to tread. Increasingly, too, it strongly piques the interests of technology and industry, inspired, as those behemoths are, by these living jewels of photonic form and function discovered along this glittering path. For these reasons and many others, I welcome and admire this timely book by my friends and colleagues Sébastien Mouchet and Olivier Deparis. These two authors are well versed in the field and have been strong contributors to it. Here, they have expertly drawn together a gamut of different parallel and often interconnecting threads that are woven through our natural photonics' learning journey. This book is designed to give the reader an honest and straightforward set of explanations about the scientific principles and physical mechanisms that underpin this field. It coherently presents the discoveries and the identification of photonic form and function across many types of fauna and flora, as well as the inspiration for applications to which these discoveries continue to give rise. I sincerely salute it and the authors for this endeavor.

Undoubtedly to this day still, as the third Lord Rayleigh acknowledged at the end of his influential paper a century ago, "much still remains to be effected towards a complete demonstration of the origin of these colours." But equally, and as this book clearly acknowledges and describes, remarkable depths of understanding of biological structural color have been reached. And so, who knows the "possibilities not hitherto contemplated"?

Only time and nature will tell.

Prof. Pete Vukusic
Professor of Biophotonics, University of Exeter
September 2021

Acknowledgments

The authors thank all their past and present colleagues and collaborators who inspired this book in various ways and produced many of the research results mentioned and referenced in this manuscript.

The authors thank their friend and former colleague Dr. Victoria L. Welch for proofreading and copyediting this manuscript.

Sébastien R. Mouchet was supported by the Belgian National Fund for Scientific Research (FRS-FNRS) (91400/1.B.309.18F) and the Maturation Fund of the Walloon Region, as a postdoctoral researcher. This book was published with the financial support of the FRS-FNRS.

Introduction

The first day the older of the two authors of this book met Jean-Pol Vigneron, he asked him this question: "Would you be keen to work on *exotic* things?". He was looking for a postdoctoral researcher in order to work with him on a European project he had just been granted (the *Biophot* project: "Complexity and evolution of photonic nanostructures in bio-organisms: templates for material sciences"). He considered my curriculum vitae (Olivier Deparis writes) as relevant for this job. As he knew, therefore, I had been working on fiber-optic communications for more than ten years and he wondered if the subject of the project were of any interest to an engineer. I answered him that I would like to study photonic crystals more deeply, and in particular, those found in the living world. I was also interested, as an engineer working in optics, in applying my experience to the fabrication of bioinspired optical devices. That was the perfect match and the beginning of a fantastic adventure with Jean-Pol in the realm of natural photonics and bioinspiration. Yet, the topic was still at its infancy at that time (2005), and the words "natural photonics," Jean-Pol liked to use to describe the research of his team, were not really in use in the optics community then. A few years later, Jean-Pol communicated with the same passion his interest for natural photonics to the younger of the two authors. It was in March 2009 and I was about the complete my bachelor's degree in physics at the University of Namur (Sébastien Mouchet writes). I was carrying out an undergrad's project in Jean-Pol's lab in the framework of a module entitled "Introduction to Scientific Research" and was flabbergasted by the research Jean-Pol was leading and describing to me. Coincidentally, it was also about time for me to take important decisions regarding a future master's degree. Thanks to Jean-Pol, I opted for a master's degree with a strong orientation in nano-optics and natural photonics and an internship within Prof. Doekele Stavenga's group, instead of the degree in climatology, to which I had briefly considered registering.

Since the first work carried out by Jean-Pol, this young branch of photonics has grown on the tree of optical science and has become an attractive interdisciplinary topic among the scientific community, continuing to gather many more physicists, biologists, chemists, and engineers. At the dawn of the twenty first-century, this "exotic" topic started to appear regularly on the program of prominent conferences, such as the SPIE annual meeting in San Diego, California. On June 24, 2013, Jean-Pol Vigneron passed away suddenly, leaving an exceptionally productive and pioneering contribution to the field of natural photonics. Serge Berthier's words, in the foreword of this book, are the most authentic testimony to Jean-Pol Vigneron, as a scientist, colleague, and friend, the authors could have ever thought of. They warmly thank him for this. In spring 2014, *Living Light*, the first conference entirely dedicated to natural photonics, was organized in Namur, Belgium, in his memory.

Other *Living Light* conferences followed in San Diego, CA (2016) and Cambridge, UK (2018), with an online edition in 2020.

The idea to gather Vigneron's pioneering work, along with those of other researcher groups around the world in a book, including our own contributions to the field, had been in the authors' minds since Jean-Pol passed away, but it did not crystalize until we were approached by the editor, Prof. Marc Wuilpart. We thank Marc for giving us this opportunity. The title of the book, *Natural Photonics and Bioinspiration*, aims to reflect the balance we intended to keep between, on one hand, fundamental aspects of physical optics and environmental biology and, on the other hand, of applied science aspects related to observations of optical effects in natural organisms. The latter comprises those observations performed in the field on natural species, numerical modeling based on morphological and optical characterizations and, eventually, design and synthesis of bioinspired photonic devices. The readership this book targets is interdisciplinary, ranging from postgraduate students and PhD students to postdoctoral researchers and academics. Optical engineers as well as physicists will be the most familiar to the core subject of the book (i.e., applied photonics). It is said that "to make a choice is to give up the other options." Hence, for the sake of the narrative, we have decided to describe the biological aspects of the field, including the biological functions, the evolution, the development, the genetics, and animal visual perception, to only a limited extent. Similarly, scientific names of natural organisms were sometimes omitted for the sake of simplicity for readers less used to taxonomy. However, we believe biologists and chemists will be interested in the very interdisciplinary nature of the subject, albeit for different reasons. In any case, we hope fellow scientists will love this book!

The first chapter, entitled "Photonics in Nature," inaugurates the reader's grand voyage in the realm of natural photonics. It recalls some basic concepts related to colors in nature and, in particular, structural colors, of course. Quantitative treatment of color vision is explained and the concept of bioinspired photonics is introduced. With the second chapter ("Diversity and Complexity of Natural Photonic Devices"), the reader is immersed into the jungle, where she or he will discover, step-by-step and in order of increasing structural complexity, those natural photonic devices, which have evolved over millions of years in order to perform, by optical means, vital functions for the species bearing them; functions such as inter- or intraspecies visual signaling, camouflage, and body thermal management. When no biological function is known, the photonic structure is considered a by-product of evolution, yet it can still be admired for its simple beauty, or be a source of inspiration for the design of photonic devices. The third chapter ("Color Additive Devices") and the fourth chapter ("Transparent Devices") will show to the reader, respectively, how sophisticated colors from natural photonic polycrystals can be in that they produce pointillist effects (such as those employed by artists) and how efficient naturally textured surfaces can be to suppress reflection from moth eyes, to make insect wings transparent or to combine transparency with other properties. Structural colors are not always static: they can be dynamic as well. For example, upon contact of certain insects' cuticles with liquids, reversible color changes can occur, with the original color reappearing when the cuticle dries. A few examples of such reversible liquid-induced color changes are discussed in detail in Chapter 5 ("Liquid-Induced Structural Color Changes"). Vapors and gases can also trigger color changes, as can other agents, such as temperature, pH, and mechanical forces. These phenomena are discussed in Chapter 6 ("Other Forms of Structural Color Changes"). Many species live their entire lives under the full glare of the Sun's illumination, as even humans sometimes do. The solar radiation they thus encounter is potentially both useful and harmful for such living organisms. Harvesting solar energy for heating body muscles is a strategy used by butterflies, as they bask under the sun before taking off. Nature has also found original solutions to provide protection against harmful ultraviolet radiation—for instance, in avian egg shells or high-altitude plants. Enhanced blackness, as is observed in some snake or insect species, results from specific light-trapping photonic structures that are designed to absorb entirely the whole spectrum of sunlight. All these fascinating effects are described and explained in Chapter 7 ("Solar Energy Harvesting Devices"). Managing light emission is a key issue for bioluminescent species. Again, some specific natural photonic structures turned out to be remarkably efficient at extracting light from a bioluminescent organ, or controlling the emission of fluorescence from embedded fluorophores. These phenomena are highlighted in Chapter 8 ("Light Emission Management Devices").

With all these lessons from nature in hand, the reader is now ready to enter the second round of adventures in the realm of natural photonics. Chapter 9 ("Design of Bioinspired Photonic Devices") opens the door to applications of nature's lessons in the technological world. Based on their own experience, within this chapter, the authors are pleased to recommend a methodology for searching for bioinspiration. Parts of that methodology are described, including the experimental and numerical tools used, respectively, for characterization and simulation of photonic structures. The chapter ends with two examples of original concepts that aim to support bioinspired design approaches. With an application in mind and a design in hand, fabrication of a bioinspired photonic device can start! In Chapter 10 ("Fabrication of Bioinspired Photonic Devices"), a selection of the methods of synthesis and fabrication techniques available for such purposes are presented, each following either a bottom-up or a top-down approach. Examples of material syntheses or device productions are provided for these fabrication strategies. The next chapter ("Ideas at Work: Bioinspired Hygrochromic Devices") addresses the specific case of synthesis of hygrochromic materials—those bioinspired materials that are able to change color according to humidity or contact with liquids in general. The last chapter ("Bioinspired Applications in Photonics") is the reward of this hopefully pleasant journey: to transform nature's ideas into human technology. As Jean-Pol Vigneron liked to say: "Nature has at its disposal only a few materials but is able to synthesize incredibly complex photonic structures, optimized through millions of years of evolution. We have at our disposal all chemical elements from Mendeleev's table. However, the photonic structures we are able to fabricate using these elements and the present technology are by far much less complex." Bioinspiration is precisely about that: using our human imaginations to produce novel materials and devices that perform as well as, or even better than, nature's wonderful templates. Sometimes even going beyond this, toward unexpected applications. This is essentially the purpose of this last chapter—to show to the reader, through just a few examples selected from the many reported in the literature, the capability of bioinspiration to meet human needs, not only in terms of technological products but also for the sake of the most noble human aspirations.

Enjoy reading!

CHAPTER 1

Photonics in Nature

Colors and their production mechanisms have always captivated humans' attention. In the early ages of humanity, colors were used for cave paintings, ceramics coloring, fabric dying, food coloring, and art. Since antiquity, humans have been describing and elucidating phenomena such as rainbows, sky colors at sunset and sunrise, tree leaf yellowing in autumn, and whiteness of clouds and auroras.

Many natural organisms, including plants, insects, birds, and mammals, exhibit striking visual appearances [1–4]. These effects are often related to inter- and intraspecific communications, such as aposematism, conspecific visual recognition, and courtship. They often have a role in thermoregulation or camouflage. Because of their significance to survival chances and reproduction in many species, they have undergone strong evolutionary selection pressures.

Whereas most natural systems are colored by pigments, some colors are produced by light emitters, such as stars, fire, or fireflies. In addition, many photonic structures give rise to some coloration and optical effects in natural organisms. Investigating these structures not only generates understanding of the optical behaviors observed in nature and of the biological role underpinning these effects, but also provides inspiration for the design and development of novel photonic materials in technological applications through a bioinspired approach [5].

In this chapter, we review the various classes of colors in nature—namely, pigments, structures, and emitters. We introduce different ways to quantify colors for human and animal vision. Finally, we describe the challenges related to bioinspiration in the field of photonics.

1.1 Colors in Nature

Colors are usually divided into three distinct categories, following the underlying physicochemical mechanism: (1) pigmentary colors, arising from the selective absorption of light by molecules called pigments; (2) structural colors, sometimes called interferential colors, originating from light wave interference within physical structures; (3) colors from light sources involving mechanisms such as incandescence, fluorescence, phosphorescence, or bioluminescence.

Pigments can be either artificial (i.e., from a chemical origin), or natural (i.e., from a mineral or biological—including plant and animal—origin) (Figure 1.1). The biological pigment category is the more widespread one among the colors of living organisms. Pigments typically act as optical filters that absorb light at some wavelengths through electron excitation processes and scatter light at other wavelengths, leading to colors. For example, melanin, pterin, bilin, ommochrome, and papiliochrome commonly color many insects [2, 6]. Melanin is usually associated with

Figure 1.1 Pigmentary colors naturally occur in minerals and biological organisms. Examples are the fur- of koalas (a), lions (b), and squirrels (c); the skin of some fish (d); the shells of snails (e); the wings of many butterflies (f); the caruncles and feathers of roosters (g); the feathers of macaws (h) and flamingos (i); the leaves of palm trees (j); the petals of *Iris* plants (k), orchids (l) and *Jacaranda* flowers (m); red sandstone in Orcombe Point, Exmouth, UK (n); minerals in volcanic rock in Death Valley National Park, United States (o); and ochre clay hills (p). (This figure was reproduced from https://commons.wikimedia.org.)

black and brown colors; bilin pigments, with red colorations; papiliochrome R, with reddish-brown hues; papliochrome II, with yellow shades; 3-hydroxykynurenine (3-OHK), with red and orange colors; and erythropterin and xanthopterin, with yellow, orange, and red colorations [2, 6–8]. Leucopterin, which absorbs in the ultraviolet (UV) range, does not lead to any color for the human visual system. However, other animal species, such as butterflies, would perceive a color [9, 10].

Structural colors (Figure 1.2) are due to simple or complex light wave interference phenomena, resulting from the interaction between incident light and nanostructures, often called photonic structures. These structures are made of components with different optical properties, the most relevant one being the refractive indices of the materials. Often, they are almost periodic, with dimensions of the order of a few hundreds of nanometres (i.e., close to visible light wavelengths).

Figure 1.2 Structural colors naturally occur in many living organisms. Examples include the skins of fish species, such as the six-finger threadfin; (a) the cardinal tetra and neon tetra (b, back and front, respectively); the blue ridges on the muzzle of the mandrill (c); the caruncles of the bare-throated bellbird (d); some feathers of the male mallard duck (e), the common pigeon (f) and the Indian peafowl (g); the chelicerae of the red-backed jumping spider (h); the integuments of some buprestids (i) and scarab beetles (j,k); the thoraces and abdomens of some flies (l); the wings of butterflies, including some lycaenids (m) and *Morpho* butterflies (n); the fruits of bastard hogberries (o); and the leaves of some *Selaginella* plants (p). (This figure was reproduced from https://commons.wikimedia.org.)

Natural structural colors have occurred in many organisms for more than 500 million years [4, 11, 12]: insects [2, 5, 13–17], arachnids [18–20], birds [21–29], fish [30, 31], ctenophores [32–34], bristle worms [35], molluscs [36, 37], diatoms [38–41], iridoviruses [42, 43], mammals [44–47], or plants [48, 49]. They are usually made of biopolymers, such as chitin in arthorpods, keratin in birds, collagen in mammals and fish, or cellulose in plants. Furthermore, these structures often have pores filled with air or liquids (usually aqueous solutions). Examples of such structures are described in the following chapter and include thin films, multilayers, diffraction gratings, photonic crystals, and disordered photonic structures. Usually, these structures give rise to iridescence; namely, change in color

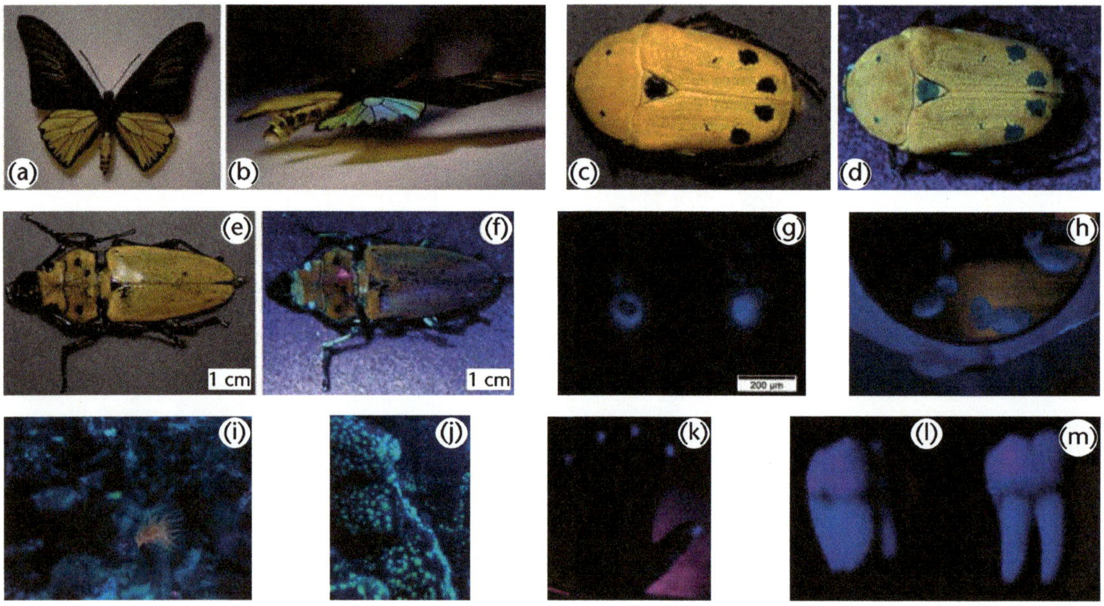

Figure 1.3 Under incident UV light, many living organisms are known to emit visible light by fluorescent decay. Insects such as the butterfly *Troïdes magellanus*(a,b), the beetle *Euchroea auripigmenta* (c,d), and the beetle *Trictenotoma childreni* are examples. They were observed here under visible incident light (a,c,e), under UV-visible light at a grazing observation angle (b), and UV incident light (d,f). The median eyes of a scorpion also fluoresce upon illumination by UV light (g). Marine organisms, including the moon jellyfish (h) and sea anemones such as the jewel anemone (i,j) emit light through fluorescent decay. White hair, nails (k), and teeth (l,m) of humans are known for being fluorescent, too. (This figure was reproduced from [60] with permission from AIP Publishing (a,b); from [61] (c,d); from [62] with permission from John Wiley and Sons (e,f); and from [63] (g). Pictures (h,j), (i), and (k-m) were taken by S. R. Mouchet at the National Marine Aquarium, Plymouth, UK on 11/19/2016, at the Birch Aquarium, La Jolla, CA, 05/05/2016 and in Namur, Belgium in September 2015, respectively.)

with variation of the incidence or observation angles. Another property of structural colors is their resilience to aging, in contrast to pigments, which bleach under incident light with time. As a result, structural colors have been observed in fossils from the plumage of paravian dinosaurs [50, 51] and birds [52, 53]; the shells of molluscs [12, 54], as well as the wings of butterflies [55], moths [56], and beetles [12,57–59].

In living organisms, two different light emission processes are typically observed, namely fluorescence (Figure 1.3) and bioluminescence (Figure 1.4). Fluorescence corresponds to long-wavelength light emission (usually in the visible range) by molecules called fluorophores, following the absorption of short-wavelength light (often in the UV and blue parts of the visible range). The emission of light results from energy transitions among electron states of the fluorophores. It takes place within the integument of many natural organisms [64, 65], including insects [60–62,66–70], arachnids [71,72], birds [73–75], reptiles [76], amphibians [77,78], marine animals [79–85], plants [86–88], and mammals [89]. Examples of fluorophores are biopterin, papiliochrome II, resilin, psittacofulvin, and the green fluorescent protein (GFP). Bioluminescence consists of light emission induced by a biochemical reaction. This reaction involves the oxydation of luciferin, fostered

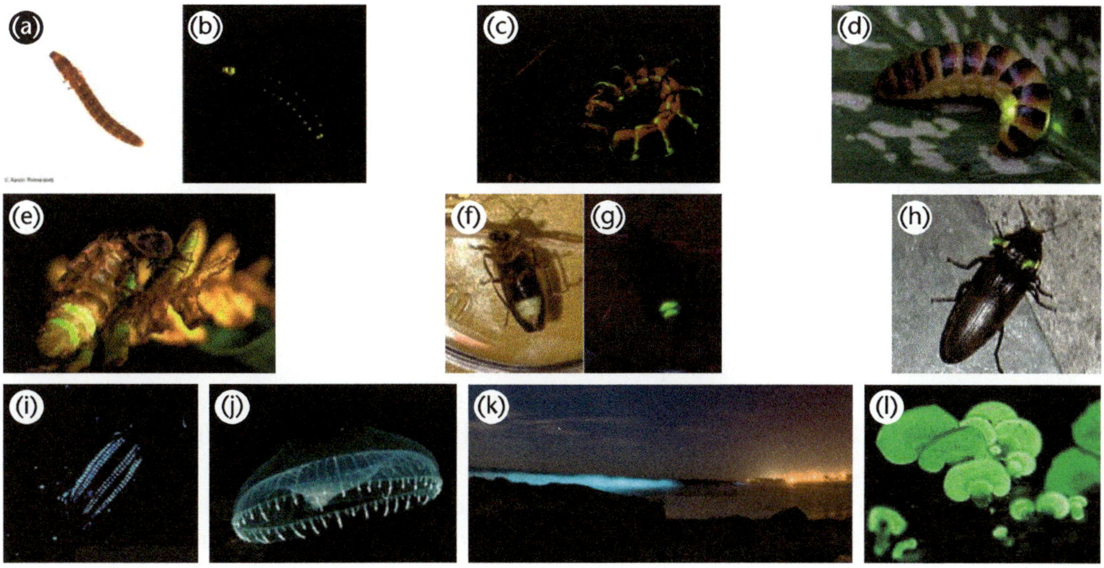

Figure 1.4 Bioluminescence can be observed from the larvae of several Elateroidea beetles (a-d) including the railroad worm, namely the larva of *Phrixothrix* sp. beetles, observed here under visible incident light (a) and in the dark (b), and the western banded glowworm (i.e., the larva of *Zarhipis integripennis*) (c); the images of many beetles, such as the female common glow-worm of Europe (*Lampyris noctiluca*) (e), a *Photuris* sp. firefly, observed here under visible incident light (f) and in the dark (g), and the headlight elater *Pyrophorus noctilucus* (h); marine animals including *Euplokamis* sp. comb jelly (i), *Aequorea victoria* crystal jellyfish (j), and dinoflagellates (k); as well as *Panellus stipticus* fungus. (This figure was reproduced from https://commons.wikimedia.org.)

by an enzyme component called luciferase. This forms a product molecule P* in an excited electron state that decays through a photon emitting relaxation [90,91]:

$$\text{luciferin} + O_2 \xrightarrow{\text{luciferase}} P^* \longrightarrow P + h\nu.$$

It occurs in various natural organisms [90] ranging from bacteria [92] and dinoflagellates [92,93], to shrimp [92,93], ctenophores [32–34,92,93], molluscs [92–94] and fish [92,93,95], to insects such as glowworms [96,97], fireflies [91], and click beetles [98]. The emission wavelengths are typically within the range of 460–560 nm [90].

In nature, organisms displaying visual appearances due to a combination of two or three color mechanisms from these categories are very common [32–34,70,99–103]. For instance, the purity of a structural color is often increased by underlying pigments [27]. Inversely, pigmentary colors may become shiny when combined with photonic structures [104]. In this book, we will mainly focus on structural colors occurring in natural organisms and related bioinspired devices. The main reason for this choice is that structuring pigment-free, colorless materials in order to produce color (or thermal properties) from them is highly attractive from a technological point of view. However, we will see that it is often relevant to understand the interplay between structures, pigments, and light emission.

1.2 Structural Colors in Nature

Throughout recent human history, many scientific fields of research have been driven by technological advancements. The investigation of structural colors in nature is no exception to this. The development that allowed the first reported scientific observation of structural color was the invention of the optical microscope in the early 17th century. Soon afterward, in 1665, Robert Hooke, who was born on the Isle of Wight and is also renowned for Hooke's law describing the compression and extension force of a spring, published his famous book *Micrographia: Or, Some Physiological Descriptions of Minute Bodies Made by Magnifying Glasses with Observations and Inquiries Thereupon* [105]. This book had an immense impact on the field of biology. In addition to magnificent illustrations, he coined the term "cell" for the biological unit of all organisms. Regarding natural structural colors, he described the "pleasing and lovely colours" of "Pearls, Mother of Pearl-shels, Oyster-shels, and almost all other kinds of stony shels whatsoever" as well as the "beauteous and vivid colours" of the feathers of peacocks and ducks (Figure 1.5). By wetting the feathers, he highlighted the structural origin of these colors: "Now, that these colours are onely fantastical ones, that is, such as arise immediately from the refractions of the light, I found by this, that water wetting these colour'd parts, destroy'd their colours, which seem'd to proceed from the alteration of the reflection and refraction. Now, though I was not able to see those hairs at all transparent by a common light, yet by looking on them against the Sun, I found them to be ting'd with a darkish red colour, nothing a-kin to the curious and lovely greens and blues they exhibited." In 1730, Newton also noted the structural origin of peacock feathers in his treatise *Opticks: Or, a Treatise of the Reflections, Refractions, Inflections, and Colours of Light* [106].

John William Strutt, 3rd Baron Rayleigh, derived a comprehensive treatment of the optical properties of multilayered media [107–109]. This followed the development of the theory of electromagnetism in the late 19th century, due to the work of James Clerk Maxwell, who demonstrated the existence of electromagnetic waves thanks to the so-called Maxwell's equations [110, 111], and Heinrich Rudolf Hertz, who experimentally confirmed the emission of such waves by electric oscillators [112]. Using his theory, Rayleigh was able to describe the interferential character of some natural colors [113]. These developments allowed a first wave of scientific studies of structural colors led by Walter [114], Michelson [115], 3rd and 4th Barons Rayleigh [113, 116–118] and Mason [44, 119–124] in various natural organisms, which were mainly insects and birds [115, 125–128]. Nevertheless, the research field was limited by the optical microscope resolution fixed by the Abbe diffraction limit. The next technological advancement that permitted researchers to elucidate natural structural colors further was the invention of the electron microscope by Ernst Ruska and Max Knoll in 1931 and its commercialization. This experimental tool allowed users to observe the micro- and nanostructures from selected organisms for the first time ever (Figure 1.6) [129–132]. As a result, researchers worldwide, mainly biologists, imaged structurally colored integuments from a large variety of animals [30, 133–145]. In parallel to these studies, scientific research allowed the development of mathematical and numerical tools in the

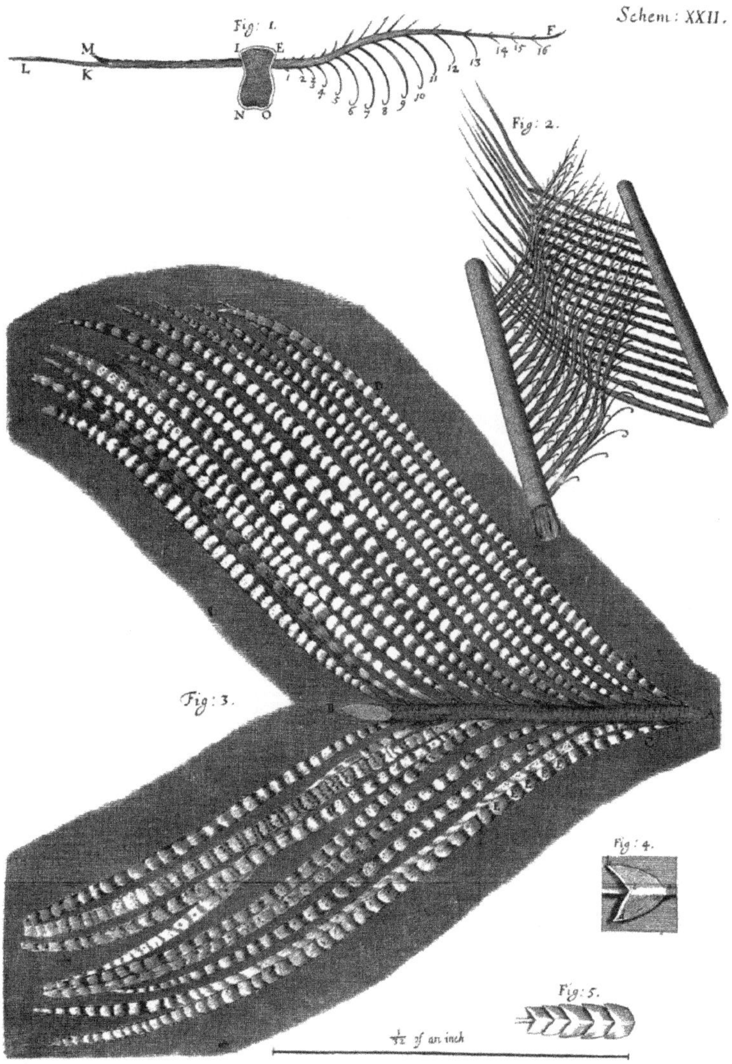

Figure 1.5 Robert Hooke made an early description of bird feathers in his book in 1665. (Reproduced from [105].)

second half of the 20th century in ways that were to prove crucial for the field of natural photonics. For instance, on the basis of Lord Rayleigh's treatment, methods allowing scientists to predict the optical response from multilayered media, such as the transfer-matrix method, or Huxley's method [146–148], or from more complex systems, such as the finite-difference time-domain (FDTD) method [149–151], the finite element method (FEM) [152, 153], or the rigorous coupled-wave analysis (RCWA) method [154–156], were suggested. These tools are still conventional methods used for research purposes today. Supercomputers, which were introduced in the 1960s, as well as computers in general, hugely improved during this period, leading to what are today powerful tools for solving Maxwell's equations.

One hundred years after Lord Rayleigh's developments, Eli Yablonovitch [157] and Sajeev John [158] investigated the optical properties of nanostructures

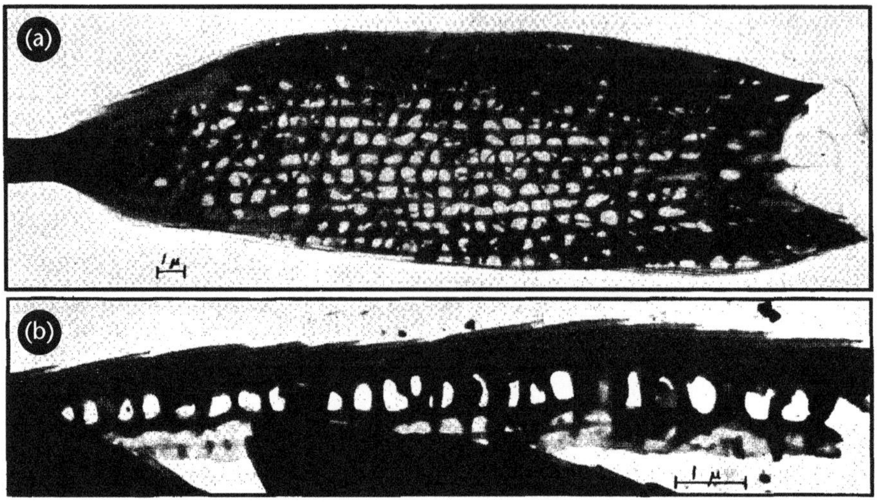

Figure 1.6 Using one of the early electron microscopes, Anderson and Richards imaged the scales covering the wings of the Cypris morpho in 1942. Their micrographs showed the ridges with top (a) and lateral views (b). (Reproduced from [131], with permission from AIP publishing.)

exhibiting a periodicity along the three dimensions of space. Such structures are now called "photonic crystals" [159] and both Eli Yablonovitch and Sajeev John are regarded as the fathers of photonics. The science of photonics started to flourish in the early 1990s [160–164]. The first synthesized photonic crystals operated in the microwave frequency range [161]. Soon afterward, researchers focused their efforts on the production of crystals active in the visible range [162]. Such crystals require their materials to be structured on the scale of visible light wavelength; namely, a few hundreds of nanometers, which is technologically challenging even today. Fortunately, this progress took place at the same time as the development of the fields of nanoscience and nanotechnology, due to, for example, the invention of the scanning tunneling microscope (STM) in 1981 [165] and the discovery of fullerenes [166] in the 1980s. By the end of the 1990s, nanoscience became a particularly trendy field within science, as well as in industry and the media. This interest increased further in the 2000s and the early 2010s. At that time, the prefix "nano-" was used to such an extent that it is now almost hackneyed through overuse.

In this context, scientists, among whom were several physicists, had their attention turned in the late 1990s and early 2000s to investigations performed by their colleagues from biology (Figure 1.7) [1,2,11,99,167–176]. The multidisciplinary backgrounds of the researchers involved in the field enabled the application of methods and techniques developed in photonics and materials science for the purpose of studying natural photonic structures [177]. This has renewed the field and helped with better understanding of the photonic effects in natural organisms. As a result, a very large number of studies have been published on this topic by many research groups worldwide since the early 2000s.

Despite this intensive research conducted over the course of decades, many natural systems display visual appearances with optical behaviors that still remain unexplained, even today. In addition, some fundamental questions related to the biological aspects, including the evolution, development, and genetics of natural

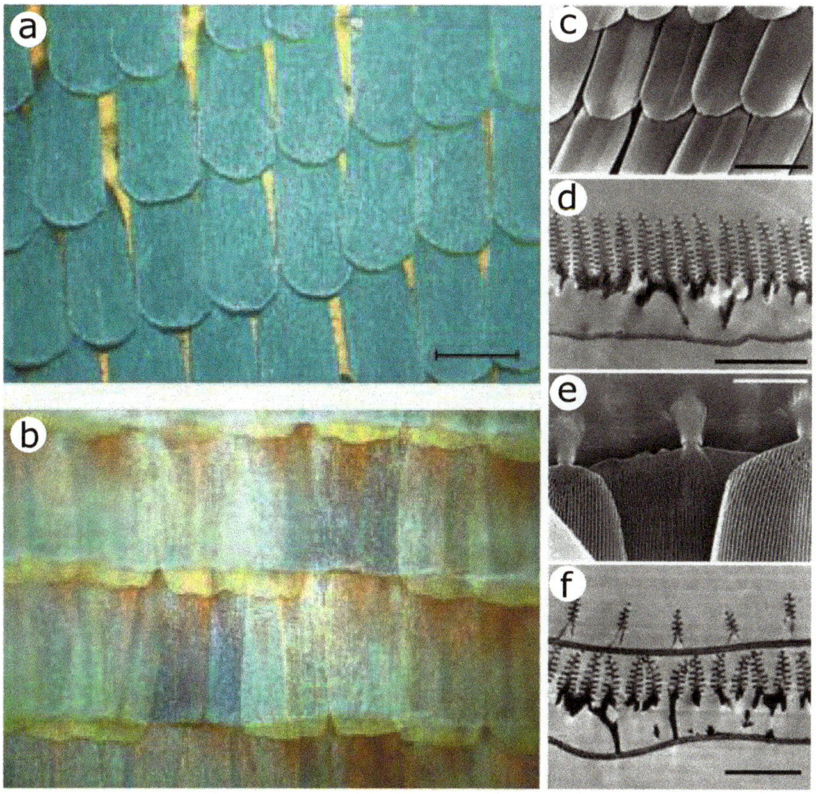

Figure 1.7 In the late 1990s and early 2000s, the emerging photonics community became interested in photonic structures occurring in nature. In their notable 1999 paper, Vukusic and colleagues investigated the photonic structures covering the wings of *Morpho rhetenor* (a,c,d) and *Morpho didius* (b,e,f), using optical (a,b) microscopy as well scanning (c,e) and transmission (d,f) electron microscopy. Scale bars: 100 μm (a,b), 120 μm (c), 3 μm (d), 30 μm (e) and 2 μm (f). (Reproduced from [99], with permission from the Royal Society.)

photonic structures are still pending. One of the leading theories regarding the evolution of structural colors in natural organisms was suggested by Andrew Parker [11, 12, 178]. According to his hypotheses, interferential structures were present in the integuments of the specimens of some species prior to the Cambrian explosion (about 541 million years ago). One could assume it was then the byproduct of another biological function. These phenotypic traits were neutrally selective, since no predator with eyes had evolved. With the Cambrian explosion, the evolution of eyes enabled a huge diversification among species and allowed predators to perceive visual images for the first time and, hence, see their prey. Thereafter, light operated a selection pressure on species and drove the evolution of both predator- and prey-coloration.

Similarly, the formation of interferential structures in natural organisms is far from fully understood. One of the systems that has been investigated the most from that perspective is the wings of some butterflies. Most of this work was led by Helen Ghiradella in the 1970s and 1980s [179–183]. The butterflies investigated bear some scales on their wings. Each scale originates from a single cell in the wing ephithelium and forms in the chrysalis, as a result of the interaction of the cell plasma membrane and the membrane of the smooth endoplasmic reticulum. Helen Ghiradella investigated the development of gyroid photonic structures (described in the following chapter) within the scales of *Callophrys* (formerly *Mitoura*) *gryneus* [182]. Whilst this research allowed science to elucidate several steps of this structure formation, questions remained. Moreover, despite further studies [184–187], the biochemical process by which chitin is produced within the cell at ambient temperature and pressure is still under debate. Furthermore, the exact development of the plasma membrane into a network of complex invaginations molding the gyroid structure, as well as the role of the smooth endoplasmic reticulum in the cell membrane invagination process, remain unexplained. The development of multilayered reflectors within the cuticle of some beetles species (see Chapter 2) was similarly described [188, 189]. In addition to insect structures, the development and evolution of bird feathers and skins (known for exhibiting structural coloration in some species, as described in Chapter 2) were also investigated [190–192]. So far, several developmental and evolutionary aspects of structural colorations in these integuments, including the formation of collagen structures and keratin polymerization, have been understood [192–196], but some underlying mechanisms remain to elucidate [195, 196]. In general, the most unclear pending question in natural photonics surely relates to genetics, with only a handful of publications reporting on possible links between structural colors and DNA [197–200] and, even then, without clear identification of any specific genes as responsible for coding structural colors in the integuments of natural organisms.

1.3 Color Vision

In general, a color cannot be scientifically described only in terms of a reflectance- or transmittance-peak wavelength and full width at half maximum. Color perception depends on the individual observers. In spite of its stunning visual sensitivity, the human eye and the human visual system in general do not allow every human being, even with healthy eyes, to perceive colors in exactly the same way. That is why, for instance, some colors are found to be green by certain people whereas they will be qualified as blue by others. This is typically the case for colors close to turquoise. In addition to the eye sensitivity and the color perception occurring in the human brain, a perceived color also depends on the incident light source. Objects lit by sodium-vapor lamps on the street at night display different colors in daylight. The same holds for the recently deployed light-emitting diode (LED) lightening.

The scientific field aiming at color quantification is called colorimetry. Several quantification systems, known as color spaces, have been developed [201,202], with different usefulness in art (e.g., photography and painting) or in industry, including the glass, metal, textile, food, cosmetics, plastics, paper, and ink industries. The most widely used color space in the study of natural organisms was developed by the International Commission on Illumination (abbreviated CIE from its French name, *Commission Internationale de l'Éclairage*) in 1931. This color space corresponds to the first international agreement on the scientific treatment of colors for human vision. Other animal species have different spectral sensitivities and are sometimes sensitive to light polarization.

1.3.1 CIE 1931 RGB Color Space

The first system defined by the CIE is a RGB color space. It was developed by color matching experiments conducted by William David Wright in 1929 [203] and John Guild in 1932 [204]. In these experiments, a few observers were asked to adjust the amount of three primary colors, red, green, and blue, related more or less to the three human visual receptors. This allowed the development of $\overline{r}$, $\overline{g}$, and $\overline{b}$ color matching functions (Figure 1.8), corresponding approximately to the chromatic sensitivities of standard human vision [201,202,205].

The red (r), green (g), and blue (b) chromaticity coordinates of the CIE 1931 RGB color space can be calculated from the color matching functions, via the R, G and B tristimulus values:

$$\begin{cases} R = \int P(\lambda)\,\overline{r}(\lambda)\,d\lambda \\ G = \int P(\lambda)\,\overline{g}(\lambda)\,d\lambda \\ B = \int P(\lambda)\,\overline{b}(\lambda)\,d\lambda \end{cases}, \tag{1.1}$$

where $P(\lambda)$ is the spectral power distribution of the light under test and λ is the wavelength [201, 202, 205]. After normalization, this leads to the chromaticity coordinates (Figure 1.8):

$$\begin{cases} r = \dfrac{R}{R+G+B} \\ g = \dfrac{G}{R+G+B} \\ b = \dfrac{B}{R+G+B} \end{cases}. \tag{1.2}$$

Due to the negative values of the $\overline{r}$, $\overline{g}$, and $\overline{b}$ functions, specifically for $\overline{r}$ between 435.8 nm and 448.1 nm (Figure 1.8), these coordinates can take negative values.

1.3.2 CIE 1931 XYZ Color Space

In order to avoid negative values of some of the chromaticity coordinates, the CIE derived another system related to the CIE 1931 RGB space by linear transformation, the CIE 1931 XYZ color space. This color space is described by new matching functions (i.e., $\overline{x}$, $\overline{y}$, and $\overline{z}$), defined by the CIE (Figure 1.9). In this space, chromaticity coordinates (x, y, z) quantify any color of a reflecting or transmitting object based on a reflectance or transmittance spectrum $S(\lambda)$, ranging from 380 to 780 nm, thanks to the formulas:

$$\begin{cases} X = k\int L(\lambda)\,S(\lambda)\,\overline{x}(\lambda)\,d\lambda \\ Y = k\int L(\lambda)\,S(\lambda)\,\overline{y}(\lambda)\,d\lambda \\ Z = k\int L(\lambda)\,S(\lambda)\,\overline{z}(\lambda)\,d\lambda \end{cases}, \tag{1.3}$$

or, in the case of the quantification of the color of a light source,

$$\begin{cases} X = \int L_\lambda(\lambda)\,\overline{x}(\lambda)\,d\lambda \\ Y = \int L_\lambda(\lambda)\,\overline{y}(\lambda)\,d\lambda \\ Z = \int L_\lambda(\lambda)\,\overline{z}(\lambda)\,d\lambda \end{cases}, \tag{1.4}$$

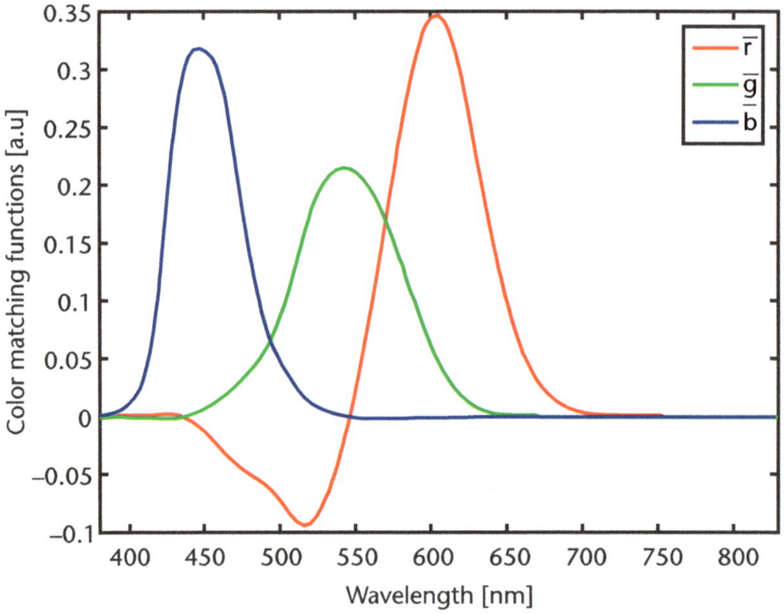

Figure 1.8 The $\overline{r}$, $\overline{g}$, and $\overline{b}$ color matching functions defined by the CIE in 1931 correspond approximately to the chromatic sensitivities of standard human vision [201, 202, 205]. (Data from [206].)

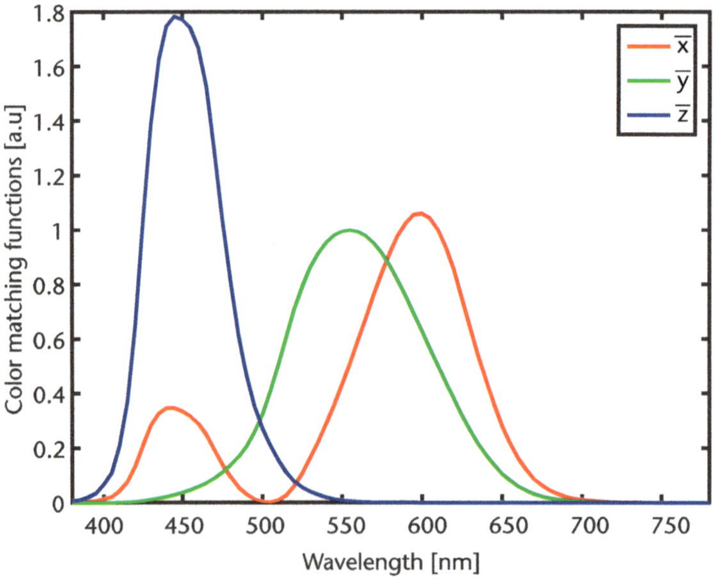

Figure 1.9 In order to avoid negative values of some of the chromaticity coordinates, the CIE derived the CIE 1931 XYZ color space relying on $\overline{x}$, $\overline{y}$, and $\overline{z}$ color matching functions. (Data from [206].)

where X, Y, and Z are termed tristimulus values; $k = 100/\int_\lambda L(\lambda)\overline{y}(\lambda)d\lambda$ normalizes tristimulus values in such a way that a hypothetical body exhibiting a flat and unitary spectrum (i.e., $S(\lambda) = 1$ for all wavelengths λ) has a Y tristimulus value equal to 100; $L(\lambda)$ corresponds to the spectral power distribution of the incident light and L_λ is the spectral radiance of the light source [201,202,205]. Usually, the

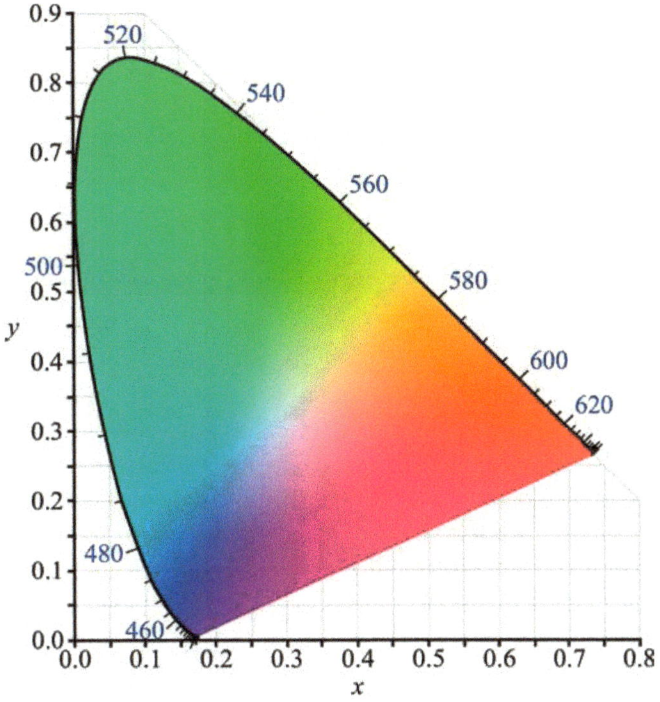

Figure 1.10 On the CIE 1931 xy diagram, (x, y) independent chromaticity coordinates define a color. (This figure was reproduced from BenRG, https://commons.wikimedia.org/wiki/File:CIE1931xy_blank.svg, public domain.)

spectrum of the incident light is assumed to be the D_{65} illuminant, corresponding to the average daylight shed on Northern Europe by a 6503.6-K black body approximating the Sun at midday. The $\overline{y}$ matching function corresponds to the average spectral sensitivity of human visual perception of brightness, setting Y as the relative luminance.

The (x, y, z) chromaticity coordinates are calculated from

$$\begin{cases} x = \dfrac{X}{X+Y+Z} \\ y = \dfrac{Y}{X+Y+Z} \\ z = \dfrac{Z}{X+Y+Z} \end{cases}, \tag{1.5}$$

where (x, y) are independent coordinates (i.e., $x + y + z = 1$) defining chromaticity on the CIE diagram (Figure 1.10). In addition to these coordinates, the luminance, Y, should be provided in order to define a color.

The CIE 1931 R, G, and B tristimulus values can be calculated from the X, Y, and Z tristimulus values, using the equation:

$$\begin{pmatrix} R \\ G \\ B \end{pmatrix} = M \begin{pmatrix} X \\ Y \\ Z \end{pmatrix}, \tag{1.6}$$

where the M matrix was defined as

$$M = \begin{pmatrix} 0,41847 & -0,15866 & -0,082835 \\ -0,091169 & 0,25243 & 0,015708 \\ 0,00092090 & -0,0025498 & 0,17860 \end{pmatrix}.$$

1.3.3 CIE 1976 L*a*b* Color Space

The major drawback of the two aforementioned color spaces lies in the distance between the coordinates of two different colors. This distance does not represent the degree of difference between the two colors. Following the investigation by David L. MacAdam in 1942 [207], the CIE suggested so-called uniform color spaces such as CIE 1964 U*V*W*, CIE 1976 L*u*v* or the CIE 1976 L*a*b* [201,202,205]. In these spaces, indistinguishable differences between perceived colors are represented by equal distances. These spaces are mathematical derivations from the X, Y, and Z tristimulus values. The CIE 1976 L*a*b* is one of the most popular and widely used in various color-related industries. a^* and b^* coordinates correspond to chromaticity coordinates and L^* is the lightness, ranging from 0 (black) to 100 (white). These coordinates are calculated from

$$\begin{aligned} L^* &= f(Y/Y_n) \\ a^* &= \tfrac{500}{116}\left[f(X/X_n) - f(Y/Y_n)\right], \\ b^* &= \tfrac{200}{116}\left[f(Y/Y_n) - f(Z/Z_n)\right] \end{aligned} \quad (1.7)$$

where

$$\begin{cases} f(w) = 116\sqrt[3]{w} - 16, & \text{if } f(w) \geq 8 \\ f(w) = \left(\tfrac{29}{3}\right)^3, & \text{if } f(w) < 8 \end{cases}, \quad (1.8)$$

X_n, Y_n, and Z_n correspond to the X, Y, and Z tristimulus values of the selected illuminant and Y_n is always equal to 100. a^* and b^* axes correspond to the green-red component and the blue-yellow component, respectively.

The CIE 1976 L*a*b* color space can be represented with cylindrical coordinates (instead of the a^* and b^* Cartesian coordinates). This leads to the CIE L*C*h°color space, with C^* representing the relative saturation, called chroma and $h°$, being the angle of the hue in the L*a*b* space, termed the hue angle. The coordinates are calculated by

$$\begin{aligned} C^* &= \sqrt{\left(a^{*2} + b^{*2}\right)} \\ h^\circ &= \arctan\left(\tfrac{b^*}{a^*}\right) \end{aligned}. \quad (1.9)$$

1.3.4 Modeling Animal Color Vision

The chromaticity diagrams described in the previous sections are only valid for human vision. Other species have photoreceptors that often exhibit different spectral sensitivities in terms of intensity distribution, as well as in terms of spectral range (Figure 1.11) [208,209]. For many animals, such as insects, fish, and birds, their

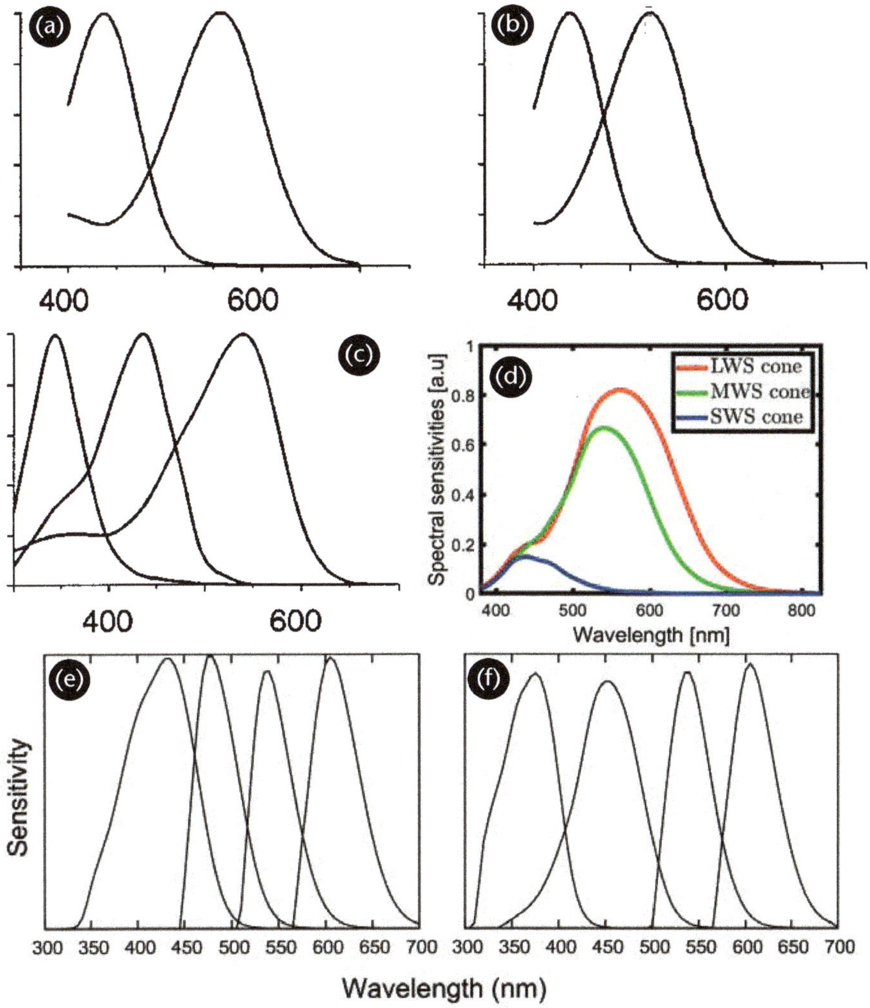

Figure 1.11 The number of cone photoreceptors and their related spectral sensitivities depends upon the species. For instance, tree shrews (a) and ground squirrels (b) have a dichromatic vision, with short- and long-wavelength-sensitive cones. Trichromatic vision is found in honeybees (c) and human beings (d), who have an extra type of medium-wavelength-sensitive cone. Bird vision is known for being tetrachromatic, with some birds exhibiting additional either violet-sensitive (e) or ultraviolet-sensitive (f) cones, such as the Indian peafowl and the Eurasian blue tit, respectively. (Reproduced from [212], with permission from the Royal Society (a-c) and [232], License CC-BY-2.0 (e,f). Data used in (d) is from [223].)

visible spectral range extends into the ultraviolet to about 320 nm. Other animals, such as butterflies and fish, have vision extending to the near infrared. Furthermore, the number of photoreceptor types depends upon the species (Figure 1.11). Human vision is trichromatic (i.e., involving three types of photoreceptors that are involved in color vision). However, some birds and some insects have tetrachromatic vision (i.e., four different types of photoreceptors). If a bare minimum of two types of photoreceptors seems to be needed in order to have color vision, some species may exceed 15 different types [210]. In addition, several animal species are able to distinguish light polarization [208,209,211].

Although developing color spaces by asking observers to match a selected color can be challenging for nonhuman species, methods allowing researchers to assess the colors experienced by other species do exist. One simple way to do this is to calculate, the relative amount of photons absorbed by each type of photoreceptor, related to the species studied [212–214]. These quantities are termed quantum catches Q_i (where i denotes the type of photoreceptor) and are calculated from

$$Q_i = k_i \int L(\lambda)\, S(\lambda)\, R_i(\lambda)\, d\lambda, \tag{1.10}$$

where $R_i(\lambda)$ represents the photoreceptor i spectral sensitivity; $k_i = 1/\int L(\lambda) R_i(\lambda) d\lambda$ is a normalization constant; $S(\lambda)$ and $L(\lambda)$ are the reflectance or transmittance spectrum and the light source spectrum, respectively. Similar equations taking into account the presence of water as the surrounding medium can be derived for marine species [213].

Often, color discrimination is the most relevant aspect of animal vision, since it underpins an animal's ability to recognize objects. In 1998, Vorobyev and Osorio introduced the receptor noise limited (RNL) model [212–215]. In this model, the discriminability of colors is assumed to arise from the noise within the photoreceptors of an animal's visual system. Equations developed on the basis of opponent mechanisms, which process the signals of the different photoreceptors and their abundances, allow scientists to calculate the color distances for di-, tri-, and tetrachromatic visual systems. This receptor-noise-limited color opponent model relies on the hypotheses that (1) in a visual system with n receptor channels, the color is coded by $n - 1$ unspecified opponent mechanisms (the achromatic signal is not taken into account), (2) opponent mechanisms give rise to zero signal for stimuli differing from the background only in intensity, and (3) the detection thresholds arise from receptor noise, and not from opponent mechanisms [212–216].

When comparing two stimuli (e.g., reflectance or luminance spectra), the chromatic distance ΔS between the stimuli only depends on the differences Δf_i in receptor signals between the stimuli for the photoreceptor type i and the standard deviation e_i of the noise in receptor mechanism [212–216]. For stimuli that are close to the achromatic background, the distance can be calculated by

$$(\Delta S)^2 = \frac{(\Delta f_{\mathrm{L}} - \Delta f_{\mathrm{S}})^2}{(e_{\mathrm{S}})^2 + (e_{\mathrm{L}})^2}, \tag{1.11}$$

for dichromatic visual systems such as those of many mammals (Figure 1.11(a,b)) [217–222], where subscripts S and L stand for short- and long-wavelength-sensitive (SWS and LWS) cones, respectively. In the case of a trichromatic visual system, such as that in human beings [223–225], some fish species [215], or honeybees [226] (Figure 1.11(c,d)), where subscript M stands for medium-wavelength-sensitive (MWS) cones, the same can be calculated by

$$(\Delta S)^2 = \frac{e_{\mathrm{S}}^2(\Delta f_{\mathrm{L}} - \Delta f_{\mathrm{M}})^2 + e_{\mathrm{M}}^2(\Delta f_{\mathrm{L}} - \Delta f_{\mathrm{S}})^2 + e_{\mathrm{L}}^2(\Delta f_{\mathrm{S}} - \Delta f_{\mathrm{M}})^2}{(e_{\mathrm{S}} e_{\mathrm{M}})^2 + (e_{\mathrm{S}} e_{\mathrm{L}})^2 + (e_{\mathrm{M}} e_{\mathrm{L}})^2}. \tag{1.12}$$

For tetrachromatic vision such as those of birds [215,227–232], the corresponding equation becomes

$$\begin{aligned}(\Delta S)^2 =&[(e_V e_S)^2(\Delta f_L - \Delta f_M)^2 + (e_V e_M)^2(\Delta f_L - \Delta f_S)^2 \\ &+ (e_V e_L)^2(\Delta f_M - \Delta f_S)^2 + (e_S e_M)^2(\Delta f_L - \Delta f_V)^2 \\ &+ (e_S e_L)^2(\Delta f_M - \Delta f_V)^2 + (e_M e_L)^2(\Delta f_S - \Delta f_V)^2] \\ &\times [(e_V e_S e_M)^2 + (e_V e_S e_L)^2 + (e_V e_M e_L)^2 + (e_S e_M e_L)^2]^{-1}, \qquad (1.13)\end{aligned}$$

where V stands for either ultraviolet- or violet-sensitive (UVS and VS, respectively) cones (Figure 1.11(e,f)).

The contrast Δf_i can be calculated for each photoreceptor type i along two models: a linear relation between the receptor signals to quantum catches $f_i = Q_i$ [212] and a log-linear relation $f_i = \ln Q_i$ [214,233,234] (where ln stands for the natural logarithm). For two close stimuli (A and B), both models are equivalent, since

$$\begin{aligned}\Delta f_i &= \ln Q_{i,A} - \ln Q_{i,B} = \ln(Q_{i,A}/Q_{i,B}) \\ &= \ln(1 + \Delta Q_i/Q_{i,B}) \simeq \Delta Q_i/Q_{i,B} \simeq \Delta Q_i. \qquad (1.14)\end{aligned}$$

The standard deviation e_i of the noise in the receptor mechanism can be associated with the signal-to-noise ratio in a single cell, which can be measured experimentally. The inverse of this ratio, called the Weber fraction, defines the threshold contrast:

$$\omega_i = \frac{\delta Q_i}{Q_i}, \qquad (1.15)$$

where δQ_i is the standard deviation of Q_i. With the log-linear relation, the standard deviation of the noise in the receptor mechanism corresponds to the Weber fraction:

$$e_i = \delta f_i = \delta \ln Q_i \simeq \delta Q_i/Q_i = \omega_i. \qquad (1.16)$$

For instance, the Weber fraction estimated in LWS cones was assessed to 0.1 for birds [233]. With the log-linear model, the receptor noise in a given photoreceptor is independent of the stimulus intensity throughout the color space. As a consequence, this model gives rise to a uniform chromaticity diagram, where ΔS is independent of the stimulus intensity. The linear model is, hence, only to be used when comparing stimuli close to the achromatic background.

1.4 Bioinspired Photonics

In addition to elucidating optical phenomena in natural integuments and explaining some related biological functions, natural photonics allows us to develop technological devices, following a bioinspired approach (Figure 1.12). Natural photonic structures are very sophisticated and ordered materials. Their optical response has been optimized through natural selection over millions of years of evolution.

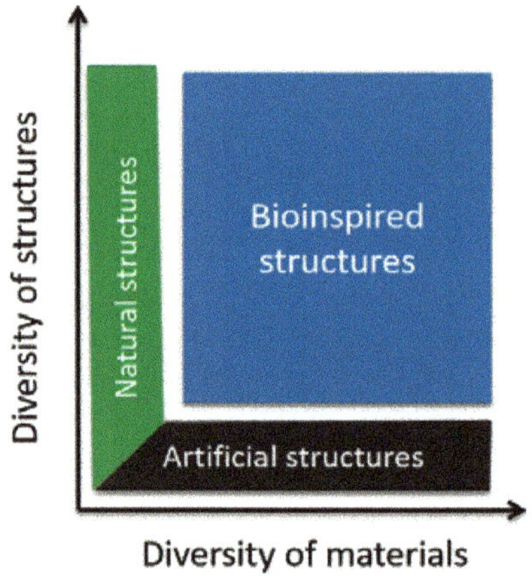

Figure 1.12 Combining the diversity of structures observed in nature with the range of materials human beings are able to synthesize permits a bioinspired approach to offer new possibilities. (Adapted from [63].)

Clearly, these structures challenge the human imagination, which has a wide range of materials at its disposal, but is limited sometimes to designing comparatively rather simple devices. Discovering such natural devices and understanding them allow us to expand upon new concepts and to envision their possible technological applications. These are exciting possibilities that have recently captured the imaginations of researchers worldwide. Such an approach is often referred to as biomimetism or bioinspiration [5]. Biomimetism and bioinspiration are two very close concepts that must be differentiated. Biomimetism is the replication of a natural model with properties of interest with the highest degree of fidelity in order to apply solutions from nature to specific human problems. Bioinspiration has a wider scope. Following observations of natural models, general concepts are retrieved and novel structures, devices, and materials are developed, thanks to inspiration from these concepts.

Outside the field of photonics, the most famous example of bioinspired application that had a commercial success is the hook-and-loop fastener, which is usually referred to by the trademark Velcro. It was invented by a Swiss electrical engineer, George de Mestral, in 1941, as he was walking in the Alps through woods with his dog after hunting. Burs of burdock clung to his coat and to his dog's fur. Thanks to microscope observation of some burs, he noticed that they each have tens of hooks that are able to attach to any sort of fiber, allowing two materials to bind reversibly. De Mestral coined the word "Velcro," which consists of a portmanteau word, blending the words "velour" and "crochet" (French for "velvet" and "hook," respectively).

Similarly, self-cleaning surfaces were conceived on the basis of the "lotus effect" [235,236], which is the superhydrophobicity displayed by the leaves of the *Nelumbo* flower, commonly known as the lotus. Due to the microstructuration of the leaf's surface that reduces its interaction potential, water is repelled from the surface [236, 237]. Water droplets that are deposited on the leaves' surfaces collect the dust as they are shed or fall under gravity, giving rise to the observed self-cleaning properties.

In photonics, some technological devices such as gas sensors, light-trapping structures, and improved optical displays have been suggested by many research groups. The last part of this book will review the major contributions in this respect. Part of the interest of taking inspiration from nature lies in the multifunctionality of natural photonic structures, which have often been optimized for various purposes, such as antireflection and hydrophobicity or color change and hydrophobicity. Another reason the subject is appealing relates to the resilience of the optical response of natural photonic structures to defects. In spite of being often described as very ordered structures, natural photonic structures usually exhibit a large amount of structural or material defects and imperfections. However, the visual appearance of structurally colored organisms is effectively reproduced notwithstanding these defects. Since the impact of defects can be crucial in photonic applications, such systems with optical responses that are tolerant to defects are beyond valuable. Finally, the fact that these systems are synthesized in nature, using natural processes at ambient pressures and temperatures, is attracting a lot of attention.

References

[1] Vukusic, P., and J. R. Sambles, "Photonic Structures in Biology," *Nature*, Vol. 424, No. 6950, 2003, pp. 852–855.

[2] Berthier, S., *Iridescences, les Couleurs Physiques des Insectes*, Paris: Springer, 2003.

[3] Welch, V. L., and J.-P. Vigneron, "Beyond Butterflies–the Diversity of Biological Photonic Crystals," *Optical and Quantum Electronics*, Vol. 39, No. 4, 2007, pp. 295–303.

[4] Kinoshita, S., *Structural Colors in the Realm of Nature*, Singapore: World Scientific Publishing Co, 2008.

[5] Biró, L. P., and J.-P. Vigneron, "Photonic Nanoarchitectures in Butterflies and Beetles: Valuable Sources for Biosinspiration," *Laser Photonics Review*, Vol. 5, No.1, 2011, pp. 27–51.

[6] Vukusic, P., and L. Chittka, "Visual Signals: Color and Light Production." In *The Insects: Structure and Function*, pp. 793–822, R. F. Chapman (ed.), Cambridge: Cambridge University Press, 2013.

[7] Morehouse, N. I., P. Vukusic, and R. Rutowski, "Pterin Pigment Granules are Responsible for Both Broadband Light Scattering and Wavelength Selective Absorption in the Wing Scales of Pierid Butterflies," *Proceedings of the Royal Society B*, Vol. 274, No. 1608, 2007, pp. 359–366.

[8] Wilts, B. D., et al., "Papiliochrome II Pigment Reduces the Angle Dependency of Structural Wing Colouration in *Nireus* Group Papilionids," *Journal of Experimental Biology*, Vol. 215, No. 5, pp. 793–805.

[9] Makino, K., et al., "Sex in *Pieris rapae* L. and the Pteridin Content of their Wings," *Nature*, Vol. 170, No. 4335, 1952, pp. 933–934.

[10] Yagi, N., "Note of Electron Microscope Research on Pterin Pigment in the Scales of Pierid Butterflies," *Annotationes Zoologicae Japonenses*, Vol. 27, 1954, pp. 113–114.

[11] Parker, A. R., "515 Million Years of Structural Colour," *Journal of Optics A: Pure and Applied Optics*, Vol. 2, No. 6, 2000, pp. R15-R28.

[12] Parker, A. R., "A Geological History of Reflecting Optics," *Journal of the Royal Society Interface*, Vol. 2, No. 2, 2005, pp. 1–17.

[13] Berthier, S., *Photonique des Morphos*, Paris: Springer, 2010.

[14] Seago, A. E., et al., "Gold Bugs and Beyond: a Review of Iridescence and Structural Colour Mechanisms in Beetles (Coleoptera)," *Journal of the Royal Society Interface*, Vol. 6, No. S2, pp. S165–S184.

[15] Vukusic, P., "Structural Colour Effects in Lepidoptera." In *Structural Colors in Biological Systems: Principles and Applications*, pp. 95–112, S. Kinoshita, and S. Yoshioka, (eds.), Osaka: Osaka University Press, 2005.

[16] Shi, N. N., et al., "Keeping Cool: Enhanced Optical Reflection and Radiative Heat Dissipation in Saharan Silver Ants," *Science*, Vol. 349, No. 6245, 2015, pp. 298–301.

[17] Mouchet, S. R., and P. Vukusic, "Structural Colours in Lepidopteran Scales," *Advances in Insect Physiology*, Vol. 54, 2018, pp. 1–53.

[18] Oxford, G. S., and R. G. Gillespie, "Evolution and Ecology of Spider Coloration," *Annual Review of Entomology*, Vol. 43, No. 1, 1998, pp. 619–643.

[19] Ingram, A. L., et al., "Characterization of the Green Iridescence on the Chelicerae of the Tube Web Spider, *Segestria florentina* (Rossi 1790) (Araneae, Segestriidae)," *Journal of Arachnology*, Vol. 37, No. 1, 2009, pp. 68–71.

[20] Ingram, A. L., et al., "Structural Origin of the Green Iridescence on the Chelicerae of the Red-Backed Jumping Spider, *Phidippus johnsoni* (Salticidae: Araneae)," *Arthropod Structure & Development*, Vol. 40, No. 1, 2011, pp. 21–25.

[21] Prum, R. O., R. L. Morrison, and G. R. Ten Eyck, "Structural Color Production by Constructive Reflection from Ordered Collagen Arrays in a Bird (*Philepitta castanea*: Eurylaimidae)," *Journal of Morphology*, Vol. 222, No. 1, 1994, pp. 61–72.

[22] Prum, R. O., et al., "Coherent Light Scattering by Blue Feather Barbs," *Nature*, Vol. 396, No. 6706, 1998, pp. 28–29.

[23] Prum, R. O., et al., "Coherent Light Scattering by Nanostructured Collagen Arrays in the Caruncles of the Malagasy Asities (Eurylaimidae: Aves)," *Journal of Experimental Biology*, Vol. 202, No. 24, 1999, pp. 3507–3522.

[24] Yoshioka, S., and S. Kinoshita, "Effect of Macroscopic Structure in Iridescent Color of the Peacock Feathers," *Forma*, Vol. 17, No. 2, pp. 169–181.

[25] Shawkey, M. D., et al., "Nanostructure Predicts Intraspecific Variation in Ultraviolet-Blue Plumage Colour," *Proceedings of the Royal Society B*, Vol. 270, No. 1523, 2003, pp. 1455–1460.

[26] Doucet, S. M., et al., "Iridescent Plumage in Satin Bowerbirds: Structure, Mechanisms and Nanostructural Predictors of Individual Variation in Colour," *Journal of Experimental Biology*, Vol. 209, No. 2, 2006, pp. 380–390.

[27] Shawkey, M. D., et al., "Significance of a Basal Melanin Layer to Production of Non-Iridescent Structural Plumage Color: Evidence from an Amelanotic Steller's Jay (*Cyanocitta stelleri*)," *Journal of Experimental Biology*, Vol. 209, No. 7, 2006, pp. 1245–1250.

[28] Vigneron, J.-P., et al., "Structural Origin of the Colored Reflections from the Black-Billed Magpie Feathers," *Physical Review E*, Vol. 73, No. 2, 2006, 021914.

[29] Yin, H., et al., "Iridescence in the Neck Feathers of Domestic Pigeons," *Physical Review E*, Vol. 74, No. 5, 2006, 051916.

[30] Denton, E., "Reflectors in Fishes," *Scientific American*, Vol. 224, No. 1, 1971, pp. 64–72.

[31] Mäthger, L. M., et al., "Rapid Colour Changes in Multilayer Reflecting Stripes in the Paradise Whiptail, *Pentapodus paradiseus*," *Journal of Experimental Biology*, Vol. 206, No. 20, 2003, pp. 3607–3613.

[32] Welch, V. L, "Structural Colouration in Jellyfish, Fish and Ctenophores," PhD Thesis, University of Oxford, UK, 2003.

[33] Welch, V. L., J.-P. Vigneron, and A. R. Parker, "The Cause of Colouration in the Ctenophore *Beroe cucumis*," *Current Biology*, Vol. 15, No. 24, 2005, pp. R985–R986.

[34] Welch, V. L., et al., "Optical Properties of the Iridescent Organ of the Comb-Jellyfish, *Beroe cucumis* (Ctenophora)," *Physical Review E*, Vol. 73, No. 4, 2006, 041916.

[35] Trzeciak, T. M., and P. Vukusic, "Photonic crystal fiber in the polychaete worm *Pherusa* sp.," *Physical Review E*, Vol. 80, No. 6, 2009, 061908.

[36] Tan, T. L., D. Wong, and P. Lee, "Iridescence of a Shell of Mollusk *Haliotis glabra*," *Optics Express*, Vol. 12, No. 20, 2004, pp. 4847–4854.

[37] Holt, A. L., et al., "Photosymbiotic Giant Clams are Transformer of Solar Flux," *Journal of the Royal Society Interface*, Vol. 11, No. 101, 2014, 20140678.

[38] Fuhrmann, T., et al., "Diatoms as Living Photonic Crystals," *Applied Physics B*, Vol. 78, No. 3, 2004, pp. 257–260.

[39] De Stefano, L., et al., "Marine Diatoms as Optical Chemical Sensors," *Applied Physics Letters*, Vol. 87, No. 23, 2005, 233902.

[40] De Stefano, L., et al., "Interfacing the Nanostructured Biosilica Microshells of the Marine Diatom *Coscinodiscus wailesii* with Biological Matter," *Acta Biomaterialia*, Vol. 4, No. 1, 2008, pp. 126–130.

[41] De Stefano, L., et al., "Nano-Biosilica from Marine Diatoms: A Brand New Material for Photonic Applications," *Superlattices and Microstructures*, Vol. 46, No. 1–2, 2009, pp. 84–89.

[42] Williams, T., and J. Cory, "DNA Restriction Fragment Polymorphism in Iridovirus Isolates from Individual Blackflies (Diptera: Simuliidae)," *Medical and Veterinary Entomology*, Vol. 7, No. 2, 1993, pp. 199–201.

[43] Williams, T., "Natural Invertebrate Hosts of Iridoviruses (Iridoviridae)," *Neotropical Entomology*, Vol. 37, No. 6, 2008, pp. 615–632.

[44] Mason, C. W., "Blue Eyes," *Journal of Physical Chemistry*, Vol. 28, No. 5, 1924, pp. 498–501.

[45] Fox D. L., *Biochromy: Natural Coloration of Living Things*, Berkeley, CA: University of California Press, 1979.

[46] Prum, R. O., and R. H. Torres, "Structural Colouration of Mammalian Skin: Convergent Evolution of Coherently Scattering Dermal Collagen Arrays," *Journal of Experimental Biology*, Vol. 207, No. 212, 2004, pp. 2157–2172.

[47] Strum, R. A., and M. Larson, "Genetics of Human Iris Colour and Patterns," *Pigment Cell Melanoma Research*, Vol. 22, No. 5, 2009, pp. 544–562.

[48] Lee, D. W., and J. B. Lowry, "Physical Basis and Ecological Significance of Iridescence in Blue Plants," *Nature*, Vol. 254, No. 5495, 1975, pp. 50–51.

[49] Vignolini, S., et al., "Pointillist Structural Color in *Pollia* Fruit," *Proceedings of the National Academy of Sciences of the United States of America*, Vol. 109, No. 39, 2012, pp. 15712–15715.

[50] Li, Q., et al., "Plumage Color Patterns of an Extinct Dinosaur," *Science*, Vol. 327, No. 5971, 2010, pp. 1369–1372.

[51] Li, Q., et al., "Reconstruction of Microraptor and the Evolution of Iridescent Plumage," *Science*, Vol. 335, No. 6073, 2012, pp. 1215–1219.

[52] Vinther, J., et al., "Structural Coloration in a Fossil Feather," *Biology Letters*, Vol. 6, No. 1, 2009, pp. 128–131.

[53] Clarke, J. A., et al., "Fossil Evidence for Evolution of the Shape and Color of Penguin Feathers," *Science*, Vol. 330, No. 6006, 2010, pp. 954–957.

[54] Parker, A. R., "Colour in Burgess Shale animals and the effect of light on evolution in the Cambrian," *Proceedings of the Royal Society B*, Vol. 265, No. 1400, 1998, pp. 967–972.

[55] Zhang, Q., et al., "Fossil Scales Illuminate the Early Evolution of Lepidopterans and Structural Colors," *Science Advances*, Vol. 4, No. 4, 2018, e1700988.

[56] McNamara, M. E., et al., "Fossilized Biophotonic Nanostructures Reveal the Original Colors of 47-Million-Year-Old Moths," *PLoS Biology*, Vol. 9, No. 11, 2011, e1001200.

[57] McNamara, M. E., et al., "The Original Colours of Fossil Beetles," *Proceedings of the Royal Society B*, Vol. 279, No. 1731, 2012, pp. 1114–1121.

[58] McNamara, M. E., et al., "Cryptic Iridescence in a Fossil Weevil Generated by Single Diamond Photonic Crystals," *Journal of the Royal Society Interface*, Vol. 11, No. 100, 2014, 20140736.

[59] McDonald, L. T., et al., "Brilliant Angle-Independent Structural Colours Preserved in Weevil Scales from the Swiss Pleistocene," *Biology Letters*, Vol. 16, No. 4, 2020, 20200063.

[60] Van Hooijdonk, E., S. Berthier, and J.-P. Vigneron, "Contribution of Both the Upperside and the Underside of the Wing on the Iridescence in the Male Butterfly *Troïdes magellanus* (Papilionidae)," *Journal of Applied Physics*, Vol. 112, No. 7, 2012, 074702.

[61] Mouchet, S. R., et al., "Colour and Fluorescence Emission of *Euchroea auripigmenta* Beetle," *Proceedings of SPIE*, Vol. 10965, 2019, pp. 72–82.

[62] Mouchet, S. R., et al., "Unveiling the Non-Linear Optical Response of *Trictenotoma childreni* Longhorn Beetle," *Journal of Biophotonics*, Vol. 12, No. 9, 2019, e201800470.

[63] Mouchet, S. R., *Mécanismes fondamentaux de la coloration structurale: addition et induction de couleurs*, PhD Thesis, University of Namur, Belgium, 2015.

[64] Lagorio, M. G., G. B. Cordon, and A. Iriel, "Reviewing the Relevance of Fluorescence in Biological Systems," *Photochemical & Photobiological Sciences*, Vol. 14, No. 9, 2015, pp. 1538–1559.

[65] Marshall, J., and S. Johnsen, "Fluorescence as a Means of Colour Signal Enhancement," *Philosophical Transactions of the Royal Society B*, Vol. 372, No. 1724, 2017, 20160335.

[66] Cockayne, E. A., "The Distribution of Fluorescent Pigments in Lepidoptera," *Transactions of the Royal Entomological Society of London*, Vol. 72, No. 1-2, 1924, pp. 1–19.

[67] Vukusic, P., and I. Hooper, "Directionally Controlled Fluorescence Emission in Butterflies," *Science*, Vol. 310, No. 5751, 2005, p. 1151.

[68] Israelowitz, M., S. H. W. Rizvi, and H. P. von Schroeder, "Fluorescence of the "Fire-Chaser" Beetle, *Melanophila acuminate*," *Journal of Luminescence*, Vol. 126, No. 1, 2007, pp. 149–154.

[69] Welch, V. L., et al., "Fluorescence in Insects," *Proceedings of SPIE*, Vol. 8480, 2012, 848004.

[70] Mouchet, S. R., et al., "Controlled Fluorescence in a Beetle's Photonic Structure and its Sensitivity to Environmentally Induced Changes," *Proceedings of the Royal Society B*, Vol. 283, No. 1845, 2016, 20162334.

[71] Lawrence, R. F., "Fluorescence in Arthropoda," *Journal of the Entomological Society of Southern Africa*, Vol. 17, No. 2, 1954, pp. 167–170.

[72] Pavan, M., and M. Vachon, "Sur l'existence d'une substance fluorescente dans les téguments des Scorpions (Arachnides)," *Comptes rendus de l'Académie des Sciences*, Vol. 239, 1954, pp. 1700–1702.

[73] Arnold, K. E., I. P. F. Owens, and N. J. Marshall, "Fluorescent Signaling in Parrots," *Science*, Vol. 295, No. 5552, 2002, p. 92.

[74] McGraw, K. J., et al., "A Description of Unique Fluorescent Yellow Pigments in Penguin Feathers," *Pigment Cell Research*, Vol. 20, No. 4, 2007, pp. 301–304.

[75] Ladouce, M., et al., "Scattering of Ultraviolet Light by Avian Eggshells," *Faraday Discussions*, Vol. 223, 2020, pp. 63–80.

[76] Mohd Top, M., et al., "First Record of Ultraviolet Fluorescence in the Bent-toed Gecko *Cyrtodactylus quadrivirgatus* Taylor, 1962 (Gekkonidae: Sauria)," *Herpetology Notes*, Vol. 13, 2020, pp. 211–212.

[77] Taboada, C., et al., "Naturally Occurring Fluorescence in Frogs," *Proceedings of the National Academy of Sciences of the United States of America*, Vol. 114, No. 14, 2017, pp. 3672–3677.

[78] Deschepper, P., B. Jonckheere, and J. Matthys, "A Light in the Dark: The Discovery of Another Fluorescent Frog in the Costa Rican Rainforests," *Wilderness & Environmental Medicine*, Vol. 29, No. 3, 2018, pp. 421–422.

[79] Phillips, C. E. S., "Fluorescence of Sea Anemones," *Nature*, Vol. 119, No. 3003, 1927, p. 747.

[80] Catala-Stucki, R., "Fluorescence Effects from Corals Irradiated with Ultra-Violet Rays," *Nature*, Vol. 183, No. 4666, 1959, p. 949.

[81] Read, K. R. H., J. M. Davidson, B. M. Twarog, "Fluorescence of Sponges and Coelenterates in Blue Light," *Comparative Biochemistry and Physiology*, Vol. 25, No. 3, 1968, pp. 873–876.

[82] Cox, G., and A. Salih, "Confocal Microscopy of GFP-Like Pigments in Corals." In *Multi-Modality Microscopy*, pp. 238–245, H. Yu, P.-C. Cheng, P.-C. Lin and F.-J. Kao (eds.), Singapore: World Scientific Publishing Co, 2006.

[83] Salih, A., "Screening Reef Corals for Novel GFP-Type Fluorescent Proteins by Confocal Imaging." In *In Vivo Cellular Imaging Using Fluorescent Proteins. Methods in Molecular Biology (Methods and Protocols)*, pp. 238–245, R. Hoffman (ed.), Totowa, NJ: Humana Press, 2012.

[84] Gruber, D. F., et al., "Adaptive Evolution of Eel Fluorescent Proteins from Fatty Acid Binding Proteins Produces Bright Fluorescence in the Marine Environment," *PLoS ONE*, Vol. 10, No. 11, 2015, e0140972.

[85] Gruber, D. F., et al., "Biofluorescence in Catsharks (Scyliorhinidae): Fundamental Description and Relevance for Elasmobranch Visual Ecology," *Scientific Reports*, Vol. 6, 2016, 24751.

[86] Goodwin, R. H., "Fluorescent substances in plants," *Annual Review of Plant Physiology*, Vol. 4, No. 1, 1953, pp. 283–304.

[87] Iriel, A. and M. G. Lagorio, "Is the Flower Fluorescence Relevant in Biocommunication?" *Naturwissenschaften*, Vol. 97, No. 10, 2010, pp. 915–924.

[88] Iriel, A. and M. G. Lagorio, "Implications of Reflectance and Fluorescence of Rhododendron Indicum Flowers in Biosignaling," *Photochemical & Photobiological Sciences*, Vol. 9, No. 3, 2010, pp. 342–348.

[89] Tani, K., et al., "Fluorescent Properties of Porcelain-Restored Teeth and Their Discrimination," *Materials Transactions*, Vol. 45, No. 4, 2004, pp. 1010–1014.

[90] Hastings, J. W., "Biological Diversity, Chemical Mechanisms, and the Evolutionary Origins of Bioluminescent Systems," *Journal of Molecular Evolution*, Vol. 19, No. 5, 1983, pp. 309–321.

[91] Inouye, S., "Firefly Luciferase: An Adenylate-Forming Enzyme for Multicatalytic Functions," *Cellular and Molecular Life Sciences*, Vol. 67, No. 3, 2010, pp. 387–404.

[92] Haddock, S. H. D., M. A. Moline, and J. F. Case, "Bioluminescence in the Sea," *Annual Review of Marine Science*, Vol. 2, No. 1, 2010, pp. 443–493.

[93] Martini, S., and S. H. D. Haddock, "Quantification of Bioluminescence from the Surface to the Deep Sea Demonstrates Its Predominance as an Ecological Trait," *Scientific Reports*, Vol. 7, 2017, 45750.

[94] Deheyn, D. D., and N. G. Wilson, "Bioluminescent Signals Spatially Amplified by Wavelength-Specific Diffusion through the Shell of a Marine Snail," *Proceedings of the Royal Society B*, Vol. 278, No. 1715, 2010, pp. 2112–2121.

[95] Davis, M. P., J. S. Sparks, W. L. Smith, "Repeated and Widespread Evolution of Bioluminescence in Marine Fishes," *PLoS One*, Vol. 11, No. 6, 2016, e0155154.

[96] Viviani, V. R., "Bioluminescence and Biological Aspects of Brazilian Railroad-Worms (Coleoptera: Phengodidae)," *Annals of the Entomological Society of America*, Vol. 90, No. 3, 1997, pp. 389–398.

[97] Meyer-Rochow, V. B., "Glowworms: a review of Arachnocampa spp. and Kin," *Luminescence*, Vol. 22, No. 3, 2007, 251–265.

[98] Harvey , E. N., and K. P. Stevens, "The Brightness of the Light of the West Indian Elaterid Beetle, *Pyrophorus*," *Journal of General Physiology*, Vol. 12, No. 2, 1928, pp. 269–272.

[99] Vukusic, P., et al., "Quantified Interference and Diffraction in Single *Morpho* Butterfly Scales," *Proceedings of the Royal Society B*, Vol. 266, No. 1427, 1999, pp. 1403–1411.

[100] Yoshioka, S., and S. Kinoshita, "Single-Scale Spectroscopy of Structurally Colored Butterflies: Measurements of Quantified Reflectance and Transmittance," *Journal of the Optical Society of America A*, Vol. 23, No. 1, 2006, pp. 134–141.

[101] Yoshioka, S., and S. Kinoshita, "Structural or Pigmentary? Origin of the Distinctive White Stripe on the Blue Wing of a *Morpho* Butterfly," *Proceedings of the Royal Society B*, Vol. 273, No. 1583, 2006, pp. 129–134.

[102] Bay, A., and J.-P. Vigneron, "Light Extraction from the Bioluminescent Organs of Fireflies," *Proceedings of SPIE*, 7401, 2009, 740108.

[103] Bay, A., et al., "Improved Light Extraction in the Bioluminescent Lantern of a *Photuris* Firefly (Lampyridae)," *Optics Express*, Vol. 21, No. 1, 2013, pp. 764–780.

[104] Vignolini, S., et al., "Directional Scattering from the Glossy Flower of *Ranunculus*: How The Buttercup Lights Up Your Chin," *Journal of the Royal Society Interface*, Vol. 9, No. 71, 2012, pp. 1295–1301.

[105] Hooke, R., *Micrographia: Or, Some Physiological Descriptions of Minute Bodies Made by Magnifying Glasses with Observations and Inquiries Thereupon*, London: J. Martyn & J. Allestry, 1665.

[106] Newton, I., *Opticks: Or, a Treatise of the Reflections, Refractions, Inflections, and Colours of Light*, London: William Innys, 1730.

[107] Lord Rayleigh, "XXVI. On the Maintenance of Vibrations by Forces of Double Frequency, and on the Propagation of Waves through a Medium Endowed with a Periodic Structure," *The London, Edinburgh, and Dublin Philosophical Magazine and Journal of Science*, Vol. 24, No. 147, 1887, pp. 145–159.

[108] Lord Rayleigh, "XXVI. On the Remarkable Phenomenon of Crystalline Reflexion described by Prof. Stokes," *The London, Edinburgh, and Dublin Philosophical Magazine and Journal of Science*, Vol. 26, No. 160, 1888, pp. 256–265.

[109] Lord Rayleigh, "On the Reflection of Light from a Regularly Stratified Medium," *Proceedings of the Royal Society A*, Vol. 93, No. 655, 1917, pp. 565–577.

[110] Maxwell, J. C., "On Physical Lines of Force," *The London, Edinburgh, and Dublin Philosophical Magazine and Journal of Science*, Vol. 23, No. 151, 1861, pp. 12–24.

[111] Maxwell, J. C., "A Dynamical Theory of the Electromagnetic Field," *Philosophical Transactions of the Royal Society of London*, Vol. 155, 1865, pp. 459–512.

[112] Hertz, H. R., "Über sehr schnelle electrische Schwingungen," *Annalen der Physik*, Vol. 267, No. 7, 1887, pp. 421–448.

[113] Lord Rayleigh, "VII. On the Optical Character of Some Brilliant Animal Colours," *The London, Edinburgh, and Dublin Philosophical Magazine and Journal of Science*, Vol. 37, No. 217, 1919, pp. 98–111.

[114] Walter, B. L. J. H., *Die Oberflächen- oder Schillerfarben*, Braunschweig: F. Vieweg und Sohn, 1895.

[115] Michelson, A. A., "LXI. On Metallic Colouring in Birds and Insects," *The London, Edinburgh, and Dublin Philosophical Magazine and Journal of Science*, Vol. 21, No. 124, 1911, pp. 554–567.

[116] Lord Rayleigh, "Studies of Iridescent Colour, and the Structure Producing It.- II. Mother-of-Pearl," *Proceedings of the Royal Society A*, Vol. 102, No. 719, 1923, pp. 674–677.

[117] Lord Rayleigh, "Studies of Iridescent Colour, and the Structure Producing It. - IV. Iridescent Beetles," *Proceedings of the Royal Society A*, Vol. 103, No. 721, 1923, pp. 233–239.

[118] Lord Rayleigh, "The Iridescent Colours of Birds and Insects," *Proceedings of the Royal Society A*, Vol. 128, No. 808, 1930, pp. 624–641.

[119] Mason, C. W., "Structural Colors in Feathers. I," *Journal of Physical Chemistry*, Vol. 27, No. 3, 1923, pp. 201–251.

[120] Mason, C. W., "Structural Colors in Feathers. II," *Journal of Physical Chemistry*, Vol. 27, No. 5, 1923, pp. 401–448.

[121] Mason, C. W., "Structural Colors in Insects. I," *Journal of Physical Chemistry*, Vol. 30, No. 3, 1926, pp. 383–395.

[122] Mason, C. W., "Structural Colors in Insects. II," *Journal of Physical Chemistry*, Vol. 31, No. 3, 1927, pp. 321–354.

[123] Mason, C. W., "Structural Colors in Insects. III," *Journal of Physical Chemistry*, Vol. 31, No. 12, 1927, pp. 1856–1872.

[124] Mason, C. W., "Transmitted Structural Blue in Microscopic Objects," *Journal of Physical Chemistry*, Vol. 35, No. 1, 1931, pp. 73–81.

[125] Bancroft, W. D., et al., "Blue Feathers," *The Auk*, Vol. 40, No. 2, 1923, pp. 275–300.

[126] Onslow, H., "On a Periodic Structure in Many Insect Scales, and the Cause of Their Iridescent Colours," *Philosophical Transactions of the Royal Society B*, Vol. 211, No. 382-390, 1923, pp. 1–74.

[127] Merritt, E., "A Spectrophotometric Study of Certain Cases of Structural Colour," *Journal of the Optical Society of America and Review of Scientific Instruments*, Vol. 11, No. 2, 1925, pp. 93–98.

[128] Raman, C. V., "The Origin of the Colours in the Plumage of Birds," *Proceedings of the Indian Academy of Sciences*, Vol. 1, 1934, pp. 1–7.

[129] Frank, F., "Die Färbung der Vogelfeder durch Pigment und Struktur," *Journal für Ornithologie*, Vol. 87, No. 3, 1939, pp. 426–523.

[130] Frank, F., and H. Ruska, "Übermikroskopische Untersuchung der Blaustruktur der Vogelfeder," *Naturwissenschaften*, Vol. 27, No. 14, 1939, pp. 229–230.

[131] Anderson, T. F., and A. G. Richards, "An Electron Microscope Study of Some Structural Colors of Insects," *Journal of Applied Physics*, Vol. 13, No. 12, 1942, pp. 748–758.

[132] Gentil, K., "Elektronenmikroskopische Untersuchung des Feinbaues schillernder Leisten von *Morpho*-Schuppen," *Zeitschrift für Morphologie und Ökologie der Tiere*, Vol. 38, No. 2, 1942, pp. 344–355.

[133] Greenewalt, C. H., W. Brandt, and D. D. Friel, "The Iridescent Colors of Hummingbird Feathers," *Proceedings of the American Philosophical Society*, Vol. 104, No. 3, 1960, pp. 249–253.

[134] Greenewalt, C. H., W. Brandt, and D. D. Friel, "Iridescent Colors of Hummingbird Feathers," *Journal of the Optical Society of America*, Vol. 50, No. 10, 1960, pp. 1005–1013.

[135] Bernhard, C. G., W. H. Miller, and A. R. Møller, "The Insect Corneal Ripple Array. A Biological Broad-Band Impedance Transformer Acts as an Antireflection Coating," *Acta Physiologica Scandinavica*, Vol. 63, Suppl. 243, 1965, pp. 1–79.

[136] Denton, E. J., and M. F. Land, "Optical Properties of the Lamellae Causing Interference Colours in Animal Reflectors," *Journal of Physiology*, Vol. 191, No. 1, 1967, pp. 23–24.

[137] Bernhard, C. G., G. Gemne, and A. R. Møller, "Modification of Specular Reflexion and Light Transmission by Biological Surface Structure - To See, to Be Seen or Not to Be Seen," *Quarterly Reviews of Biophysics*, Vol. 1, No. 1, 1968, pp. 89–105.

[138] Hinton, H. E., "Some Structures of Insects as Seen with the Scanning Electron Microscope," *Micron*, Vol. 1, No. 2, 1969, pp. 84–108.

[139] Neville, A. C., and S. Caveney, "Scarabaeid Beetle Exocuticle as an Optical Analogue of Cholesteric Liquid Crystals," *Biological Reviews*, Vol. 44, No. 4, 1969, pp. 531–562.

[140] Durrer, H., and W. Villiger, "Schillerradien des Goldkuckucks (*Chrysococcyx cupreus* (Shaw)) im Elektronenmikroskop," *Zeitschrift für Zellforschung und Mikroskopische Anatomie*, Vol. 109, No. 3, 1970, pp. 407–413.

[141] Durrer, H., and W. Villiger, "Schillerfarben der Stare (*Sturnidae*) - Elektronenmikroskopischer Beitrag zur Färbung und Systematik der Farmilie," *Journal für Ornithologie*, Vol. 111, No. 2, 1970, pp. 133–153.

[142] Denton, E. J., and M. F. Land, "Mechanism of Reflexion in Silvery Layers of Fish and Cephalopods," *Proceedings of the Royal Society B*, Vol. 178, No. 1050, 1971, pp. 43–61.

[143] Durrer, H., and W. Villiger, "Schillerfarben von *Euchroma gigantea* (L.): (Coleoptera: Buprestidae): Elektronenmikroskopische Untersuchung der Elytra," *International Journal of Insect Morphology and Embryology*, Vol. 1, No. 3, 1972, pp. 233–240.

[144] Ghiradella, H., et al., "Ultraviolet Reflection of a Male Butterfly: Interference Color Caused by Thin-Layer Elaboration of Wing Scales," *Science*, Vol. 178, No. 4066, 1972, pp. 1214–1217.

[145] Land, M. F., "The Physics and Biology of Animal Reflectors," *Progress in Biophysics and Molecular Biology*, Vol. 24, 1972, pp. 75–106.

[146] Abelès, F., "La détermination de l'indice et de l'épaisseur des couches minces transparentes," *Le Journal de Physique et le Radium*, Vol. 11, No. 7, 1950, pp. 310–314.

[147] Born, M., and E. Wolf, *Principles of Optics: Electromagnetic Theory of Propagation, Interference and Diffraction of Light*, Oxford, UK: Pergamon Press, 1964.

[148] Huxley, A. F., "A Theoretical Treatment of the Reflexion of Light by Multilayer Structures," *Journal of Experimental Biology*, Vol. 48, No. 2, 1968, pp. 227–245.

[149] Yee, K., "Numerical Solution of Initial Boundary Value Problems Involving Maxwell's Equations in Isotropic Media," *IEEE Transactions on Antennas and Propagation*, Vol. 14, No. 3, 1966, pp. 302–307.

[150] Taflove, A., "Application of the Finite-Difference Time-Domain Method to Sinusoidal Steady State Electromagnetic Penetration Problems," *IEEE Transactions on Electromagnetic Compatibility*, Vol. 22, No. 3, 1980, pp. 191–202.

[151] Taflove, A., and S. C. Hagness, *Computational Electrodynamics: The Finite-Difference Time-Domain Method*, Norwood, MA: Artech House, 2005.

[152] Bathe, K. J., *Finite Element Procedures*, Cambridge: Pearson Education, 2006.

[153] Zienkiewicz, O. C., R. L. Taylor, J. Z. Zhu, *The Finite Element Method: Its Basis and Fundamentals*, Oxford: Butterworth-Heinemann, 2013.

[154] Moharam, M. G., and T. K. Gaylord, "Rigorous Coupled-Wave Analysis of Planar-Grating Diffraction," *Journal of the Optical Society of America*, Vol. 71, No. 7, 1981, pp. 811–818.

[155] Pendry, J. B., "Photonic Band Structures," *Journal of Modern Optics*, Vol. 41, No. 2, 1994, pp. 209–229.

[156] Vigneron, J.-P., et al., "Theory of Electromagnetic Energy Transfer in Three-Dimensional Structures," *Ultramicroscopy*, Vol. 61, No. 1-4, 1995, pp. 21–27.

[157] Yablonovitch, E., "Inhibited Spontaneous Emission in Solid-State Physics and Electronics," *Physical Review Letters*, Vol. 58, No. 20, 1987, pp. 2059–2062.

[158] John, S., "Strong Localization of Photons in Certain Disordered Dielectric Superlattices," *Physical Review Letters*, Vol. 58, No. 23, 1987, pp. 2486–2489.

[159] Joannopoulos, J. D., et al., *Photonic Crystals: Molding the Flow of Light*, Princeton, NJ: Princeton University Press, 2008.

[160] Ho, K. M., C. T. Chan, and C. M. Soukoulis, "Existence of a Photonic Gap in Periodic Dielectric Structures," *Physical Review Letters*, Vol. 65, No. 25, 1990, pp. 3152–3155.

[161] Yablonovitch, E., T. J. Gmitter, and K. M. Leung, "Photonic Band Structure: The Face-Centered-Cubic Case Employing Nonspherical Atoms," *Physical Review Letters*, Vol. 67, No. 17, 1991, pp. 2295–2298.

[162] Krauss, T. F., R. M. De La Rue, and S. Brand, "Two-Dimensional Photonic-Bandgap Structures Operating at Near-Infrared Wavelengths," *Nature*, Vol. 383, No. 6602, 1996, pp. 699–702.

[163] Foresi, J. S., et al., "Photonic-Bandgap Microcavities in Optical Waveguides," *Nature*, Vol. 390, No. 6656, 1997, pp. 143–145.

[164] Boroditsky, M., et al., "Light Extraction from Optically Pumped Light-Emitting Diode by Thin-Slab Photonic Crystals," *Applied Physics Letters*, Vol. 75, No. 8, 1999, pp. 1036–1038.

[165] Binnig, G., and H. Rohrer, "Scanning Tunneling Microscopy," *IBM Journal of Research and Development*, Vol. 30, No. 4, 1986, pp. 355–369.

[166] Kroto, H. W., et al., "C60: Buckminsterfullerene," *Nature*, Vol. 318, No. 6042, 1985, pp. 162–163.

[167] Parker, A. R., "Discovery of Functional Iridescence and Its Coevolution with Eyes in the Phylogeny of Ostracoda (Crustacea)," *Proceedings of the Royal Society B*, Vol. 262, No. 1365, 1995, pp. 349–355.

[168] Parker, A. R., D. R. Mckenzie, and M. C. J. Large, "Multilayer Reflectors in Animals Using Green and Gold Beetles as Contrasting Examples," *Journal of Experimental Biology*, Vol. 201, No. 9, 1998, pp. 1307–1313.

[169] Parker, A. R., "The Diversity and Implications of Animal Structural Colours," *Journal of Experimental Biology*, Vol. 201, No. 16, 1998, pp. 2343–2347.

[170] Vukusic, P., J. R. Sambles, and C. R. Lawrence, "Colour Mixing in Wing Scales of a Butterfly," *Nature*, Vol. 404, No. 6777, 2000, p. 457.

[171] Large, M. C. J., et al., "The Mechanism of Light Reflectance in Silverfish," *Proceedings of the Royal Society A*, Vol. 457, No. 2007, 2001, pp. 511–518.

[172] Parker, A. R., et al., "Aphrodite's Iridescence," *Nature*, Vol. 409, No. 6816, 2001, pp. 36–37.

[173] Stavenga, D. G., "Colour in the Eyes of Insects," *Journal of Comparative Physiology A*, Vol. 188, No. 5, 2002, pp. 337–348.

[174] Biró, L. P., et al., "Role of Photonic-Crystal-Type Structures in the Thermal Regulation of a Lycaenid Butterfly Sister Species Pair," *Physical Review E*, Vol. 67, No. 2, 2003, 021907.

[175] Stavenga, D. G., et al., "Butterfly Wing Colours: Scale Beads Make White Pierid Wings Brighter," *Proceedings of the Royal Society B*, Vol. 271, No. 1548, 2004, pp. 1577–1584.

[176] Vigneron, J.-P., et al., "Optical Structure and Function of the White Filamentary Hair Covering the Edelweiss Bracts," *Physical Review E*, Vol. 67, No. 1, 2003, 021907.

[177] Vukusic, P., and D. G. Stavenga, "Physical Methods for Investigating Structural Colours in Biological Systems," *Journal of the Royal Society Interface*, Vol. 6, Suppl. 2, 2009, pp. S133–S148.

[178] Parker, A. R., and D. R. McKenzie "The Cause of 50 Million-Year-Old Colour," *Proceedings of the Royal Society B*, Vol. 270, Suppl. 2, 2003, S151–S153.

[179] Ghiradella, H., and W. Radican, "Development of Butterfly Scales II. Struts, Lattices and Surface Tension," *Journal of Morphology*, Vol. 150, No. 2, 1976, pp. 279–298.

[180] Ghiradella, H. "Development of Ultraviolet-Reflecting Butterfly Scales: How to Make an Interference Filter," *Journal of Morphology*, Vol. 142, No. 4, 1978, pp. 395–410.

[181] Ghiradella, H., "Structure and Development of Iridescent Lepidopteran Scales: the Papilionidae as a Showcase Family," *Annals of the Entomological Society of America*, Vol. 78, No. 2, 1985, pp. 252–264.

[182] Ghiradella, H., "Structure and Development of Iridescent Butterfly Scales: Lattices and Laminae," *Journal of Morphology*, Vol. 202, No. 1, 1989, pp. 69–88.

[183] Ghiradella, H., "Hairs, Bristles and Scales." In *Microscopic Anatomy of Invertebrates*, pp. 257–287, F. W. Harrison, and M. Locke, (eds.), New York: John Wiley & Sons, 1998.

[184] Merzendorfer, H., and L. Zimoch, "Chitin Metabolism in Insects: Structure, Function and Regulation of Chitinsynthases and Chitinases," *Journal of Experimental Biology*, Vol. 206, No. 24, 2003, pp. 4393–4412.

[185] Saranathan, V., et al., "Structure, Function, and Self-Assembly of Single Network Gyroid ($I4_132$) Photonic Crystals in Butterfly Wing Scales," *Proceedings of the National Academy of Sciences of the United States of America*, Vol. 107, No. 26, 2010, pp. 11676–11681.

[186] Dinwiddie, A., et al., "Dynamics of F-Actin Prefigure the Structure of Butterfly Wing Scales," *Developmental Biology*, Vol. 392, No. 2, 2014, pp. 404–418.

[187] Pomerantz, A. F., et al., "Developmental, Cellular, and Biochemical Basis of Transparency in the Glasswing Butterfly *Greta oto*," 2020, (hal-03012452).

[188] Schultz, T. D., and M. A. Rankin, "Developmental Changes in the Interference Reflectors and Colorations of Tiger Beetles (*Cicindela*)," *Journal of Experimental Biology*, Vol. 117, No. 1, 1985, pp. 111–117.

[189] Onelli, O. D., et al., "Development of Structural Colour in Leaf Beetle," *Scientific Reports*, Vol. 7, 2017, 1373.

[190] Prum, R. O., "Development and Evolutionary Origin of Feathers," *Journal of Experimental Zoology*, Vol. 285, No. 4, 1999, pp. 291–306.

[191] Prum, R. O., and R. H. Torres, "Structural Colouration of Avian Skin: Convergent Evolution of Coherently Scattering Dermal Collagen Arrays," *Journal of Experimental Biology*, Vol. 206, No. 14, 2003, pp. 2409–2429.

[192] Prum, R. O., "Evolution of the Morphological Innovations of Feathers," Vol. 304B, No. 6, 2005, pp. 570–579.

[193] Prum, R. O., et al., "Development of Colour-Producing β-Keratin Nanostructures in Avian Feather Barbs," *Journal of the Royal Society Interface*, Vol. 6, No. S2, 2009, pp. S253–S265.

[194] Maia, R., R. H. F. Macedo, and M. D. Shawkey, "Nanostructural Self-Assembly of Iridescent Feather Barbules through Depletion Attraction of Melanosomes During Keratinization," *Journal of the Royal Society Interface*, Vol. 9, No. 69, 2012, pp. 734–743.

[195] Burg, S. L., and A. J. Parnell, "Self-Assembling Structural Colour in Nature," *Journal of Physics: Condensed Matter*, Vol. 30, No. 41, 2018, 413001.

[196] Saranathan, V., and C. Finet, "Cellular and Developmental Basis of Avian Structural Coloration," *Current Opinion in Genetics & Development*, Vol. 69, 2021, pp. 56–64.

[197] Zhang, L., A. Mazo-Vargas, and R. D. Reed, "Single Master Regulatory Gene Coordinates the Evolution and Development of Butterfly Color and Iridescence," *Proceedings of the National Academy of Sciences of the United States of America*, Vol. 114, No. 40, 2017, pp. 10707–10712.

[198] Brien, M. N., et al., "Phenotypic Variation in *Heliconius erato* Crosses Shows that Iridescent Structural Colour is Sex-Linked and Controlled by Multiple Genes," *Interface Focus*, Vol. 9, No. 1, 2018, 20180047.

[199] Johansen, V. E., et al., "Genetic Manipulation of Structural Color in Bacterial Colonies," *Proceedings of the National Academy of Sciences of the United States of America*, Vol. 115, No. 11, 2018, pp. 2652–2657.

[200] Thayer, R. C., et al., "Structural Color in *Junonia* Butterflies Evolves by Tuning Scale Lamina Thickness," *eLife*, Vol. 9, 2020, e52187.

[201] Judd, D. B., and G. Wyszecki, *Color in Business, Science and Industry*, New York: John Wiley & Sons, 1975.

[202] Chamberlin, G. J., and D. G. Chamberlin, *Colour: Its Measurement, Computation and Application*, London: Heyden International Topics in Science, 1980.

[203] Wright, W. D., "A Re-Determination of the Trichromatic Coefficients of the Spectral Colors," *Transactions of the Optical Society*, Vol. 30, No. 4, 1929, pp. 141–164.

[204] Guild, J., "The Colorimetric Properties of the Spectrum," *Philosophical Transactions of the Royal Society A*, Vol. 230, No. 681-693, 1932, pp. 149–187.

[205] Elias, M., and J. Lafait, *La Couleur : Lumière, Vision et Matériaux*, Paris: Éditions Belin, 2006.

[206] http://www.cie.co.at/, website of *Commission Internationale de l'Éclairage.*

[207] MacAdam, D. L., "Visual Sensitivities to Color Differences in Daylight," *Journal of the Optical Society of America*, Vol. 32, No. 5, 1942, pp. 247–274.

[208] Land, M. F., and D. E. Nilsson, *Animal Eyes*, Oxford: Oxford University Press, 2002.

[209] Barta, A., and G. Horváth, "Why Is It Advantageous for Animals to Detect Celestial Polarization in the Ultraviolet? Skylight Polarization Under Clouds and Canopies is Strongest in the UV," *Journal of Theoretical Biology*, Vol. 226, No. 4, 2004, pp. 429–437.

[210] Marshall, N. J., and J. Oberwinkler, "The Colour World of the Mantis Shrimp," *Nature*, Vol. 401, No. 6756, 1999, pp. 873–874.

[211] Marshall N. J., et al., "Behavioural Evidence for Polarization Vision in Stomatopods Reveals a Potential Channel for Communication," *Current Biology*, Vol. 9, No. 14, 1999, pp. 755–758.

[212] Vorobyev M., and D. Osorio, "Receptor Noise as a Determinant of Colour Thresholds," *Proceedings of the Royal Society B*, Vol. 265, No. 1394, 1998, pp. 351–358.

[213] Vorobyev M., et al., "Colorful Objects through Animal Eyes," *Color Research and Application*, Vol. 26, No. S1, 2001, pp. S214–S217.

[214] Vorobyev M., et al., "Colour Thresholds and Receptor Noise: Behaviour and Physiology Compared," *Vision Research*, Vol. 41, No. 5, 2001, pp. 639–653.

[215] Vorobyev, M., et al., "Colourful Objects Through Animal Eyes," *Color Research and Application*, Vol. 26, No. S1, 2001, pp. S214–S217.

[216] Kelber, A., M. Vorobyev, and D. Osorio, "Animal Colour Vision – Behavioural Tests and Physiological Concepts," *Biological Reviews*, Vol. 78, No. 1, 2003, pp. 81–118.

[217] Petry, H. M., and F. I. Harosi, "Visual Pigments of the Tree Shrew (*Tupaia belangeri*) and Greater Galago (*Galago crassicaudatus*): A microspectrophotometric Investigation," Vision Research, Vol. 30, No. 6, 1990, pp. 839–851.

[218] Jacobs, G. H., J. Neitz, and M. Crognale, "Spectral Sensitivity of Ground Squirrel Cones Measured with ERG Flicker Photometry," *Journal of Comparative Physiology A*, Vol. 156, No. 4, 1985, pp. 503–509.

[219] Bowmaker, J. K., "Evolution of Colour Vision in Vertebrates," *Eye*, Vol. 12, No. 3, 1998, pp. 541–547.

[220] Hunt, D. M., "Cone Visual Pigments in Two Species of South American Marsupials," *Gene*, Vol. 433, No. 1-2, 2009, pp. 50–55.

[221] Palacios, A. G., "Retinal Photoreceptor Arrangement, SWS1 and LWS Opsin Sequence, and Electroretinography in the South American Marsupial *Thylamys elegans* (Waterhouse, 1839)," *Journal of Comparative Neurology*, Vol. 518, No. 9, 2009, pp. 1589–1602.

[222] Gutierrez, E. de A., et al., "Behavioural Evidence of Dichromacy in a Species of South American Marsupial," *Animal Behaviour*, Vol. 81, No. 5, 2011, pp. 1049–1054.

[223] Smith V. C., and J. Pokorny, "Spectral Sensitivity of the Foveal Cone Photopigments between 400 and 500 nm," *Vision Research*, Vol. 15, No. 2, 1975, pp. 161–171.

[224] Stockman, A., D. I. A. MacLeod, and N. E. Johnson, "Spectral Sensitivities of the Human Cones," *Journal of the Optical Society of America A*, Vol. 10, No. 12, 1993, pp. 2491–2521.

[225] Stockman, A., and L. T. Sharpe, "The Spectral Sensitivities of the Middle- and Long-Wavelength-Sensitive Cones Derived from Measurements in Observers of Known Genotype," *Vision Research*, Vol. 40, No. 13, 2000, pp. 1711–1737.

[226] Menzel, R., and W. Backhaus, "Color Vision in Insects." In *Vision and Visual Dysfunction. VI. Perception of Color*, pp. 262–293, P. Gouras, (ed), Houndmills, UK: Macmillan Press, 1991.

[227] Maier, E. J., and J. K. Bowmaker, "Colour Vision in the Passeriform Bird, *Leiothrix lutea*: Correlation of Visual Pigment Absorbance and Oil Droplet Transmission with Spectral Sensitivity," *Journal of Comparative Physiology A*, Vol. 172, No. 3, 1993, pp. 295–301.

[228] Bowmaker, J. K., et al., "Visual Pigments and Oil Droplets from Six Classes of Photoreceptor in the Retina of Birds," *Vision Research*, Vol. 37, No. 16, 1997, pp. 2183–2194.

[229] Hart, N. S., et al., "Visual Pigments, Oil Droplets, Ocular Media and Cone Photoreceptor Distribution in Two Species of Passerine Bird: The Blue Tit (*Parus caeruleus* L.) and the Blackbird (*Turdus merula* L.)," *Journal of Comparative Physiology A*, Vol. 186, No. 4, 2000, pp. 375–387.

[230] Hart, N. S., "The Visual Ecology of Avian Photoreceptors," *Progress in Retinal and Eye Research*, Vol. 20, No. 5, 2001, pp. 675–703.

[231] Hart, N. S., "Vision in the Peafowl (Aves: *Pavo cristatus*)," *Journal of Experimental Biology*, Vol. 205, No. 24, 2002, pp. 3925–3935.

[232] Ödeen, A., O. Håstad, and P. Alström, "Evolution of Ultraviolet Vision in the Largest Avian Radiation - the Passerines," *BMC Evolutionary Biology*, Vol. 11, No. 1, 2011, 313.

[233] Vorobyev M., et al., "Tetrachromacy, Oil Droplets and Bird Plumage Colours," *Journal of Comparative Physiology A*, Vol. 183, 1998, pp. 621–633.

[234] Siddiqi, A., et al., "Interspecific and Intraspecific Views of Color Signals in the Strawberry Poison Frog *Dendrobates pumilio*," *Journal of Experimental Biology*, Vol. 207, No. 14, 2004, pp. 2471–2485.

[235] Guo, Z., et al., "Stable Biomimetic Super-Hydrophobic Engineering Materials," *Journal of the American Chemical Society*, Vol. 127, No. 45, 2005, pp. 15670–15671.

[236] Solga, A., et al., "The Dream of Staying Clean: Lotus and Biomimetic Surfaces," *Bioinspiration & Biomimetics*, Vol. 2, No. 4, 2007, pp. S126–S134.

[237] Lafuma, A., and D. Quere, "Superhydrophobic States," *Nature Materials*, Vol. 2, No. 7, 2003, pp. 457–460.

CHAPTER 2

Diversity and Complexity of Natural Photonic Devices

The scattering of visible light by an object determines its visual appearance, and, in particular, its color. The very first obstacle light encounters is the surface of the object. When the object can be assimilated to a semi-infinite, homogeneous, and isotropic medium, the Fresnel formulas allow us to determine the amount of light that is reflected (as well as transmitted) at the dioptre as a function of the wavelength, angle of incidence, and state of polarization of the light. The simplest extension of Fresnel's treatment is the case of a double interface; namely, a thin film. Thin films of appropriate refractive index and thickness, such as in soap bubbles or oil drops, are well-known to produce rainbow colors as a result of light wave interference. Since cuticle layers in some insect species have evolved as the simplest natural color-producing devices, this chapter will first recall basic concepts of thin film optics and then extend the analysis to the case of multilayers. The particular case of the periodic multilayer stack, also called Bragg mirror or one-dimensional photonic crystal, will be discussed, since it is the second natural color-producing device, by increasing order of complexity, and, probably, by the number of its occurrence among species. With a similar complexity, diffraction gratings, which can take the form of 1D or 2D surface relief gratings, will be described. We will end the chapter by increasing the complexity with two- and three-dimensional photonic crystals and consider disorder in natural photonic structures.

2.1 Thin Films: The Simple Beauty

It is always amazing for children to discover how easy it is to produce a rainbow from water and soap. Even if we have replaced in our adult's mind the mystery of the soap bubble's rainbow colors by their physical explanation[1], as humble disciples of Hooke, Newton, and later Fresnel, we may still stare at the simple beauty of thin film colors.

2.1.1 Light Wave Interference Pattern and Film Color

The key concept behind thin film colors is light wave interference. We will start this section by recalling the basic results of thin film theory [2, 3]. Let us consider a homogeneous thin film of thickness d and refractive index n, which is either suspended in air or deposited on a semi-infinite substrate (Figure 2.1). First, the

1. Actually, an accurate description of interference colors of soap bubbles may turn out to be quite involved [1]. For instance, the film thickness is not uniform along the sphere because of the gravitational effect.

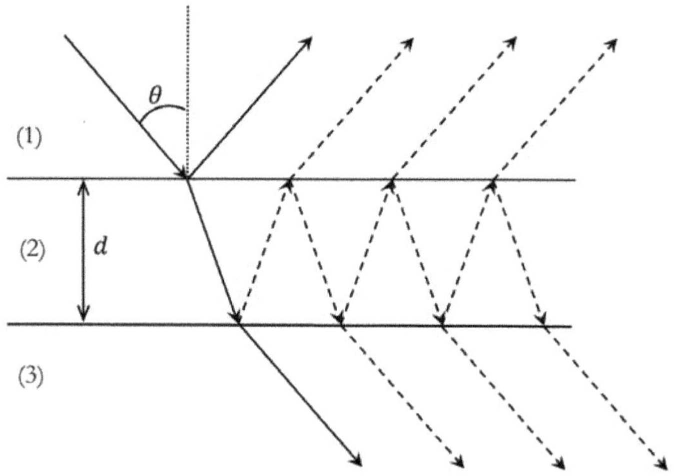

Figure 2.1 Ray-tracing sketch of light wave interference in a thin film. Solid-line rays indicate the shortest pathways followed by incoming light as it reflects or transmits across the film, whereas dashed-line rays indicate the infinite set of increasingly longer pathways, due to multiple bounces at both interfaces. The incidence angle is denoted by θ. The film thickness and refractive index are denoted by d and n, respectively. The labels (1), (2), and (3) refer to the semi-infinite incidence medium, the film, and the semi-infinite emergence medium, respectively.

reader may wonder what does "thin" mean precisely. In fact, the film thickness must be small enough that light waves retain a sufficiently high level of coherence along their propagation path.[2] In this case, interference phenomenon can occur, for instance, between the wave that is reflected at the first interface and the waves that pass through that interface after multiple bounces inside the film (Figure 2.1).

Depending on the film thickness, the wavelength and the angle of incidence of the light, these multiple reflected waves interfere either constructively or destructively. This interference manifests itself as maxima and minima in the spectra of the reflected and transmitted light. These maxima and minima are regularly spaced when the reflectance and transmittance are plotted as a function of the angular frequency ω or the wavenumber $k = \frac{\omega}{c}$ (where c stands for the speed of light in free space) (Figure 2.2(a)). If the wavenumbers corresponding to these extrema are plotted as a function of their sequence number (Figure 2.2(c)), they all align along a straight line and the film thickness can be deduced from the slope once the refractive index is known [3]. The state of polarization of the incident light matters as soon as it impinges at non-zero angles θ from the direction perpendicular to the film. In this situation, we must distinguish between the cases of s-polarized light and p-polarized light.[3] The peculiar case of an incidence close to the Brewster's angle, where the

2. We will follow Stenzel's definition [3] of a *thin* film by the condition imposed on its thickness: $d \ll \frac{\lambda^2}{2\pi n \Delta\lambda}$, where $\Delta\lambda$ is the spectral width of the light whose central wavelength is λ and n is the film refractive index. When applied to interferential colors ($\Delta\lambda = 700 - 400 = 300$ nm, $\lambda = (700 + 400)/2 = 550$ nm), this definition leads to a rather stringent condition on the film thickness: typically $d < 107$ nm for $n = 1.5$. In natural "thin" film devices, the thickness is of the order of several hundreds of nanometres, which means that decoherence effects cannot be neglected at visible light wavelengths [4].
3. The notations s and p refers to, respectively, transverse electric (TE) and transverse magnetic (TM) waves, meaning that the electric field and the magnetic field, respectively, oscillate perpendicularly and parallelly to the plane of incidence. These letters stand for the German *senkrechte und parallele Polarisationen*.

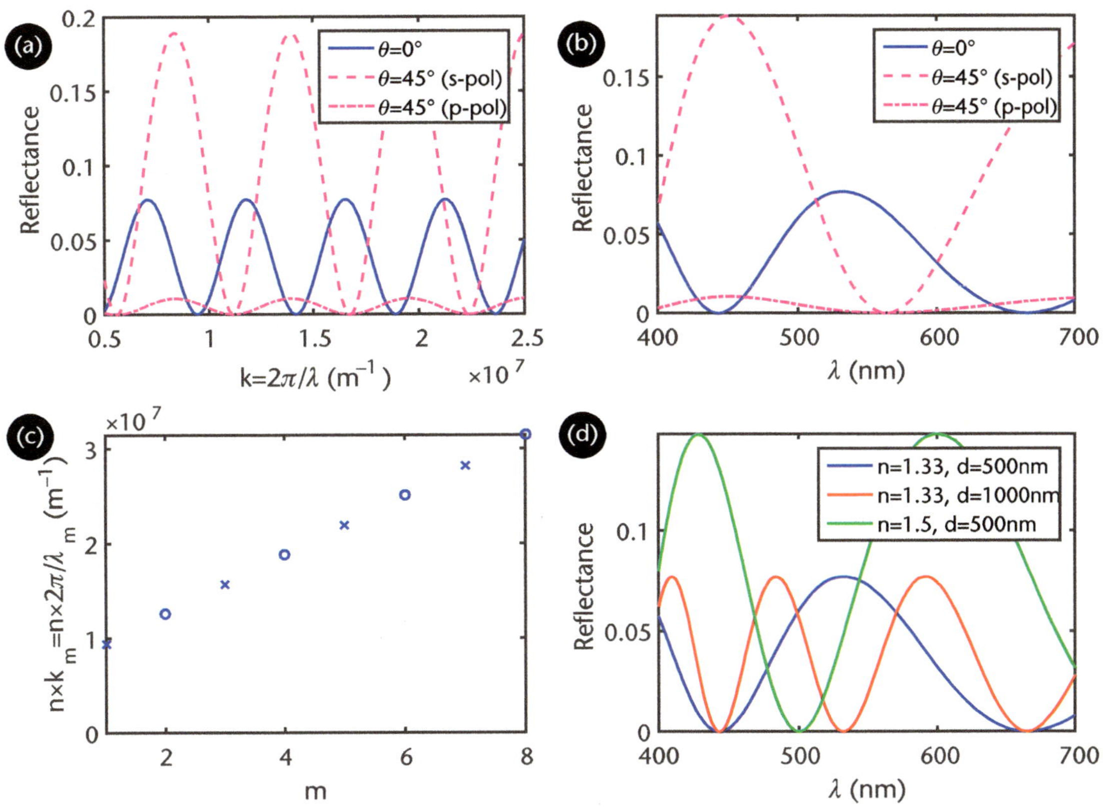

Figure 2.2 Reflectance spectra of a thin film. (a) Reflectance as a function of the free-space wavevector k in the case of a thin film ($d = 500$ nm, $n = 1.33$) suspended in air, at normal ($\theta = 0°$) incidence (solid curve) and oblique ($\theta = 45°$) incidence for s-polarized light (dashed curve) and p-polarized light (dashed-dotted curve). The Brewster's angle is equal to $\theta_B = \arctan(n) = 56.3°$. (b) The same spectra as in (a) but plotted as a function of the free-space wavelength λ in the visible range of the electromagnetic spectrum. (c) Wavevector values corresponding to maxima (crosses) and minima (circles) of the reflectance spectrum at normal incidence displayed in (a), as a function of their sequence number m. (d) Reflectance spectra calculated at normal incidence for thin films differing by their refractive index or thickness (blue curve: $n = 1.33, d = 500$ nm, green curve: $n = 1.33, d = 1000$ nm, red curve: $n = 1.5, d = 500$ nm). Refractive index values shown are typical of water (1.33) and a biopolymer, such as chitin (1.5).

reflectance tends to vanish for p-polarized light, is illustrated in Figure 2.2(a,b). Also notice the shift of the interference pattern toward higher frequencies (wavevectors) with increasing incidence angle: its effect on the film color will be described hereafter. At normal incidence ($\theta = 0°$), however, it is no longer relevant to distinguish between polarizations, since the s- and p-states are degenerated. When the reflectance is plotted as a function of the wavelength ($\lambda = \frac{2\pi}{k}$), the distance between successive peaks increases with increasing wavelength (Figure 2.2b). The number of peaks and their positions within the visible range ($\lambda \in$ [400–700 nm]) determines the thin film color. In the case shown in Figure 2.2(b), the peak around 550 nm gives the film a green color at normal incidence. Increasing the angle of incidence shifts the interference pattern to shorter wavelengths and, therefore, changes the film color, as the peaks shift to the visible range: this effect accounts for the iridescence behavior introduced in the previous chapter. A point must be made here about determining the film's optical response from calculated refectance spectra. Colors in nature are observed under sunlight, whose state of polarization is mostly random. In order to take this important aspect into account, reflectance spectra must be calculated for both *s*-polarization and *p*-polarization (except at normal incidence, where they are identical), and the arithmetic mean of these spectra must be considered to determine the color.[4] Thin film theory also predicts that the interference pattern, hence the film color, changes as the film thickness or refractive index are varied (Figure 2.2(d)).

The mathematical treatment of thin film interference is indebted to Sir Georges B. Airy [5] and is based on Fresnel formulas. Taking into account the successive reflections and refractions the light wave experiences at the first and second interfaces, the infinite set of multiple reflected waves, whose pathways are depicted in Figure 2.1 is summed; the converging series gives the complex reflection coefficient for the electric field r_{123}. The subscript 123 denotes the system that is formed by the incidence medium (1), the film (2), and the emergence medium (3). For either *s*-polarized or *p*-polarized light, the reflectance is obtained by taking the square modulus of the corresponding coefficient: $R = |r_{123}|^2$. In the case of nonabsorbing materials,[5] the reflectance is expressed by [3]:

$$R = |r_{123}|^2 = \frac{r_{12}^2 + r_{23}^2 + 2r_{12}r_{23}\cos 2\delta}{1 + r_{12}^2 r_{23}^2 + 2r_{12}r_{23}\cos 2\delta}, \tag{2.1}$$

where r_{ij} is the Fresnel reflection coefficient for light impinging from medium (i) onto medium (j), and $\delta = \frac{2\pi}{\lambda} nd\sqrt{n^2 - \sin^2\theta}$ is the phase shift due to a single-pass propagation across the film. This formula was used to produce the spectra shown in Figure 2.2. If light propagates normally to the interfaces, ($\theta = 0°$), completely constructive (destructive) interference takes place when the double-pass phase shift (2δ) is an even (odd) integer multiple of π, which corresponds to maximal (minimal) reflectance, as it can be easily checked from (2.1). The periodicity introduced by $\cos 2\delta$ in the frequency (wavevector) domain response explains the regular spacing of the reflectance extrema (Figure 2.2(a,c)). Furthermore, the dependence of δ on the film parameters (n, d) as well as on the incidence angle (θ) easily explains the corresponding modifications of the interference pattern (Figure 2.2(d)).

2.1.2 Thin Film on Thick Slab

The above discussion pertained to a thin film suspended in air. Of course, the same formula, (2.1), can be applied to calculate the reflectance of a thin film deposited on a substrate simply by using the appropriate refractive index (n_3). However, this substrate was assumed to be semi-finite. An actual substrate has indeed a finite thickness and reflection at the back side that modifies the reflectance of the film-substrate system. This effect can be taken into account by extending Fresnel's treatment to

4. This averaging procedure is justified by the fact that sunlight is, on average, half of the time polarized in the *s*-state and half of the time polarized in the *p*-state.
5. For nonabsorbing materials, all Fresnel coefficients r_{ij} and the phase shift δ are real. In the case of absorbing materials, these are complex numbers and the reflectance must be calculated using the most general formula $R = |r_{123}|^2$.

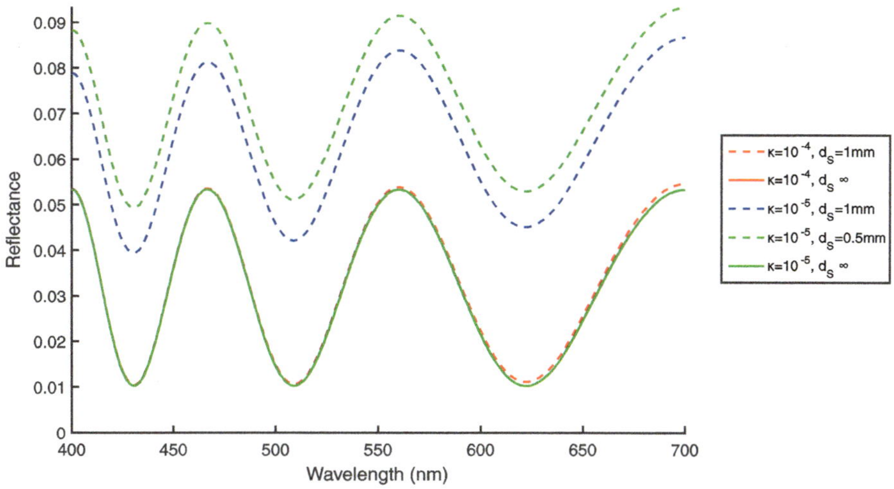

Figure 2.3 Reflectance spectra (at normal incidence) of a thin film ($n = 1.4$, $d = 1000$ nm) lying on a substrate that exhibits absorption ($n_S \equiv \Re(n_S) = 1.6$, red curves: $\kappa \equiv \Im(n_S) = 10^{-4}$, blue and green curves: $\kappa \equiv \Im(n_S) = 10^{-5}$). Solid curves were calculated, assuming a semi-infinite substrate, whereas dashed curves were caculated for a finite substrate thickness (red and blue curves: $d_S = 1$ mm, green curve: $d_S = 0.5$ mm) and assuming incoherent light wave superposition. For both values of κ chosen here, the reflectance spectrum is almost unchanged if the substrate is assumed to be infinitely thick.

a two-layer system made of the thin film on top of a thick slab. The first layer is treated as before (waves superimpose coherently in the thin film). The second one, however, is assumed to be so thick that waves superimpose incoherently in the substrate. Allowing both the film and substrate materials to be absorbing,[6] the reflectance is now expressed by [3]:

$$R = |r_{123}|^2 + \frac{|t_{123}|^2|r_{31}|^2|t_{321}|^2 e^{-4\Im(\delta_S)}}{1 - |r_{321}|^2|r_{31}|^2 e^{-4\Im(\delta_S)}}, \tag{2.2}$$

where $\Im$ denotes the imaginary part, δ_S is associated with the substrate of finite thickness d_S, and the other parameters have the same meaning as earlier.

Let us consider a thick substrate made of a material that is not completely transparent. As long as the product of the absorption coefficient by the substrate thickness is large enough, we can expect that the reflectance spectra will not differ significantly when calculated, either by assuming a semi-infinite substrate or a finite substrate. This is indeed true, as illustrated in Figure 2.3 for $\Im(n_S) = 10^{-4}$ with $d_S = 1$ mm. When the substrate is either less absorbing or less thick, its effect is to increase the reflectance level through the contribution of the reflection at the back side, as illustrated in Figure 2.3. Moreover, since the substrate is absorbing and treated incoherently, the thinner it is, the higher the contribution of the back side to the reflectance (compare curves for $\Im(n_S) = 10^{-5}$ with $d_S = 1$ mm or $d_S = 0.5$ mm).

6. Material absorption is accounted for by adding an imaginary part to the refractive index, which becomes a complex quantity: $\tilde{n} = n + i\kappa$. The absorption coefficient (m^{-1}) is defined by: $\alpha = \frac{4\pi}{\lambda}\kappa$.

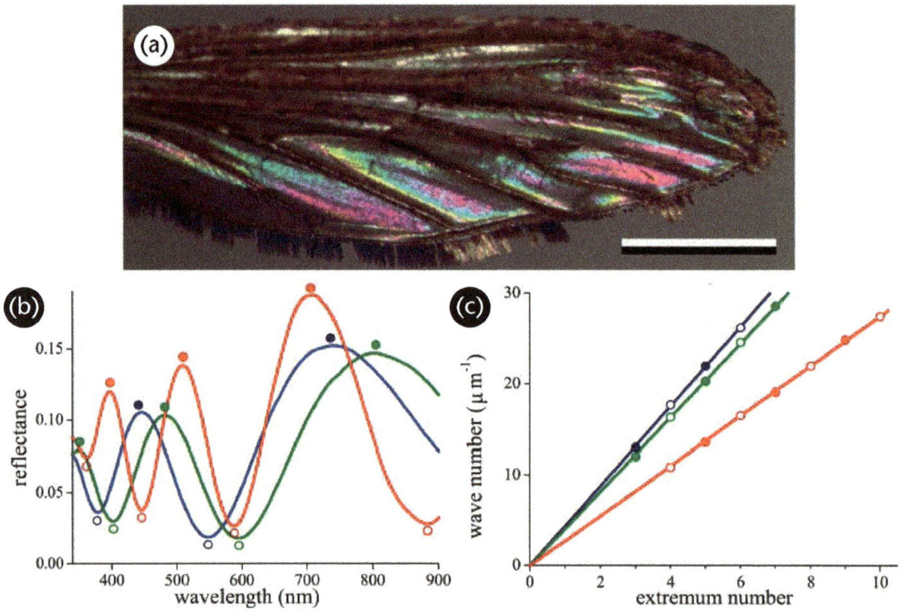

Figure 2.4 Thin film structure in the wing of a mosquito. The wing exhibits distinct interference colors (a). Reflectance spectra measured at three different locations on the wing (b). Plot of wavenumbers ($\frac{1}{\lambda}$) of the reflectance extrema (closed symbols: maxima, open symbols: minima) as a function of their sequence number, together with a linear fit to data (c). Scale bar: 1 mm. (Reproduced from [6], with permission from Elsevier.)

2.1.3 Natural Thin Film Devices

Although examples of interference colors are numerous in insects and birds, it is not easy to isolate the contribution of a thin film, since it may rest upon a layered substrate, or come together with a diffraction grating, for instance.

The wing of a mosquito is quite a rare example of a thin film suspended in air (Figure 2.4). The wing membrane does not contain any pigments and acts as a transparent thin film. Because the membrane has a variable thickness according to its position on the wing, the typical oscillations in the reflectance spectra shift accordingly and distinct colors are observed. Thickness variations (from about 360 nm to about 570 nm) can be estimated from linear fits to the wavenumbers of reflectance extrema [6].

The opaque and iridescent wing of the giant tropical wasp *Megascolia procer javanensis* is a striking example of a thin film lying on a thick absorbing substrate (Figure 2.5) [7]. The iridescent bluish-green color arises from a uniformly thin ($\simeq$ 300 nm) cuticle layer covering the bulk of the wing, which contains the melanin pigment responsible for opacity. Although melanin absorption does not play a role in determining the hue here, it does provide the wing with a dark background, which makes the perception of the iridescent color easier. The cuticle layer and the wing substrate both consist of chitin, but the former is assumed to lack melanin, unlike the latter. Given uncertainties in the refractive indices, it is worth noting the good agreement between measurements and simulations of the reflectance spectra at incident angles ranging from 0° to 60°.

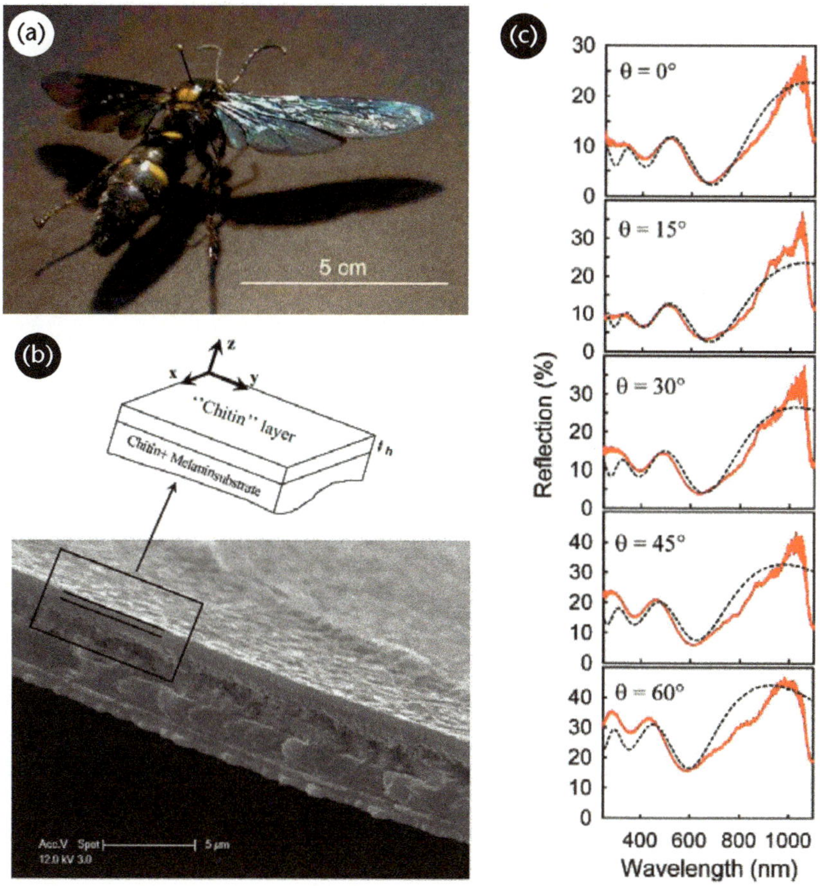

Figure 2.5 Thin film device in the wing of a wasp. The dark wing displays iridescent bluish-green colors at oblique viewing angles (a). Thin-film model used for simulations ($h = 286$ nm) and built from cross-sectional SEM image of the wing (b). Reflectance spectra at various incidence angles (c): measurements (solid red curve) and simulations (dashed black curves). (Reproduced from [7], with permission from the American Physical Society.)

In butterflies, structural colors are usually produced by the scales covering the wing. Therefore, the wing membrane is usually not considered to elucidate the coloration. However, the swordtail butterfly *Graphium sarpedon* is an exception, with its glass scales (Figure 2.6). These scales actually consist of an upper and lower lamina joined together to form a thin film, the thickness of which is about 400 nm, leading to interference in the visible range of the electromagnetic spectrum.

2.2 Multilayers: More Layers, More Effects

A stack of planar layers is a step further toward complexity beyond the thin film. If we return to the ray-tracing figure considered above (Figure 2.1), additional interfaces must now be considered, each creating both reflected and refracted (transmitted) waves, according to Snell-Descartes laws. In principle, the same treatment used above can be applied in order to find the complex reflection coefficient of a

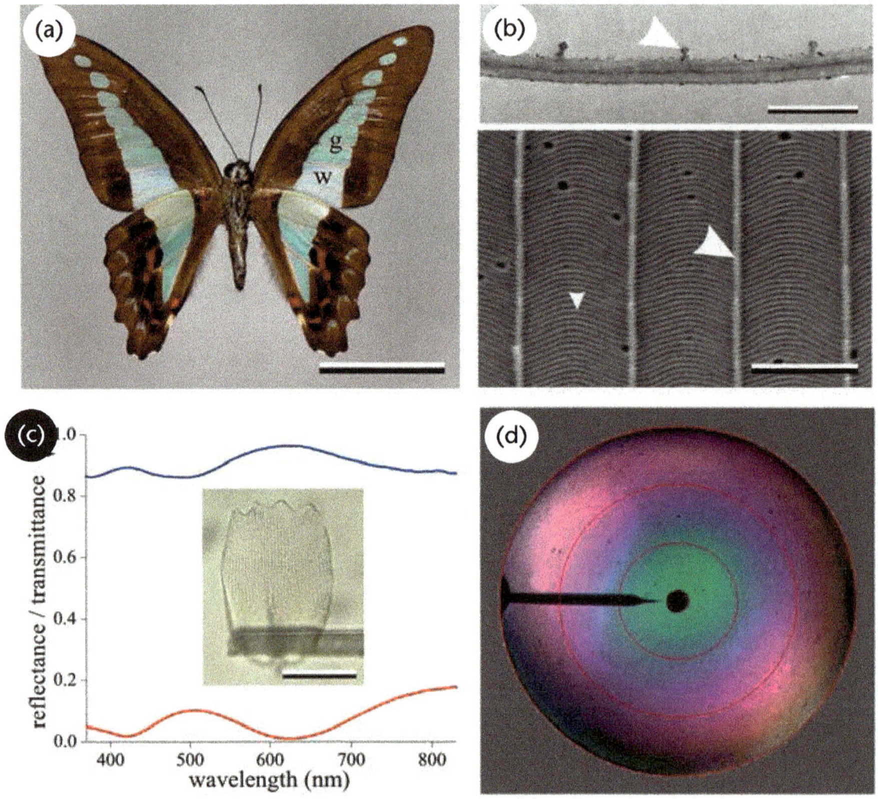

Figure 2.6 Thin film device in the wing of a butterfly. Ventral wing of the swordtail butterfly *Graphium sarpedon* (a), with colored patches of either nonpigmented glass (labeled *g*) scales or white (labeled *w*) scales. TEM and SEM images of glass scales, which consist of an upper lamina with small ridges (large arrowhead) that is interconnected by cross ribs (small arrowhead) and an opposed lower lamina; together they form a thin film (b). Transmittance and reflectance spectra (c) of a glass scale (inset). Scatterogram of a glass scale showing interference colors (d). Scale bars not provided in the original work. (Reproduced from [6], with permission from Elsevier.)

N-layer system from which the reflectance is deduced; that is, $R = |r_{12...N}|^2$. This approach becomes very tedious as the number of interfaces increases, since more and more multiple bounces and pathways must be considered in the summation leading to the reflected electric field expression. In his 1949 thesis, Florin Abelès introduced the concept of optical impedance and used matrix algebra for solving this problem [8]. This transfer matrix formalism has been widely used subsequently [9]. Essentially, each layer is characterized by a four-by-four matrix, which connects electric fields of forward- and backward-propagating waves at the layer interfaces. Reflectance R, transmittance T, and absorptance $A = 1 - R - T$ are deduced from the matrix of the multilayer system, which is obtained by matrix multiplication in the order imposed by the layer stack. An alternative method that is indebted to our colleagues Vigneron and Lambin, derives a continued fraction expression of R and T by exploiting Abelès' impedance concept [10]. Both methods solve Maxwell's equations exactly and rely on the expression of electromagnetic boundary conditions at interfaces. When applied to a single layer, they, of course, lead to the same result as the Airy's formula; see (2.1). In general, they can be applied to any arbitrarily stratified medium ($0 < z < z_N$), provided it is divided into layers within which the permittivity profile $\varepsilon(z)$ and the permeability profile $\mu(z)$ of refractive index: $\tilde{n} = \sqrt{\varepsilon\mu}$ can be considered to be constant across the layer thicknesses $z \in [z_j, z_{j+1}]$. The incidence medium ($z < 0$) is usually air and the emergence medium ($z > z_N$) is either air or any substrate material—either transparent, or absorbing at the wavelengths of interest. The calculated quantities depend upon the state of polarization of the incident light, as well as on the incidence angle, except at normal incidence, for which they are identical, as already mentioned for the case of a thin film.

A particular and important instance of stratified media is the periodic multilayer system—the Bragg stack, which is obtained by stacking two distinct layers (thicknesses: d_1, d_2, refractive indices: n_1, n_2) according to a repeated sequence of N bilayers [9]. The bilayer thickness, $a = d_1 + d_2$, defines the spatial period of the Bragg stack, and the ratio $c = \frac{n_2}{n_1} = \sqrt{\frac{\varepsilon_2}{\varepsilon_1}}$ denotes the refractive index contrast. Because of the periodicity along the longitudinal direction (i.e., perpendicular to the interfaces) and the spatial invariance along the lateral direction, the Bragg stack can be regarded as a finite-size realization of a one-dimensional photonic crystal [11]: a concept that we will use hereafter to derive the asymptotic value ($N \to \infty$) of the Bragg wavelength. This rephrasing is not just fancy terminology, but brings with it fruitful interpretations such as those related to Brillouin zone, photonic bandgap, and density of optical states (DOS). Along these lines, Vigneron introduced in 2006 the useful asymptotical concept of *dominant reflected wavelength*, λ_B, in order to capture the essential characteristic of natural periodic multilayer devices (i.e., iridescence) [12]. The dominant reflected wavelength is derived from a long-wavelength (nonretarded) approximation of the asymptotic 1D photonic crystal [12] and expressed by the simple formula

$$\lambda_B = \frac{2a\sqrt{\bar{n}^2 - \sin^2\theta}}{m}, \tag{2.3}$$

where a is the spatial period of the refractive index perturbation (contrast assumed to be small), θ is the incidence angle, m is an integer (defining the fundamental Bragg peak at $m = 1$ and its harmonics) and $\bar{n}$ is the average real refractive index in the long-wavelength approximation.[7] The latter quantity depends on the incident electric field orientation (polarization), since it results from an effective medium approximation of a bilayer system that is macroscopically anisotropic (form birefringence). At normal incidence and for s-polarized light at all angles, the expression of $\bar{\varepsilon} = \bar{n}^2$ is reminiscent of parallel capacitance assembly:

$$\bar{n}^2 = \frac{d_1 n_1^2 + d_2 n_2^2}{d_1 + d_2}. \tag{2.4}$$

As the angle becomes large and for p-polarized light, the expression of $\bar{\varepsilon}$ is, however, reminiscent of series capacitance assembly

$$\frac{1}{\bar{n}^2} = \frac{d_1 \frac{1}{n_1^2} + d_2 \frac{1}{n_2^2}}{d_1 + d_2}. \tag{2.5}$$

7. Actually, any valid approximation can be used to calculate $\bar{n}$, such as the Bruggeman effective medium approximation [13, 14]. The choice of the approximation must be evaluated with care in each case.

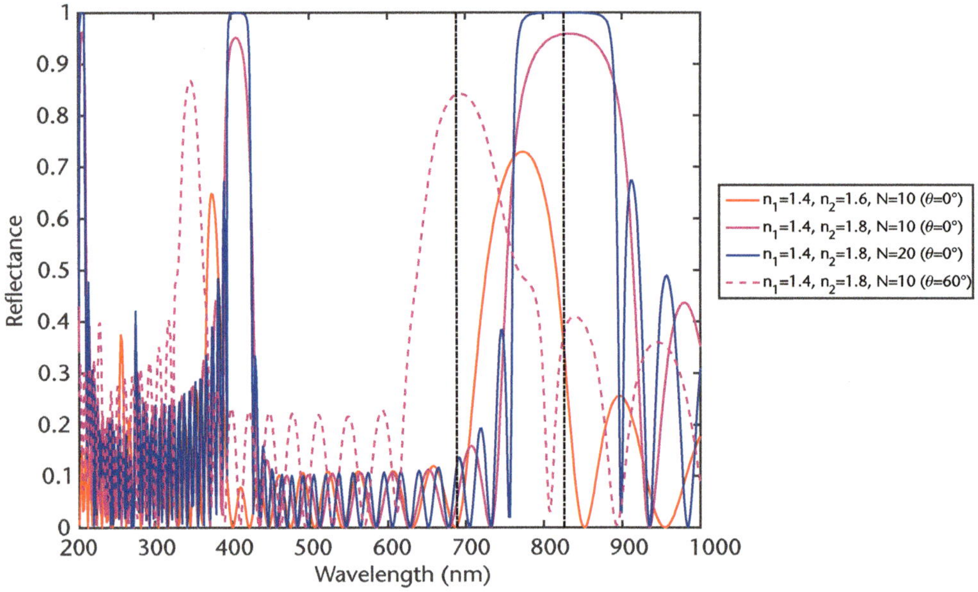

Figure 2.7 Reflectance spectra of periodic multilayer stacks ($d_1 = 100$ nm, $d_2 = 150$ nm). Spectra were calculated for a refractive index contrast equal to $c = \frac{n_2}{n_1} = \frac{1.8}{1.4}$ or $c = \frac{n_2}{n_1} = \frac{1.6}{1.4}$ and for $N = 10$ or $N = 20$ periods, at normal ($\theta = 0°$) or oblique ($\theta = 60°$) incidences. At oblique incidence, the reflectance is the average between the *s*-polarization and *p*-polarization responses. Incidence and emergence media were air. The dominant reflected wavelength (vertical dashed-dotted lines) was predicted using (2.3). At normal incidence, the average refractive index was obtained from (2.4) ($\bar{n} = 1.65$); whereas, at oblique incidence, it was obtained by taking the arithemic mean between the values obtained from (2.4) and (2.5) ($\bar{n} = 1.62$).

The dominant reflected wavelength predicts the actual position of the Bragg reflectance peak quite well, with increasing accuracy as the number of periods increases, of course (Figure 2.7). Formula (2.3) is therefore useful to predict the iridescence of Bragg reflectors (hues at various incidence angles).

The calculation of the reflectance allows us to analyze in more detail the spectral response of periodic stacks [9, 11]. We will briefly summarize hereafter their well-known properties with a few examples (Figure 2.7). In the absence of material absorption and for a given refractive index contrast, the reflectance peak value tends to unite when the number of periods becomes sufficiently large. In other words, the stack tends to behave as a perfect reflector around λ_B, which justifies the widespread use of wavelength-selective Bragg mirrors. Note that perfect reflection ($R = 100\%$) is the limit case that results from the existence of a photonic bandgap—namely, a frequency range where the DOS is vanishing and light cannot propagate in the infinite crystal. As the refractive index contrast increases, the almost unitary reflectance is reached with fewer periods and the reflectance acquires more rapidly a wide and flat-topped spectrum. The iridescence is explained by the shift of λ_B (photonic bandgap) toward shorter wavelengths (higher frequencies) as the incidence angle increases. Reminiscent of a thin film effect are the oscillations between the Bragg resonance harmonics, which are due to the effective Fabry-Pérot cavity formed by the finite size of the stack.

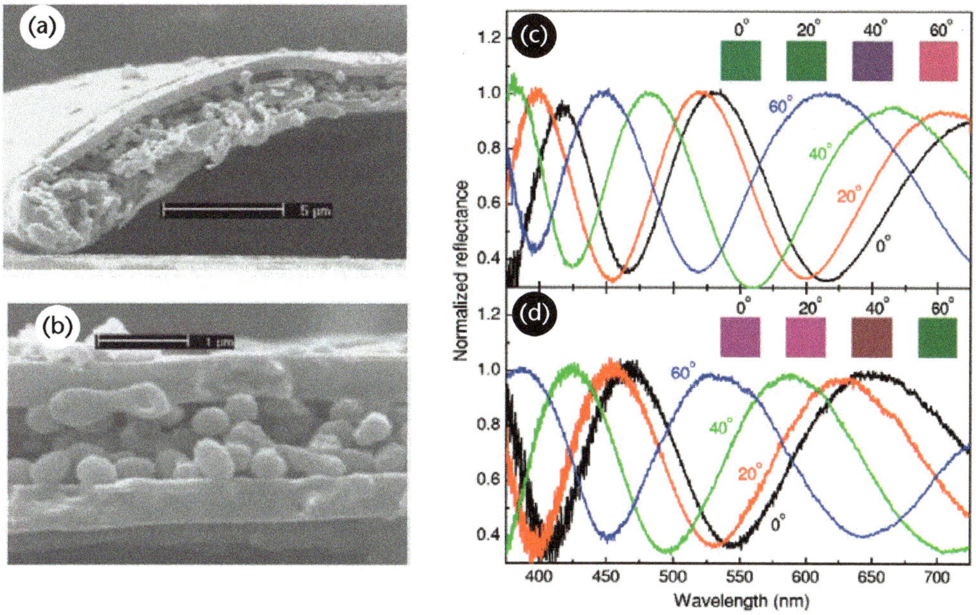

Figure 2.8 Thin-film interference colors in the neck feathers of domestic pigeons. SEM images of cross sections of a green barbule (a) and a purple barbule (b). Reflectance spectra measured on green (c) and purple (d) feathers at different incidence angles. The corresponding RGB colors are shown in insets. (Reproduced from [16], with permission from the American Physical Society.)

2.2.1 Few-Layer Systems in Natural Organisms

There are cases of natural multilayer devices made of only a few (two or three) layers, which behave roughly as an effective thin film in terms of interference color production. The iridescence of green and purple neck feathers of domestic pigeons, observed by Merritt a century ago [15], is a good example of this kind of configuration. The barbules, which are made of a smooth keratin cortex surrounding a medullary layer (melanin granules), reportedly form a three-layer system [16]. However, the absorbing medullary layer acts as a poor mirror: the color is primarily due to thin-film interference in the top cortex layer, the mean thickness of which is about 595 nm for green barbules and 530 nm for purple barbules (Figure 2.8). The resulting reflectance fringe patterns are such that, with increasing angles from normal to grazing incidence, the purple color changes to green in one barbule type and vice versa in the other type, producing a kind of opposite iridescence in both type of barbules (Figure 2.8).

Flowers of many buttercups (*Ranunculus* species) exhibit a glossy yellow coloration that is attributed to a flat and smooth pigmented epidermal layer acting as a mirror. In 2011, Vignolini and coworkers performed anatomical observations and optical measurements on *Ranunculus repens*. They found a multilayer structure within the adaxial epidermis of the petals that consisted of a 5-μm-thick upper epidermal layer containing carotenoid pigment and a 7-μm-thick starch layer. These two layers are separated by an approximately 0.5-μm-thick discontinuous air gap [17]. As an explanation for the directional scattering, they proposed that the upper layer was responsible for a mirrorlike reflection, while the underneath inhomogeneous layers scattered light, hence producing a more diffused yellow color. They further suggested that the air gap could provide a second planar interface for specular reflection. Extending the study to several buttercup species, Stavenga's group confirmed the previous findings through anatomical observations and optical measurements, as well as through detailed simulations of the whole multilayer system [18]. They clearly identified thin-film oscillations due to the upper epidermis layer and confirmed that this air-suspended thin film served as a filter for light backscattered by the starch and mesophyll layers. Glossy buttercup flowers give us a nice example of combined structural and pigmentary coloration. It also shows that detailed optical simulations are able to isolate the predominant effect arising from an apparent multilayer system, here the thin-film effect and its filtering action on the backscattered light.

There are cases, however, where interferential colors in a few-layer system cannot be explained solely by a thin-film effect. Stavenga and coworkers reported a seven-layer device in the blue scales of the peacock spider *Maratus splendens* [18]. In these unpigmented scales, two chitinous layers, with ridges on their outer sides and an array of chitinous filaments on their inner sides, are separated by an air gap. The ridges were treated as graded-index layers and the filaments (filling fraction of approximately 0.25) as thin homogenous effective layers. Modeling revealed that the ridge layers did not affect the reflectance spectrum significantly, whereas the filament layers had a major spectral effect in narrowing the blue reflectance band and reducing the far-red band.

2.2.2 Periodic Multilayer Systems

There are numerous examples of periodic multilayer systems in nature that produce metallic colors [12, 19–21]. The adjective "metallic" stands for the strong, highly directional (specular) reflection such systems possess and, although they are only made of dielectric layers, it recalls their visual assimilation to mirrors. Due to their inherent wavelength-selective Bragg-type reflection, however, these systems produce iridescent colors, unlike real metallic mirrors.

The case of the metallic woodboring beetle *Chrysochroa vittata* (Figure 2.9) is noteworthy in the sense that the periodic layer stack found in its cuticle is constructed from thin, dense chitin layers and much thinner and irregular air gaps [12]. Numerical simulations have been conducted on the structure observed on the ventral side of the insect's abdomen (chitin layer thickness: about 194 nm, air gap thickness: approximately 10 nm, $N = 20$ periods). These employed a refractive index contrast of $c = 1.56$, which is the largest contrast available, considering that the device is built from a single material (assumed to be chitin) and voids (air). The simulations predicted a strong reflectance (approximately 60%), peaking at 632 nm at normal incidence, which explains the observed metallic red color on the insect's ventral side. At grazing incidence ($\theta = 75°$), the reflectance is slightly higher and shifted to about 488 nm, in agreement with the observed metallic bluish-green color. The dominant reflected wavelength *predictions* are consistent with simulations at different incidence angles (less than 5% deviation). This consistency between both predictions consolidates the interpretation of the cuticle structure as a

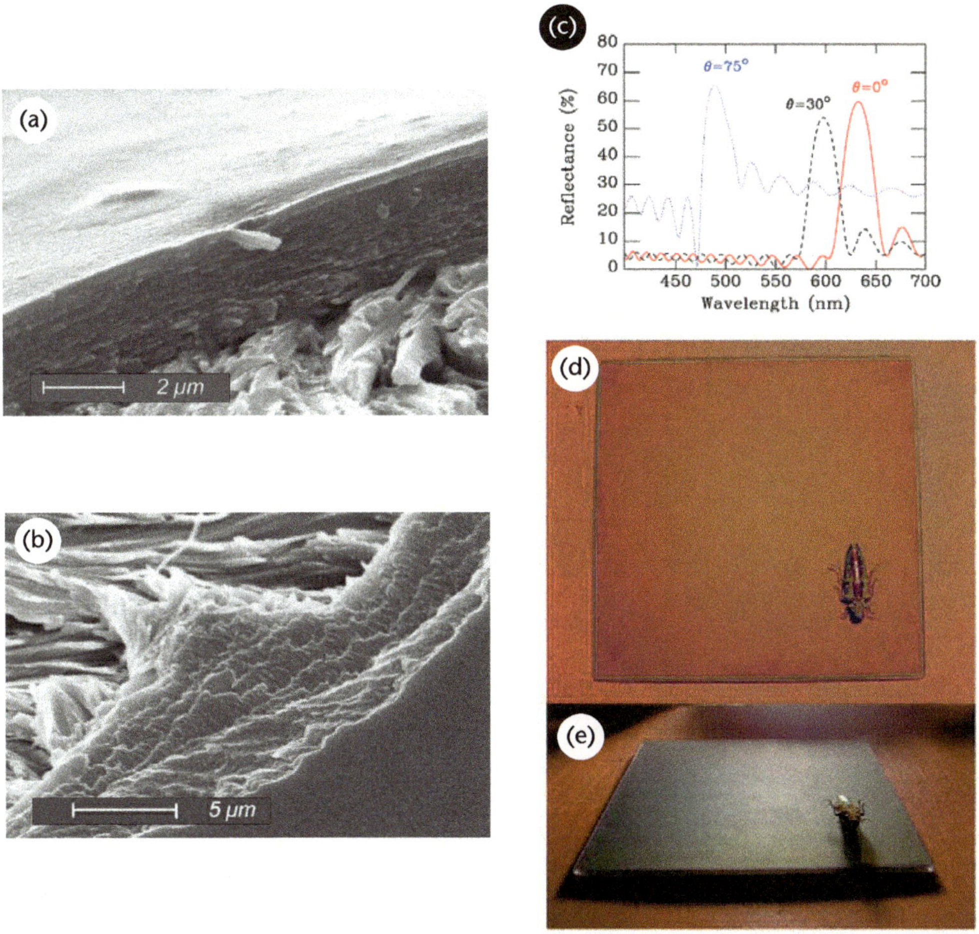

Figure 2.9 Dissymmetric periodic multilayer device in the cuticle of the metallic woodboring beetle *Chrysochroa vittata*. SEM images of cross sections of the cuticle on the ventral side of the insect's abdomen (a,b), showing the stack of chitin layers, the surface roughness of which creates very thin air gaps at periodic positions. Simulated reflectance spectra at various incidence angles (c) and photographs of an artificial SiO/Ni multilayer coating at normal (d) and grazing (e) incidence and observation. The design of the coating was inspired by the structure of the iridescent abdomen of the beetle. (Reproduced from [12], with permission from the American Physical Society.)

homogenous piece of chitin, in which a series of ultrathin air gaps are introduced as equidistant index-perturbing planes, in order to open a photonic bandgap. A salient conclusion about such a dissymmetric dielectric mirror design[8] is that its iridescent colors depend only on the distance separating the perturbing planes, not on their material composition (here, air). As long as the perturbing layers are kept thin, similar coloration could be obtained with any other material in lieu of air in the gaps. This idea, which is contained implicitly in the dominant reflected wavelength

8. Classical dielectric mirror designs are characterized by similar optical paths (i.e., products of thickness and refractive index) in both layers forming the Bragg period.

formula[9], has inspired the fabrication of a SiO/Ni coating on a glass substrate that reproduced precisely the visual appearance of the beetle (Figure 2.9d,e) [12]. The use of ultrathin (10 nm) *metallic* (nickel) layers demonstrated the robustness of the design with respect to the perturbing material. Other examples of bioinspired iridescent coatings will be discussed in subsequent chapters.

2.2.3 Curved Periodic Multilayer Systems

When a Bragg mirror lies on a curved surface with a curvature radius much larger than the stack thickness, its local orientation can be such that an incident light ray bounces several times on the surface. The reflected light may hence be scattered in many directions and color mixing may occur, as the iridescent periodic multilayer stack is lit locally at various incidence angles.

Let us consider, for instance, a periodic multilayer stack covering the surface of a hemispherical concave cavity (Figure 2.10). The global incidence angle θ is defined with respect to the normal to the top plane of the cavity. Meanwhile, the local incidence angle γ is defined with respect to the local normal to the surface at the point where the incident light reflects. The case of double reflection by a pair of orthogonal surface elements leading to retro-reflection of the incident light is illustrated in Figure 2.10. This symmetric configuration occurs when incident light interacts with the first surface element with a local incidence angle equal to $\gamma = \frac{\pi}{4} - \frac{\theta}{2}$. When $\theta = 0°$ (global normal incidence), two pathways contribute to the backscattering of light: the first one passes through the center of the cavity, whereas the second one goes through the inclined sides of the cavity, where $\gamma = \frac{\pi}{4}$. Since the local incidence angles are not the same at both points, the Bragg stack reflects two distinct iridescent colors, which produce a mixed color when observed in the far field.

This color mixing effect was reported for the first time by Vukusic and coworkers in the wing scales of the male *Papilio palinurus* butterfly [22]. A regular 2D array of hemispherical concavities covered the scale surface. Their size (diameter: 4-6 μm, depth: 0.5-3 μm) was comparable with the multilayer stack thickness. The flat bottom of the concavity reflected yellow light, whereas its inclined sides reflected blue light. The double reflection mechanism was confirmed by the observation of polarization-conversion in the retro-reflected blue light.

The dorsal side of the scutellum of the African shield bug *Calidea panethiopica* exhibits yellow, green, and blue iridescent areas arising from a myriad of hemispherical pits (cavities) [23]. A similar color mixing effect to that in the butterfly *P. palinurus* was observed; but, in this case, the mean diameter of the pits (88 μm) was much larger than the Bragg stack dimensions (period: $a = 190$ nm, $N = 15$ periods). In each pit, a yellow central spot was observed, which was surrounded by an annular blue ring (Figure 2.11). Accordingly, a broad peak (approximately 560 nm) and a sideband (about 500 nm) were observed in the reflectance spectrum

9. Indeed, at a fixed incidence angle, the dominant reflected wavelength depends only on the spatial period and the average refractive index. As long as one of the layers forming the period is thin enough, the average refractive index is about the same as the refractive index of the other layer. Therefore, the actual composition of the thinner layer does not matter anymore. This exact principle underpins the design idea of strongly dissymmetry mirrors, which will be developed further in Chapter 9, Section 9.3.1.

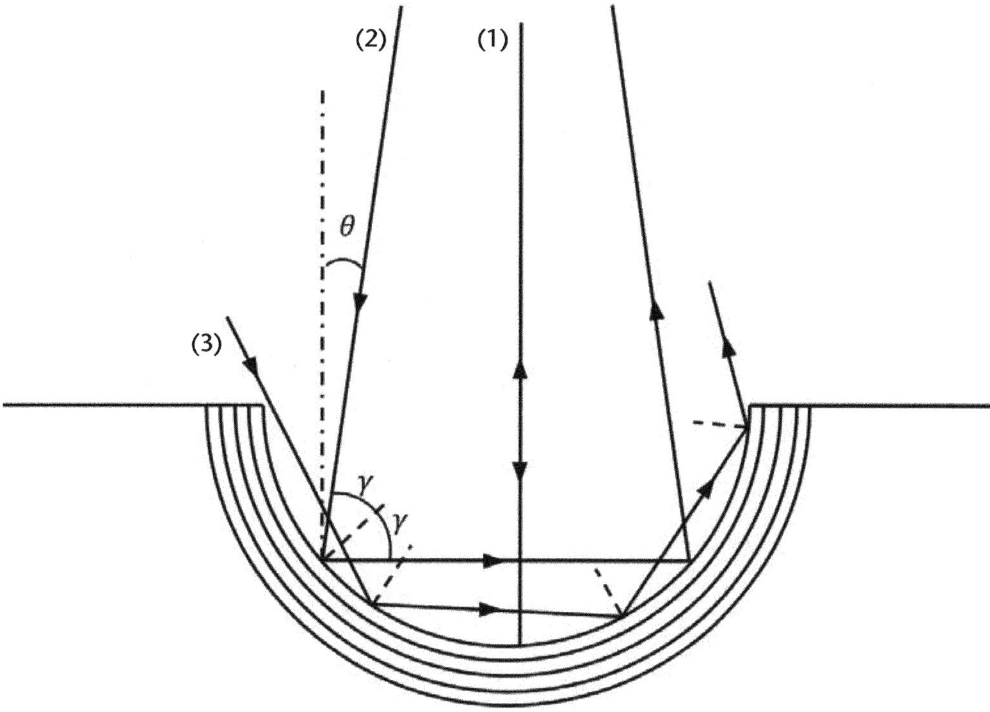

Figure 2.10 Hemispherical concave cavity covered by a periodic multilayer stack. θ is the global incidence angle with respect to the normal (dashed-dotted line) to the top plane of the cavity. Pathways are sketched for a few light rays: a ray reflecting at the center of the cavity along the vertical direction (1), a ray reflecting at a point on the inclined side of the cavity at $\gamma = \frac{\pi}{4} - \frac{\theta}{2}$ with respect to the local normal (dotted line) and exiting the cavity after a double reflection in a symmetric direction (2), a ray reflecting first on a point on the inclined side of the cavity and exiting the cavity after three reflections in a dissymmetric direction (3).

near normal incidence. A Monte Carlo calculation of the reflectance spectrum, using a computational code that combined ray tracing (at the scale of the surface curvature) and wave optics (at the scale of the Bragg stack), reproduced the measured spectrum very well and highlighted the difference between the curved multilayer stack and its flattened counterpart (Figure 2.11). At an oblique incidence ($\theta = 45°$), the main peak shifted toward shorter wavelengths, as expected for reflection from the flat bottom of the cavity; however, the sideband shifted in the opposite way. This intriguing *reverse* iridescence effect for the double reflection from the inclined sides of the cavity can be explained easily by adapting the dominant wavelength formula to a curved surface. Indeed, substituting $\gamma = \frac{\pi}{4} - \frac{\theta}{2}$ to θ in (2.3), makes it immediately apparent that λ_B becomes an increasing function of the global incidence angle θ.

2.2.4 Multilayer Systems with Variable Layer Thicknesses

Besides Bragg mirrors, nonperiodic multilayer systems can be found in nature, too. If the layer thickness varies progressively across the depth of the stack, the multilayer system can be approximated by a superposition of partial Bragg reflectors, with each of them having a progressively shorter or longer period and a small number of periods, depending on the approximation. This *chirped* multilayer system exhibits

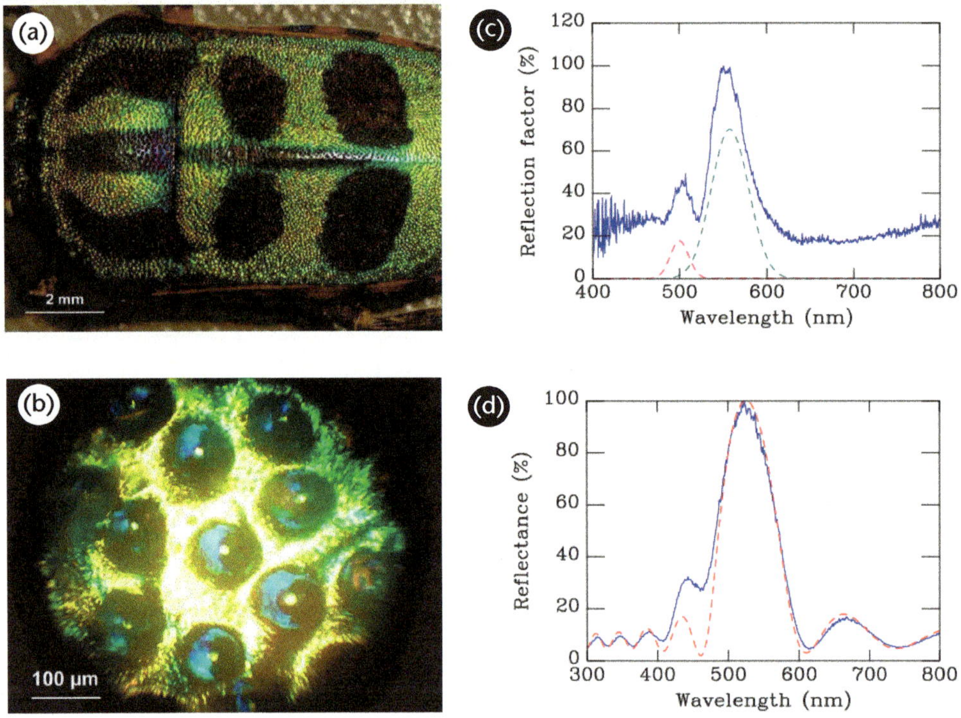

Figure 2.11 Periodic multilayer concavities on the scutellum of the African shield bug *Calidea panethiopica*. Microscopic overview of the scutellum's dorsal side (a), showing myriad of yellow, green, and blue iridescent spots. Optical microscope image of pits with a yellow central spot surrounded by an annular blue ring and a yellow flat surface between them (b). Reflection factor, measured in specular configuration at near normal incidence (dashed curves are components of a best-fit Gaussian band decomposition of the measured spectrum) (c). Reflectance spectrum simulated by a Monte Carlo computer code using a combined ray-tracing and wave-optics multiscale approach (d). The source and detector axes are set up in a specular geometry, at near normal incidence (the dashed line is the reflectance of the flattened multilayer system). (Reproduced from [23], with permission from the American Physical Society.)

a broadband reflectance spectrum, which can be interpreted as the collection of narrowband Bragg peaks associated with each sampled period.

In some beetle species, chirped multilayer devices have evolved to produce a metallic golden coloration by reflecting strongly all wavelengths in the visible range, except in the blue part of the spectrum. This is the case, for instance, in *Charidotella ambita*, where the chirped gold reflector is located in the endocuticle, which is the inner layer of the exoskeleton of an insect just under the exocuticle and above the epidermis (Figure 2.12) [24]. The multilayer system comprises up to 50 bilayers alternating high and low refractive indices, with a progressive decrease in layer thickness across its depth (from approximately 150 nm to approximately 100 nm). In a related species, *Charidotella egregia*, a chirped gold reflector was also observed, but it is located in the exocuticle (i.e., the intermediate layer of a cuticle, between the epicuticle and the endocuticle) [25].

A gradient of layer thickness across the depth is not the only way to achieve broadband reflection. Another design exists, for instance, in silvery fish species, using random thickness variation. The Atlantic herring *Clupea harengus* and the

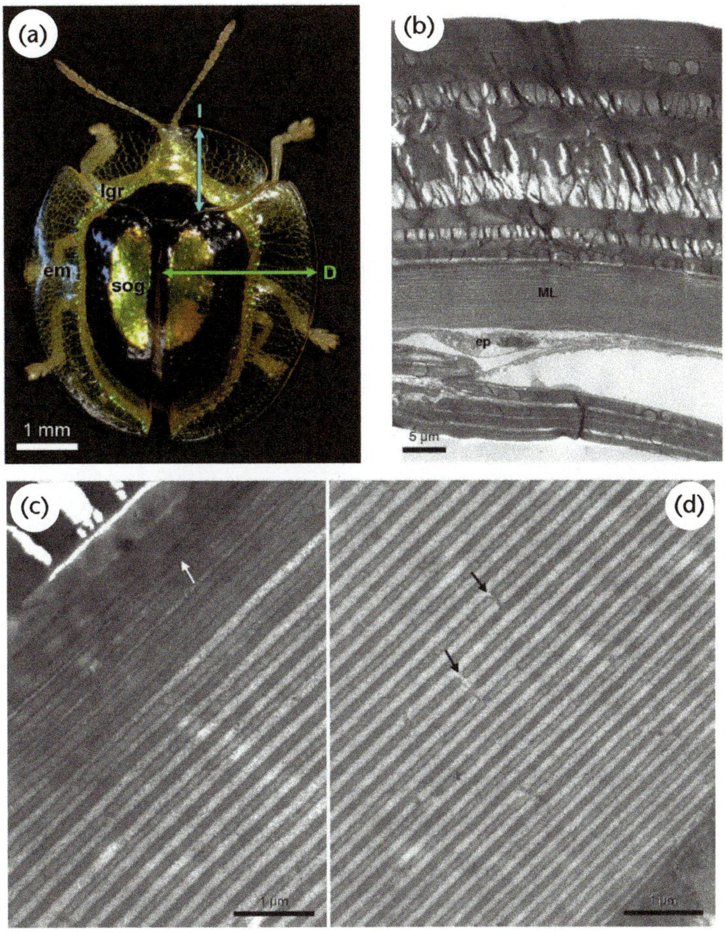

Figure 2.12 Chirped gold reflector in the endocuticle of *Charidotella ambita*. Overview of the beetle elytra showing several golden regions (a). TEM image of an elytral cross section, taken in the semi ovoid gold region (ML: multilayer, ep: epidermis) (b). TEM images of the endocuticular multilayer stack (c,d). (Reproduced from [24], with permission from Elsevier.)

European sardine *Sardina pilchardus* possess a multilayer system made of alternating isotropic layers (cytoplasm) and birefringent crystal (guanine) layers with random thicknesses that produces a silver coloration. In addition, these broadband reflectors exhibit an almost polarization-independent response, which is not encountered in either isotropic multilayer systems or anisotropic multilayer systems with a single crystal axis orientation [26]. With the support of anisotropic transfer matrix modeling, such a polarization-insensitivity was shown to arise from the random mixing of two types of uniaxial crystal layers that have their optical axis either parallel or perpendicular to the layer stacking direction. This arrangement tends to cancel the effects of interfacial Brewster angles on the reflectance [26].

2.2.5 The Twists of Bouligand Structures

An interesting case of Bragg mirror is that of Bouligand structures. They consist of incrementally rotated birefringent thin layers that are stacked upon one another

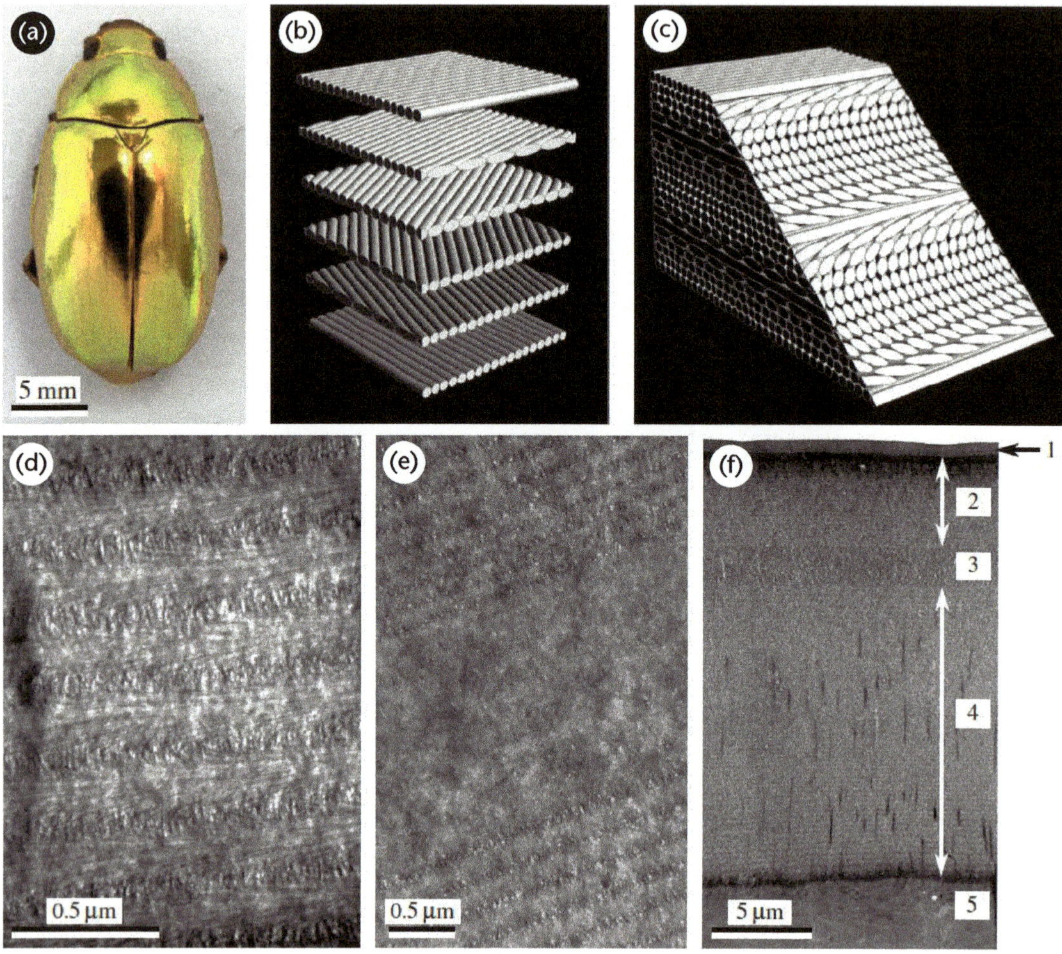

Figure 2.13 *Chrysina resplendens* exhibits a golden appearance (a) associated with both left-handed circularly polarized and right-handed circularly polarized light, when it is illuminated by left-handed circularly polarized light and right-handed circularly polarized light, respectively. This optical behavior is due to a pair of chirped left-handed Bouligand structures (b-d) separated by a half-wave plate (e). A Bouligand structure consists of incrementally rotated birefringent thin layers stacked upon each other (b). It can be identified in electron micrographs of oblique sections of such integuments as periodic lamellar arrangements of Bouligand curves (c,d). (Reproduced from [31], with permission from Elsevier.)

(Figure 2.13(b)). These helicoidal structures are usually identified in electron micrographs of oblique sections of such integuments by periodic lamellar arrangements of Bouligand curves or Bouligand arcs Figure 2.13(c,d) [27]. The helicoidal morphology leads to circularly light polarizing effects [20,28–33]. A Fresnel-type reflection on the surface of an isotropic medium gives rise to a cross-circularly polarized reflection, which means that, a left-handed circularly polarized light reflects as right-handed circularly polarized and vice versa. A Boulingand structure, by contrast, allows cocircularly polarized reflection (i.e., left- or right-handed circularly polarized light reflecting as left- or right-handed circularly polarized light) depending upon the structure's handedness. The period of the helicoid, also known as the pitch p, is defined as the length over which the layers have rotated over 2π rad.

If the pitch is comparable to the wavelength of visible light, Bragg-like reflection takes place, with a peak wavelength $\lambda = \bar{n} p \cos(\theta - \phi)$, where $\bar{n}$ is the average in-plane refractive index of the cuticle material; θ, the incidence angle with respect to the normal direction to the sample, and ϕ, the axis of the helicoid, with respect to the normal [34–36]. The spectral range $\Delta\lambda$ over which the Bragg peak spreads is a function of the material birefringence: $\Delta\lambda = p(n_e - n_o)$, where n_e and n_o, respectively, represent the refractive indices of the polarization parallel (known as *extraordinary*) and perpendicular (known as *ordinary*) to the axis of anisotropy.

Such a structure was reported for the first time in nature by Michelson [37] in the case of the beetle *Plustiotis resplendens*, known today as *Chrysina resplendens*. He described its visual appearance "as if coated with an electrolytic deposit of metal, with a lustre resembling brass." Using a Babinet compensator, he observed that the light reflected by this animal was circularly polarized at an incidence normal to its cuticle, with both polarized and nonpolarized incident light. Michelson concluded that this behavior must originate from a "screw structure" at an "ultra microscopic" scale. Subsequently, many researchers have investigated the circularly polarized light reflection in other beetles [30–33,38–42]. Soon after Michelson's observations, similarities between the optical behaviors of these beetles and the ones of cholesteric liquid crystals were noticed [43–51].

Bouligand structures with quasi-constant pitch lengths across the helicoid axes give rise to single hues, such as in the case of *Chrysina gloriosa*. However, the metallic visual appearances of, for example, *C. resplendens* and *C. optima*, are often due to Bouligand structures with a higher number of repetitions of the pitch, as well as a higher variation in the pitch length across the depth of the beetle's cuticle. This variation gives rise to a chirped structure in a way similar to the ones described in Section 2.2.4.

Most of these beetles display a strong left-handed circularly polarized light reflection when illuminated by left-handed circularly polarized light [31, 52, 53]. One remarkable case is *C. resplendens*, since it also reflects right-handed circularly polarized light under right-handed circularly polarized light [31, 47]. This beetle exhibits two chirped left-handed helicoids separated by a birefringent retarder, acting as a half-wave plate in its integument (Figure 2.13). Incident right-handed circularly polarized light is transmitted through the first Bouligand structure. The wave plate converts this light polarization into a left-handed circularly polarization that is reflected by the second Bouligand structure. The light eventually propagates back into the incident medium with a right-handed circularly polarization [31].

Cellulose-based materials also give rise to Bouligand structures, thanks to helicoidal arrangements of cellulose microfibrils, with pitch lengths ranging from hundreds of nanometers to several microns. However, exclusively structural colors in plants are scarcer than in insects. Thus, only a few examples of Bouligand structures that are active in the visible range have been noted in plant leaves. Such optical devices have been recorded in neotropical ferns, including *Selaginella willdenowii* and *Danaea nodosa* [54–59], flowering plants [60, 61], and sedges [62]. For a handful of these plants, a selective left-circularly polarized light reflection was observed [59, 62]. In addition, Bouligand structures have been found in the cell walls of the tissues of some fruits such as *Margaritaria nobilis* [63, 64], a tree, also known as the bastard hogberry, and *Pollia condensata* [65]. In the latter case, Vignolini and coworkers observed both left- and right-handed Bouligand structures, reflecting either left- or right-handed circularly polarized light, respectively [65].

In general, Bouligand structures have been described in many crustaceans, such as stomatopods [66], crabs [67–69], and lobsters [70, 71]. However, the pitches of these structures are too long (about 1 μm) to generate any response in the visible range. In these instances, the role of these structures lies in their high stiffness properties [72].

Mueller matrix spectroscopic ellipsometry is a perfect tool for investigating such samples with polarization effects [73–75]. Mueller matrix formalism is a method allowing scientists to treat light propagation in structured media, while taking into account arbitrary states of polarization of the light. In the field of natural photonics, Hans Arwin's group has made an extensive contribution thanks to this method [30, 33, 75–81].

2.3 Diffraction Gratings: Playing with the Second Dimension

Diffraction gratings, which often take the form of periodically arranged surface protuberances or corrugations, for example, on butterfly wings (Figure 2.14), scatter light in specific directions. For a one-dimensional grating, such as an array of grooves or ridges, these directions are given by the well-known equation:

$$\sin\phi_m = \sin\theta + m\frac{\lambda}{b}, \tag{2.6}$$

where θ and ϕ_m are, respectively, the incidence and diffraction angles, b is the spatial period, λ is the wavelength, and the integer m is the diffraction order. Under white light illumination, red wavelength components are diffracted at larger angles ϕ_m than blue wavelength components. By changing the incidence angle, diffraction angles are modified and some new diffraction orders may appear while others may disappear.[10] Where the diffraction directions are determined by (2.6), the diffraction efficiencies[11] are determined by the grating spatial profile. Gratings with a slanted profile (*blazed* gratings) allow the diffracted intensity to be concentrated into a single diffraction order by making all other orders evanescent.

A quite intriguing dual diffraction effect was observed by Ingram and coworkers in the male *Lampronelis nitida* butterfly [82]. When its hindwing was illuminated in a certain direction (Figure 2.14(h,i)), green to red iridescence was observed in backscattering. Following an azimuthal rotation of both the incidence and viewing directions by 180°, a fainter blue to violet iridescence was observed at the *same* location. The absence of specular reflection suggested the presence of two separate blazed gratings, which were optimized to concentrate the diffracted light in the $m = -1$ backscattered order. Indeed, two structures were identified in the wing cross ribs and ridges, respectively. The former consisted of slanted plates (2-μm wide,

10. Disappearance of the diffracted wave (in the far field) occurs when the wave becomes evanescent; that is, when ϕ_m no longer gives a real solution of (2.6).
11. For a given order, the diffraction efficiency is defined as the relative intensity of the corresponding diffractive wave.

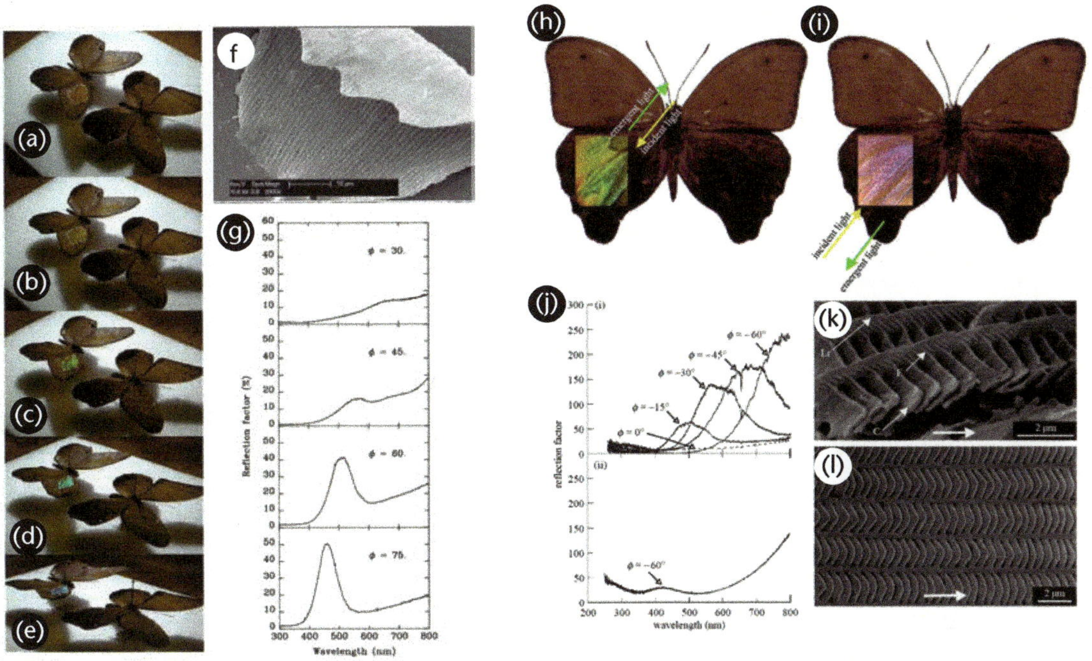

Figure 2.14 Diffraction gratings in butterfly wing scales. *Pierella luna* female and male (behind) viewed at different angles (a–e). Only the male displays rainbow colors and these occur in reverse order compared with that expected for a flat diffraction grating. SEM image of a curved scale, the cross ribs of which form an almost vertical grating that operates in transmission (f). Measured reflectance spectra at various viewing angles ϕ for illumination at $\theta = 45°$ (g). Iridescent colors of the hindwings of the male *Lampronelis nitida* observed in the backscattering direction at a grazing angle for anteroposterior (h) and posteroanterior (i) illuminations. Backscattered reflection factor spectra (j) for anteroposterior ("i") and posteroanterior ("ii") illuminations ($\theta = 75°$ and $\theta = 60°$, respectively) observed at various emerging angles ϕ. SEM images of the wing for an oblique view (k,l), showing the blazed gratings formed by the slanted cross ribs (c) and the inclined lateral proturberances (f) on the ridges (Lr). (Reproduced from [83] (a–g) and [82] (h–l), with permission from the American Physical Society and the Royal Society, respectively.)

100-nm thick) with a periodicity of about 582 nm and a tilt angle of 30° toward the costal wing margin, whereas the latter were tiny lateral ridge protuberances (100-nm thick), with a periodicity of about 205 nm and a tilt angle of 45° toward the *opposite* direction (Figure 2.14(k,l)). These concomitant and opposite blazed gratings forming a natural dual grating device are so far unique. However, they explain the observed dual diffraction well. In particular, the diffracted directions in both cases were correctly predicted by the grating equation (2.6) that was modified to account for the scale tilt γ with respect to the wing surface (i.e., $\sin(\phi_m + \gamma) = \sin(\theta - \gamma) + m\frac{\lambda}{b}$).

Another puzzling effect is the reverse rainbow color sequence in the diffraction from the forewing of the male *Pierella luna* [83]. When illuminated with white light at oblique incidence ($\theta = 45°$) toward the wing tip, a sequence of blue, cyan, green, yellow, and red colors is observed in the forward direction as the viewing angle is varied from grazing to normal inclination (Figure 2.14(a–e)). As a matter of fact, there is no way a surface grating, lying on the wing, could explain this reverse color sequence; that is, one opposed to that predicted by (2.6). In the iridescent area and only there, curled scales were found, with their distal ends protruding high above the wing surface (Figure 2.14(f)). Their cross ribs form a diffraction grating with a much smaller period ($b = 440$ nm) than the scale curvature radius (approximately 25 μm). This exceptional upright grating, which is more or less perpendicular to the wing surface, operates in transmission rather than in reflection and obeys the grating equation [83]:

$$\cos(\phi_m + \gamma) = -\cos(\theta - \gamma) + m\frac{\lambda}{b}, \tag{2.7}$$

with γ a tilt angle that accounts for a slightly slanted position of the grating with respect to the vertical direction ($\gamma \simeq 20°$). At $\theta = 45°$ incidence, given both the grating period and its orientation, the specularly reflected or transmitted light is absorbed by the pigmented wing membrane and only the $m = 1$ diffraction order produces the iridescence. The emergence angles of the observed reverse sequence colors are satisfactorily predicted by (2.6). The influence of the slight grating curvature on the reflectance spectra was also found to be well described by Fraunhofer diffraction simulations using a curved scale model with regularly spaced point scatterers [83].

2.4 Photonics Crystals: Playing with Order at One, Two, or Three Dimensions

Photonics crystals have been the subject of research for more than three decades and are well described in the literature [11]. Typically, they are formed by a periodic arrangement, in one, two, or three dimensions, of two regions of distinct dielectric material properties (refractive indices), the size of these regions and the periodicity being of the order of the wavelength of interest. Provided the refractive contrast is large enough, coherent multiple scattering from the spatially ordered interfaces may lead to a photonic bandgap. The resulting forbidden light propagation depends on the direction, the crystal lattice class, and the light polarization state. In a photonic crystal slab of finite size, even with a relatively low number of periods, a strong reflection of light occurs at the slab surface for wavevector directions where a photonic bandgap exists for any wavelength within the gap. This therefore explains how such structures, a large variety of which are found in nature, can produce bright iridescent colors.

2.4.1 One-Dimensional Photonic Crystals

A periodic multilayer stack is simply a slab of one-dimensional photonic crystal, the surface termination of which is perpendicular to the direction of discrete translational invariance; namely, the direction along the dielectric permittivity (refractive index) varies periodically as $\varepsilon(z) = \varepsilon(z + a)$, where a is the Bragg period. The shift in the polarization-dependent photonic bandgap induced by changing the wavevector direction (incidence angle) explains the iridescence. We will not reconsider the aforementioned natural multilayer devices further in the framework of photonic crystal theory. Rather, we will take the opportunity to discuss the approximation of a natural structure by a 1D photonic crystal. Often, in multilayer stacks made, for example, from chitinous material, the refractive index contrast is achieved by porosity, which lowers the cuticle bulk refractive index by the corresponding volume filling fraction of the pores. Porosity can take various forms. For instance, in the cuticle of *C. vittata*, porosity consists of very thin void spaces left by rough chitin plates stacked together [12], as described earlier. By contrast, in the scales of the beetles *Hoplia coerulea* or *Tmesisternus isabellae*, the porosity is due to pores between chitin rods forming a complex network that still has short-range, parallel, and equidistant alignment [84–86]. In these cases, the pore mean size and lateral variations are much smaller than visible wavelengths, meaning that the porous layers can be treated as an effective medium in first approximation. As a result, the porous multilayer can be assimilated to an effective 1D photonic crystal.

2.4.2 Two-Dimensional Photonic Crystals

Peacock feathers are striking examples of structural coloration arising from natural 2D photonic crystal devices [87–89]. In the cortex of blue, green, yellow, and brown barbules of the tail feathers of the male *Pavo muticus*, Zi and coworkers discovered quite similar 2D photonic crystal structures; they explained the coloration palette through photonic band structure calculations (Figure 2.15). The cortex structures, enclosing a disordered medullar core, consist of rectangular lattices of melanin rods (approximately 0.7 μm long) connected by a keratin matrix and interspersed with gaps. In blue, green, and yellow barbules, the lattice constant[12] determines mainly the hue, in order to produce the correct balance for brown coloration,[13] as predicted by photonic band structure calculations showing a partial photonic bandgap along the Γ-X direction (i.e., the normal to the surface). The number of periods, however, varies from one barbule type to another and influences the color brightness. It is as low as $N = 4$ in brown barbules.

Zi's group studied the coloration of brown barbules in detail [88]. This case is particularly interesting for two reasons. Firstly, the brown color, which is usually due to pigments, arises here from the mixing of blue and orange-red structural colors, instead. Secondly, the partial photonic bandgap of the infinite crystal explains only the orange-red band in the refectance spectrum of the finite photonic crystal slab. The blue band, however, originates from a Fabry-Pérot effect related to the finite crystal size. The contribution of the blue component to the color mixing is critically adjusted by the number of periods, which needs to be relatively small.

The blue feathers of the neck and breast of the peacock *Pavo cristatus* have similar 2D photonic crystal structures in their barbules. In this case, an effective medium approximation of the two-dimensional structure by a multilayer stack was found to be sufficient to encapsulate the blue band features in the reflectance spectrum [89].

12. In blue, green, and yellow barbules, the 2D photonic crystal has a square array, with a lattice constant equal to approximately 140 nm, about 150 nm, and around 165 nm, respectively, whereas in the brown barbules, it has a rectangular array with lattice constants along the directions parallel and perpendicular to the cortex surface equal to $a_{\parallel} \simeq 150$ nm and $a_{\perp} \simeq 185$ nm [87].
13. The yellow barbule is actually an intermediate case. The yellow color is not due to a single band in the reflectance spectrum, as for blue and green barbules, but to a mix of blue and green-orange. However, the blue band is less pronounced than in the brown barbule.

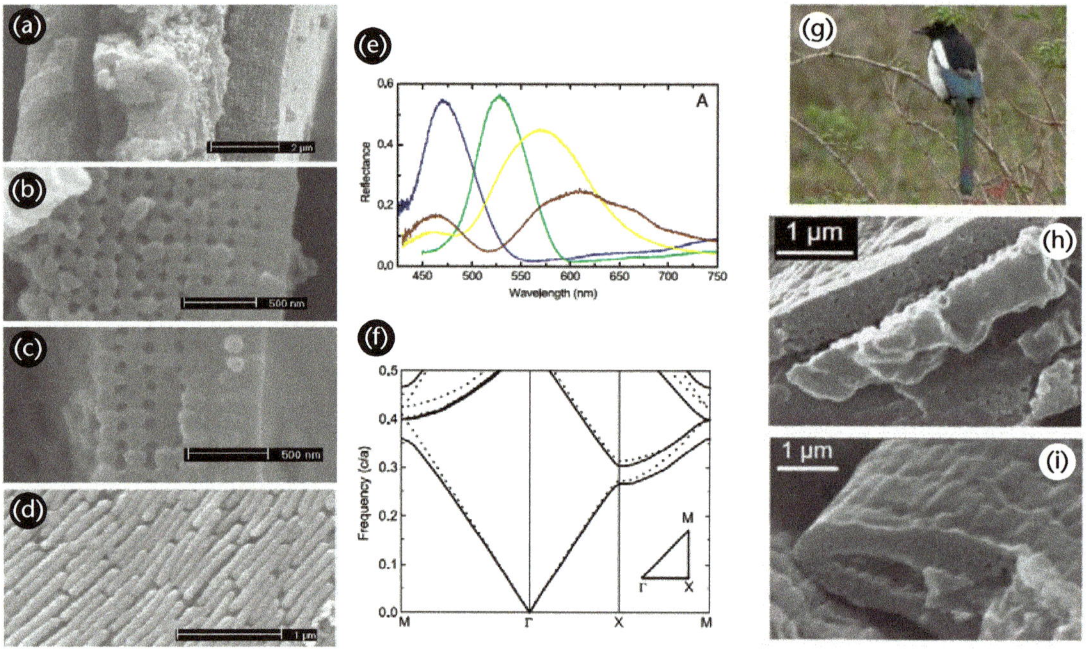

Figure 2.15 Two-dimensional photonic crystal structures in bird feathers. SEM images of barbules of peacock tail feathers (a–d), showing transverse cross sections of a green barbule (a,b), transverse cross section of a brown barbule, (c) and longitudinal cross section of a green barbule with the thin keratin surface layer removed (d). Reflectance spectra of blue, green, yellow, and brown barbules measured at normal incidence with light polarization parallel to the rods (e). Photonic band structure in normalized frequency units of an infinite 2D crystal (square lattice constant a) calculated for polarizations parallel (solid lines) and perpendicular (dashed lines) to the rods (f). Photograph of a black-billed magpie (g), showing yellow-green and bluish iridescence in tail feathers and wing feathers, respectively. SEM images of cross sections of a yellow-green barbule (h) and a bluish barbule (i). (Reproduced from [87] (a–f) and [94] (g–i), with permission from PNAS and the American Physical Society, respectively.)

Welch, Parker, and colleagues also found two-dimensionally periodic photonic crystals among the comb jellyfish (phylum Ctenophora) [90–93].

The back-billed magpie *Pica pica* looks like a black-and-white bird at first glance. However, under appropriate illumination and viewing conditions, weak yellow-green and bluish iridescence can be observed from the otherwise dark tail and wing feathers, respectively (Figure 2.15). Research by Jean-Pol Vigneron's group revealed that two-dimensional photonic crystals found in the barbules were again responsible for the coloration [94]. However, they featured another type of geometry; namely, a hexagonal lattice of cylindrical holes. The modeling with a 2D crystal was justified by the large aspect ratio between the mean cylinder length (1.5 μm) and the inter-hole distance (a = 180 nm in tail feathers, a = 270 nm in wing feathers). The volume filling fraction of air holes (radius: 50 nm) in the keratin-melanin matrix led to an average refractive index estimated as $\bar{n}$ = 1.9. According to the photonic crystal theory, the first two photonic bandgaps, which determine, respectively the lowest forbidden frequency range and the one immediately above it, correspond to wavevectors reaching the first and second Brillouin zones, respectively. From such considerations, Vigneron extended the concept of dominant reflected wavelength to the case of 2D photonic crystals and derived a formula for a hexagonal lattice:

$$\lambda_B = \gamma \frac{a\bar{n}}{m}, \tag{2.8}$$

where a is the nearest neighbor distance between nodes of the hexagonal lattice, $\bar{n}$ is the average refractive index over the 2-D crystal unit cell. The integer m is the bandgap order and $\gamma = (\frac{\pi\sqrt{3}}{2})^{\frac{1}{2}}$ is a geometrical factor. This formula gave an accurate prediction of the yellow-green color for $m = 1$, $a = 180$ nm, and $\bar{n} = 1.9$, which means that the color arises from the first-order bandgap. Because of the *larger* period found in the wing feathers, the formula failed to predict the bluish color with a first-order bandgap but did succeed with a second-order bandgap ($\lambda_B = 423$ nm for $m = 2$, $a = 270$ nm and $\bar{n} = 1.9$). This second-gap coloration mechanism contrasts with the peacock coloration strategy, where blue color is obtained by shortening the period of the crystal.

2.4.3 Three-Dimensional Photonic Crystals

Mineral opals are natural occurrences of three-dimensional photonic crystals consisting of close-packed hexagonal or cubic lattices of silica spheres (150–300 nm diameter). Some of them exhibit partial photonic bandgaps within the visible wavelength range that give rise to spectacular iridescence due to strong reflections in specific crystallographic directions. In 2003, Parker and coworkers observed 3D photonic structures in the lumen of the scales of the weevil *Pachyrrhynchus argus* that were described as biological analog of opal, in the form of a close-packed hexagonal lattice of chitinous spheres [95]. From the observed sphere diameter (250 nm) and the chitin refractive index ($n = 1.56$), a crude estimation of the reflectance peak wavelength matched the yellow-green color of the weevil quite well.

The orange scales located in the annular spots covering the thorax and abdomen of the weevil *Pachyrrhynchus congestus pavonius* drew the attention of Welch, Parker, and Vigneron, who noted that orange structural coloration is not very common in insects [91, 96, 97]. In SEM images of fractured orange scales (Figure 2.16), the structure looked like stacks of perforated chitin plates with cylindrical protrusions and holes forming hexagonal arrays. After examination of stacking possibilities, the structure was assimilated to a 3D photonic crystal having a faced-centered cubic lattice. SEM images also revealed several domains, corresponding to different crystal orientations, which make the actual structure a photonic *polycrystal*. Transfer-matrix simulations confirmed that orientational averaging was responsible for the observed angular-independence of the orange reflection (i.e., polycrystallinity suppressed the iridescence). This explanation of the reflection angular broadening was confirmed in a later study by Wilts and Saranathan, who extended the analysis to adjacent scales, which form radial rainbow color sequences in annular rings [98]. Using small-angle X-ray scattering for detailed structural characterization, they found that the 3D photonic structure was better described by a *diamond* lattice similar to the one developed 6 years earlier for another weevil [99, 100] (see next paragraph). This symmetry is shared by all the differently colored scales of *P. congestus pavonius*. They confirmed the presence of small crystalline domains (average size: 2.5 μm) with (111) preferential orientation

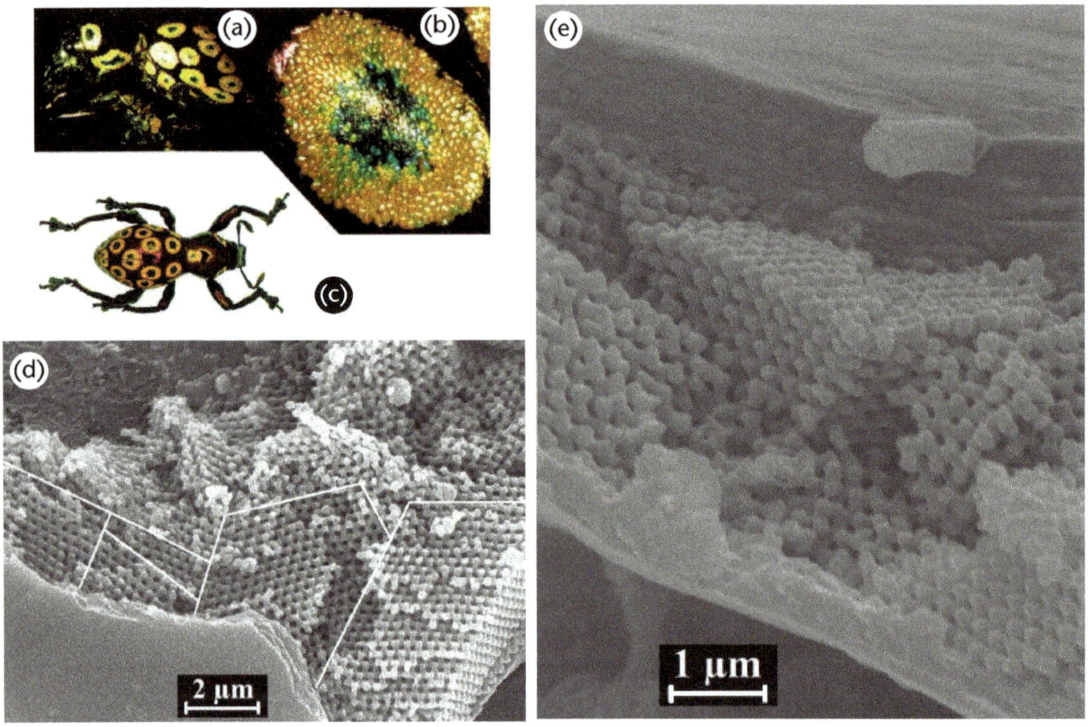

Figure 2.16 Three-dimensional photonic crystal structures in weevils. The weevil *Pachyrrhynus congestus pavonius* (a-c) with its annular ring patches (a) of orange scales (b) on the exoskeleton (a). SEM images of a fractured scale: top view with cortex layer removed (d), cross-sectional side view (e) showing polycrystal domains with different orientations. (Reproduced from [96], with permission from the American Physical Society.)

in the direction normal to the scale surface. Interestingly, the rainbow palette was due to scale-by-scale variations of the lattice constant, as well as changes in chitin filling fraction f_c. The influence of this crystal parameter on the purity of the hue[14] was highlighted by calculations of the gap to mid-gap ratios, which turned out to be close to optimal values ($f_c \simeq 0.35 - 0.40$) for yellow, orange, and red scales, but lower than optimal ($f_c < 0.30$) for blue and green scales.

Although diamond is the champion lattice for achieving a complete photonic bandgap [11], the low refractive index contrast available in biological structures (typically 1.56 for chitin/air) produces only partial photonic bandgaps in certain crystallographic directions. In the iridescent scales of the *Lamprocyphus augustus* weevil, all partial gaps of the diamond photonic crystal (lattice constant: a = 450 nm) are within the green region (525–625 nm) and the macroscopically angle-independent green color arises from three different polycrystal domain orientations (Figure 2.17) [101, 102]. The Brazilian weevil *Entimus imperialis* features scales with a few (three to five) *large* domains[15] (20–30 μm, the colors—ranging

14. The purity of the hue increases with decreasing reflection optical bandwidth. The optical bandwidth $\Delta\lambda$ is in turn determined by the size of the photonic bandgap in the frequency domain $\Delta\nu$: $|\Delta\lambda| = \lambda\frac{|\Delta\nu|}{\nu}$ ($\frac{|\Delta\nu|}{\nu}$: gap to mid-gap ratio).
15. Roughly domains extend from a quarter to a half of the scale dimensions (100 μm × 50 μm).

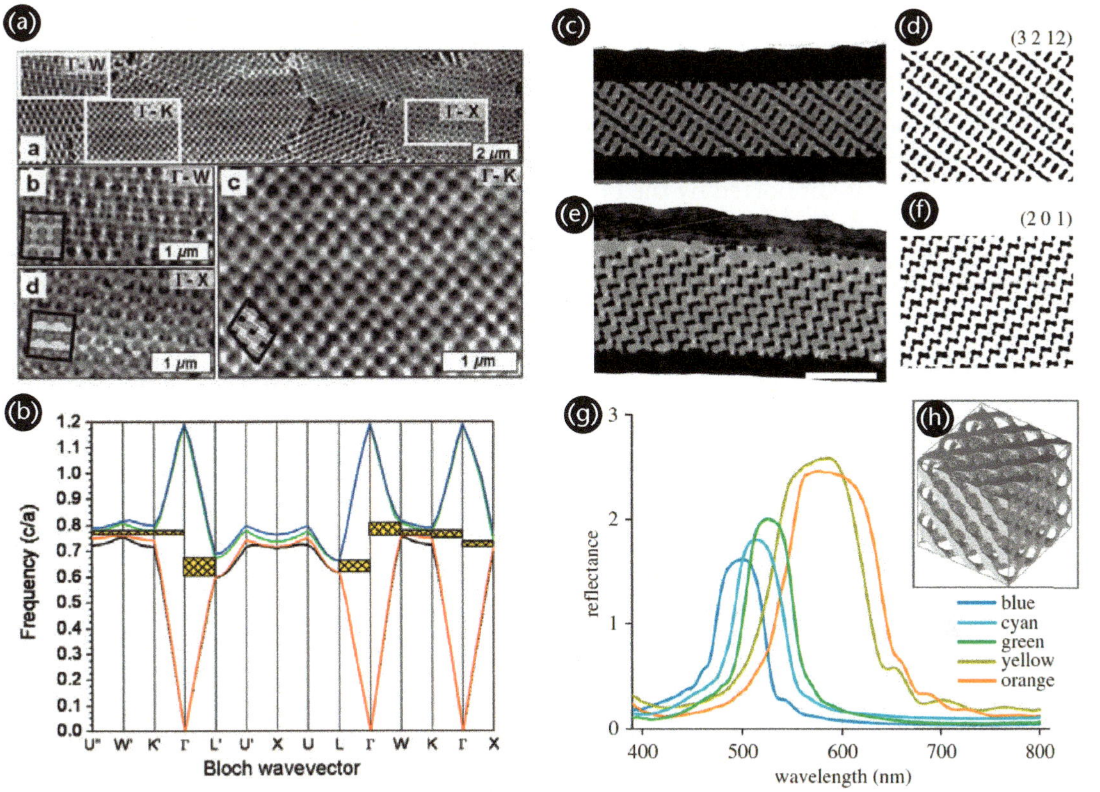

Figure 2.17 Diamond-type photonic crystal structures in weevils. SEM images of focused-ion-beam (FIB) milled top surface of a scale of *Lamprocyphus augustus* (a), showing three different polycrystal domains (highlighted by white frames) and their corresponding computer-generated projections, based on a diamond lattice model (black frames). Photonic band diagram (b) of the diamond lattice model, showing three overlapping partial bandgaps along the $\Gamma - W$, $\Gamma - K$, and $\Gamma - X$ directions (Miller indices: (210), (110) and (100)). TEM images of a scale of *Entimus imperialis* (c,e), showing two different polycrystal domains and their corresponding computer-generated projections (d,f) based on a diamond lattice model. Measured reflectance spectra of single, differently colored domains (g) and 3D rendering of the diamond lattice model (h). (Reproduced from [101] (a,b) and [99] (c–h), with permission from the American Physical Society and the Royal Society, respectively.)

from blue to orange—of which are associated with different orientations of the same photonic crystal of diamond lattice constant of $a = 445$ nm and chitin filling fraction of $f_c \simeq 0.3$ (Figure 2.17) [99, 100, 103–107]. The matte green color observed in the far field results from orientational averaging, with an additional contribution due to the scales' arrangement in concave pits [99, 106].

The four-fold coordinated diamond lattice is one of the three fundamental associated with triply periodic minimal surfaces[16] [108]. The two others are associated with the six-fold coordinated cubic (primitive) lattice and the three-fold coordinated gyroid lattice. In cell biology and coblock polymer chemistry, such surfaces are known to divide, for example, a biphasic system consisting of two continuous interpenetrated regions of rather complex geometries. Detailed comparisons between SEM or TEM images and 2D projections of computer-generated 3D bicontinuous

16. Triply periodic minimal surfaces are surfaces that are periodic along the three directions of space and exhibit a mean curvature equal to zero at every point when averaging over all directions.

volumes have shown that the gyroid photonic crystal structure is frequently encountered in butterfly wing scales, for instance in the papilionids *Parides sesostris* and *Teinopalpus imperialis*, as well as the lycaenids *Callophrys rubi*, *Callophrys dumetorum*, *Callophrys* (formerly *Mitoura*) *gryneus*, and *Cyanophrys remus* [108–111]. The idea of mesoscopic self-assembly of bicontinuous gyroid structures in butterfly wing scales is supported by the developmental phases of scale cells which, after cell death, eventually result in the formation of a single gyroid network of chitin and air [110].

2.5 Disordered Structures: Imperfections May Be Helpful

Natural photonic structures always come with disorder at different length scales. Polycrystallinity is a mesoscopic disorder that arises from the finite size of naturally grown photonic crystal grains. Multiple scattering at grain boundaries causes a decorrelation between light incidence and emergence directions, which leads to a diffuse reflection. It also causes randomization of the incidence direction on the crystal grains, resulting in angle averaging of the individual iridescence spectra. The effect of polycrystalline disorder is therefore to cancel iridescence and to produce a matte coloration, as aforementioned. The resulting broadband and diffuse character of the reflection is expected to be reinforced with decreasing grain sizes. The long greenish-white bristles (elongated scales) of *Prosopocera lactator* are a typical example of strong color desaturation, due to a polycrystalline disorder featuring many small (approximately 2 μm) grains of a single crystal [112]. The photonic crystal was modeled as a 3D array of spheres connected by short rods (Figure 2.18). A detailed analysis of SEM images concluded that a face-centered cubic lattice was consistent with the appearance displayed (100), (110), and (111) the crystallite orientations, as well as with X-ray tomography observations. An advantage of this noninvasive approach for analyzing the photonic crystallography is the ability to rotate the 3D-reconstructed image around a fixed point, in order to expose different reticular planes and to compare their patterns with SEM images taken from different viewing directions.

Three-dimensional structures that lack long-range order can hardly be regarded 3D photonic crystals. Filamentary structures belong to this loose category of randomly disordered media, in which multiple scattering of visible light produces a matte white. A network of interconnected, rod-like chitin filaments was found in the scales of a species of the white beetle *Cyphochilus* sp., and they were more or less oriented parallel to the scale surface (Figure 2.18). The anisotropy of the structure turned out to be the key to explaining the exceptionally short value of the mean free length of the photon transport (estimated at approximately 1.5 μm) in a material with such a low refractive index ($n \simeq 1.5$) and high filling factor ($f_c \simeq 0.6$) in regard to the small thickness (8 μm) of the scattering layer which is, a priori, not favorable for multiple scattering [113].

Butterfly scales with their ribs and cross ribs are never perfectly regular. Additional disorder can be present in the form of oblong granules that are randomly arranged between the upper and the lower lamina, as in pierid butterflies [114]. In the small white butterfly *Pieris rapae* (Figure 2.17), Stavenga and coworkers showed

Figure 2.18 Disorder in natural photonic structures. The longhorn beetle *Prosopocera lactator* featuring greenish-white patches (a) and SEM image of the polycrystalline structure in a greenish-white scale (b). The white beetle *Cyphochilus* (c) and SEM image of the scale anisotropic filamentary structure (d). The white pierid *Pieris rapae* (e) and SEM image of the scale structure featuring randomly dispersed granules between upper and lower lamina (f). (Reproduced from [112] (a,b), [113] (c,d), and [115] (e,f), with permission from the American Physical Society, under License CC-BY-4.0 and with permission from the Royal Society, respectively.)

that beads of typical sizes ranging from 100 nm to 500 nm, which are hollow and do not contain visible-light-absorbing pterin pigments, helped with producing a bright white aspect through random multiple scattering [115]. In a later study on the pierid *Ponti protodice*, Morehouse, Vukusic, and Rutowski removed selectively UV-absorbing petrin granules by using a chemical solvent. As a consequence, they observed not only an expected increase of reflectance in the UV, but also a reduction of reflectance at visible wavelengths [116]. The latter result, together with a positive correlation found between the granule density and the reflectance intensity, confirmed the role played by granules in bright white coloration. Finally, a more macroscopic type of disorder exists in butterfly wings, due to the overlapping between ground scales and cover scales. Using a simple optical model based on multiple incoherent scattering in a stack of plates, Stavenga and coworkers explained the higher reflectance of a wing with overlapping scales as compared to that of an isolated scale [114].

References

[1] Jaszkowski, D., and J. Rzeszut, "Interference Colours of Soap Bubbles," *The Visual Computer*, Vol. 19, 2003, pp. 252–270.

[2] Born, M., and E. Wolf, *Principles of Optics*, Cambridge: Cambridge University Press, 2003.

[3] Stenzel, O., *The Physics of Thin Film Optical Spectra*, Berlin: Springer, 2005.

[4] Dellieu, L., et al., "Light Coherence Time Modifies Color Perception of Living Beings," *Materials Today: Proceedings*, Vol. 4, No. 4, 2017, pp. 4952–4958.

[5] Airy, G. B., "On the Phænomena of Newton's Rings when Formed between Two Transparent Substances of Different Refractive Powers," *The London, Edinburgh, and Dublin Philosophical Magazine and Journal of Science*, Vol. 2, No. 7, 1833, pp. 20–30.

[6] Stavenga, D. G., "Thin Film and Multilayer Optics Cause Structural Colors of Many Insects and Birds," *Materials Today: Proceedings*, Vol. 1S, 2014, pp. 109–121.

[7] Sarrazin, M., et al., "Nanomorphology of the Blue Iridescent Wings of a Giant Tropical Wasp *Megascolia procer javanensi*," *Physical Review E*, Vol. 78, No. 5, 2008, 051902.

[8] Abelès, F., "La théorie générale des couches minces," *Journal de Physique et le Radium*, Vol. 11, No. 7, 1950, pp. 307–309.

[9] Yeh, P., *Optical Waves in Layered Media*, Hoboken, NJ: Wiley, 2005.

[10] Dereux, A., et al., "Polaritons in Semiconductor Multilayered Materials," *Physical Review E*, Vol. 38, No. 8, 1988, pp. 5438–5452.

[11] Joannopoulos, J. D., et al., *Photonic Crystals Molding the Flow of Light*, Princeton: Princeton University Press, 2008.

[12] Vigneron, J.-P., et al., "Spectral Filtering of Visible Light by the Cuticle of Metallic Wood-boring Beetles and Microfabrication of a Matching Bioinspired Material," *Physical Review E*, Vol. 73, No. 4, 2006, 041905.

[13] Weiglhofer, W. S., A. Lakhtakia, and B. Michel, "Maxwell Garnett and Bruggeman Formalisms for a Particulate Composite with Bianisotropic Host Medium," *Microwave and Optical Technology Letters*, Vol. 15, No. 4, 1998, pp. 263–266.

[14] Choy, T. C., *Effective Medium Theory: Principles and Applications*, Oxford: Oxford University Press, 1999.

[15] Merritt, E., "A Spectrophotometric Study of Certain Cases of Structural Color," *Journal of Optical Society of America*, Vol. 11, No. 2, 1925, pp. 93–98.

[16] Yin, H., et al., "Iridescence in the Neck Feathers of Domestic Pigeons," *Physical Review E*, Vol. 74, No. 5, 2006, 051916.

[17] Vignolini, S., et al. "Directional Scattering from the Glossy Flower of *Ranunculus*: How the Buttercup Lights Up Your Chin," *Journal of the Royal Society Interface*, Vol. 9, 2012, pp. 1295–1301.

[18] van der Kooi, C. J., et al., "Functional Optics of Glossy Buttercup Flowers," *Journal of the Royal Society Interface*, Vol. 14, No. 127, 2017, 20160933.

[19] Land, M. F., "The Physics and Biology of Animal Reflectors," *Progress in Biophysics and Molecular Biology*, Vol. 24,1972, pp. 75–106.

[20] Seago, A. E., et al., "Gold Bugs and Beyond: A Review of Iridescence and Structural Colour Mechanisms in Beetles (Coleoptera)," *Journal of the Royal Society Interface*, Vol. 6, 2008, pp. S165–S184.

[21] Stavenga, D. G., et al., "Polarized Iridescence of the Multilayered Elytra of the Japanese Jewel Beetle, *Chrysochroa fulgidissima*," *Philosophical Transactions of Royal Society B*, Vol. 366, No. 1565, 2011, pp. 709–723.

[22] Vukusic, P., J. R. Sambles, and C. R. Lawrence, "Colour Mixing in Wing Scales of a Butterfly," *Nature*, Vol. 404, 2000, p. 457.

[23] Vigneron, J.-P., et al., "Spectral Sideband Produced by a Hemispherical Concave Multilayer on the African Shield-Bug *Calidea panaethiopica* (Scutelleridae)," *Physical Review E*, Vol 79, No. 21, 2009, 21907.

[24] Pasteels, J., et al., "Structural and Physical Evidence for an Endocuticular Gold Reflector in the Tortoise Beetle, *Charidotella ambita*," *Arthropod Structure and Development*, Vol. 45, 2016, pp. 509–518.

[25] Vigneron, J.-P., et al., "Switchable Reflector in the Panamanian Tortoise Beetle *Charidotella egregia* (Chrysomelidae: Cassidinae)," *Physical Review E*, Vol 76, No. 3, 2007, 031907.

[26] Jordan, T. M., J. C. Partridge, and N. W. Roberts, "Non-Polarizing Broadband Multilayer Reflectors in Fish," *Nature Photonics*, Vol. 6, 2012, pp. 759–763.

[27] Bouligand Y., "Sur une architecture torsadée répandue dans la nombreuses cuticules d'Arthropodes," *Comptes Rendus Hebdomadaires des Séances de l'Académie des Sciences*, Vol. 261, 1965, pp. 3665–3668.

[28] Bouligand Y., "Sur l'existence de "Pseudomorphoses cholestériques" chez divers organismes vivants," *Journal de Physique Colloques*, Vol. 30, No. C4, 1969, pp. 90–103.

[29] Neville A. C., and S. Levy, "The Helicoidal Concept in Plant Cell Wall Ultrastructure and Morphogenesis," In *Biochemistry of Plant Cell Walls*, pp. 99–124, C. T. Brett and J. R. Hillman (eds.), Cambridge: Cambridge University Press, 1985.

[30] Fernandez del Río, L., H. Arwin, and K. Järrendahl, "Polarizing Properties and Structure of the Cuticle of Scarab Beetles from the *Chrysina* genus," *Physical Review E*, Vol. 94, No. 1, 2016, 012409.

[31] Finlayson E. D., L. T. McDonald, and P. Vukusic, "Optically Ambidextrous Circularly Polarized Reflection from the Chiral Cuticle of the Scarab Beetle *Chrysina resplendens*," *Journal of the Royal Society Interface*, Vol. 14, No. 131, 2017, 20170129.

[32] McDonald L. T., et al., "Circularly Polarized Reflection from the Scarab Beetle *Chalcothea smaragdina*: Light Scattering by a Dual Photonic Structure," *Interface Focus*, Vol.7, No. 4, 2017, 20160129.

[33] Mendoza-Galván, A., et al., "Graded Pitch Profile for the Helicoidal Broadband Reflector and Left-Handed Circularly Polarizing Cuticle of the Scarab Beetle *Chrysina chrysargyrea*," *Scientific Reports*, Vol. 8, 2018, 6456.

[34] Belyakov, V. A., and V. E. Dmitrienko, *Optics of Chiral Liquid Crystals*, Chur: Harwood Academic Publishers, 1989.

[35] de Gennes, P. G., and J. Prost, *The Physics of Liquid Crystals*, Oxford, UK: Oxford Science Publications, 1995.

[36] Oswald, P., and P. Pieranski, *Nematic and Cholesteric Liquid Crystals: Concepts and Physical Properties Illustrated by Experiments*, Boca Raton, FL: Taylor and Francis, 2005.

[37] Michelson A. A., et al., "On Metallic Colouring in Birds and Insects," *Philosophical Magazine*, Vol. 21, No. 124, 1911, pp. 554–567.

[38] Sharma, V., et al., "Structural Origin of Circularly Polarized Iridescence in Jeweled Beetles," *Science*, Vol. 325, No. 5939, 2009, pp. 449–451.

[39] Hernández-Jiménez, M., et al., "Qualitative Correlation between Structural Chirality through the Cuticle of *Chrysina aurigans* scarabs and Left-Handed Circular Polarization of the Reflected Light," *Optical Materials Express*, Vol. 4, No. 12, 2014, pp. 2632–2645.

[40] Sharma V., et al., "Bouligand Structures Underlie Circularly Polarized Iridescence of Scarab Beetles: A Closer View," *Materials Today: Proceedings*, Vol. 1S, 2014, pp. 161–171.

[41] Azofeifa, D. E., et al., "A Quantitative Assessment Approach of Feasible Optical Mechanisms Contributing to Structural Color of Golden-like *Chrysina aurigans* Scarab Beetles," *Journal of Quantitative Spectroscopy and Radiative Transfer*, Vol. 160, 2015, pp. 63–74.

[42] Carter I. E., et al., "Variation in the Circularly Polarised Light Reflection of *Lomaptera* (Scarabaeidae) Beetles," *Journal of the Royal Society Interface*, Vol. 13, No. 120, 2016, 20160015.

[43] Gaubert, P., "Sur la polarisation circulaire de la lumière réfléchie par les Insectes," *Comptes Rendus Hebdomadaires des Séances de l'Académie des Sciences*, Vol. 179, 1924, pp. 1148–1150.

[44] Mathieu, J. P. and, N. Faraggi, "Etude de la lumière polarisée circulairement réfléchie par certains Coléoptères," *Comptes Rendus Hebdomadaires des Séances de l'Académie des Sciences*, Vol. 205, 1937, pp. 1378–1380.

[45] Robinson, C., "The Cholesteric Phase in Polypeptide Solutions and Biological Structures," *Molecular Crystals and Liquid Crystals*, Vol. 1, No. 4, 1966, pp. 467–494.

[46] Neville, A. C. and, S. Caveney, "Scarabaeid Beetle Exocuticle as an Optical Analogue of Cholesteric Liquid Crystals," *Biological Reviews of the Cambridge Philosophical Society*, Vol. 44, No. 4, 1969, 531–562.

[47] Caveney, S., "Cuticle Reflectivity and Optical Activity in Scarab Beetles: The Role of Uric Acid," *Proceedings of the Royal Society B*, Vol. 178, No. 1051, 1971, pp. 205–225.

[48] Bouligand, Y., "Recherches sur les textures des états mésomorphes – 2. – Les champs polygonaux dans les cholestériques," *Journal de Physique*, Vol. 33, No. 7, 1972, pp. 715–736.

[49] Bouligand, Y., "Presentation of a Film Entitled 'Textures of Nematic and Cholesteric Liquid Crystals"', *Journal de Physique Colloques*, Vol. 36, No. C1, 1975, pp. 173–176.

[50] Bouligand, Y., "Some Geometrical Problems in Liquid Crystals," In *Bifurcation Theory, Mechanics, and Physics*, pp. 357–381, C. P. Bruter, A. Arangol, and A. Lichenrowicz (eds.), Dordrecht: Reidel Publishing Company, 1983.

[51] Neville, A. C., *Biology of Fibrous Composites: Development Beyond the Cell Membrane*, Cambridge, UK: Cambridge University Press, 1993.

[52] Jewell, S. A., P. Vukusic, and N. W. Roberts, "Circularly Polarized Colour Reflection from Helicoidal Structures in the Beetle *Plusiotis boucardi*," *New Journal of Physics*, Vol. 9, No. 99, 2007, pp. 1–10.

[53] Pye J. D., "The Distribution of Circularly Polarized Light Reflection in the Scarabaeoidea (Coleoptera)," *Biological Journal of the Linnean Society*, Vol. 100, No. 3, 2010, pp. 585–596.

[54] Hebant, C., and D. W. Lee, "Ultrastructural Basis and Developmental Control of Blue Iridescence in Selaginella Leaves," *American Journal of Botany*, Vol. 71, No. 2, 1984, pp. 216–219.

[55] Graham, R.M., D. W. Lee, and K. Norstog, "Physical and Ultrastructural Basis of Blue Leaf Iridescence in Two Neotropical Ferns," *American Journal of Botany*, Vol. 80, No. 2, 1993, pp. 198–203.

[56] Lee, D. W., *Nature's Palette: The Science of Plant Color*, Chicago: University of Chicago Press, 2007.

[57] Thomas, K. R., et al., "Function of Blue Iridescence in Tropical Understorey Plants," *Journal of the Royal Society Interface*, Vol. 7, 2010, pp. 1699–1707.

[58] Vignolini, S., et al., "Analysing Photonic Structures in Plants," *Journal of the Royal Society Interface*, Vol. 10, No. 87, 2013, 20130394.

[59] Steiner L. M., et al., "Structural Colours in the Frond of *Microsorum thailandicum*," *Interface Focus*, Vol. 9, No. 1, 2018, 20180055.

[60] Gould, K. S., and D. W. Lee, "Physical and Ultrastructural Basis of Blue Leaf Iridescence in Four Malaysian Understory Plants," *American Journal of Botany*, Vol. 83, No. 1, 1996, pp. 45–50.

[61] Lee, D. W., G. T. Taylor, and A. K. Irvine, "Structural Fruit Coloration in *Delarbrea michieana* (Araliaceae)," *International Journal of Plant Sciences*, Vol. 161, No. 2, 2000, pp. 297–300.

[62] Strout, G., et al., "Silica Nanoparticles Aid in Structural Leaf Coloration in the Malaysian Tropical Rainforest Understorey Herb *Mapania caudata*," *Annals of Botany*, Vol. 12, 2013, pp. 1141–1148.

[63] Kolle, M., et al., "Bio-Inspired Band-Gap Tunable Elastic Optical Multilayer Fibers," *Advanced Materials*, Vol. 25, No. 15, 2013, pp. 2239–2245.

[64] Vignolini, S., et al., "Structural Colour from Helicoidal Cell-Wall Architecture in Fruits of *Margaritaria nobilis*," *Journal of the Royal Society Interface*, Vol. 13, No. 124, 2016, 20160645.

[65] Vignolini, S., et al., "Pointillist Structural Color in *Pollia fruit*," *Proceedings of the National Academy of Sciences of the United States of America*, Vol. 109, No. 39, 2012, pp. 15712–15715.

[66] Chiou, T. H., et al., "Circular Polarization Vision in a Stomatopod Crustacean," *Current Biology*, Vol. 18, No. 6, 2008, pp. 429–434.

[67] Neville, A. C., and B. M. Luke, "Form Optical Activity in Crustacean Cuticle," *Journal of Insect Physiology*, Vol. 17, no. 3, 1971, pp. 519–523.

[68] Giraud-Guille, M. M., "Fine Structure of the Chitin-Protein System in the Crab Cuticle," *Tissue and Cell*, Vol. 16, No. 1, 1984, pp. 75–92.

[69] Raabe, D., C. Sachs, and P. Romano, "The Crustacean Exoskeleton as an Example of a Structurally and Mechanically Graded Biological Nanocomposite Material," *Acta Materialia*, Vol. 53, No. 15, 2005, pp. 4281–4292.

[70] Raabe, D., et al., "Microstructure and Crystallographic Texture of the Chitin-Protein Network in the Biological Composite Material of the Exoskeleton of the Lobster *Homarus americanus*," *Materials Science and Engineering: A*, Vol. 421, No. 1–2, 2006, pp. 143–153.

[71] Raabe, D., et al., "Preferred Crystallographic Texture of Alpha-Chitin as a Microscopic and Macroscopic Design Principle of the Exoskeleton of the Lobster *Homarus americanus*," *Acta Biomaterialia*, Vol. 3, No. 6, 2007, pp. 882–895.

[72] Apichattrabrut, T., and K. Ravi-Chandar, "Helicoidal Composites," *Mechanics of Advanced Materials and Structures*, Vol. 13, No. 1, 2006, pp. 61–76.

[73] Goldstein, D. H., "Reflection Properties of Scarabaeidae," *Proceedings of SPIE*, Vol. 5888, 2005, 58880T.

[74] Goldstein, D. H., "Polarization Properties of Scarabaeidae," *Applied Optics*, Vol. 45, No. 30, 2006, pp. 7944–7950.

[75] Arwin, H., "Application of Ellipsometry Techniques to Biological Materials," *Thin Solid Films*, Vol. 519, No. 9, 2011, pp. 2589–2592.

[76] Arwin, H., et al., "Cuticle Structure of the Scarab Beetle *Cetonia aurata* Analyzed by Regression Analysis of Mueller Matrix Ellipsometric Data," *Optics Express*, Vol. 21, No. 19, 2013, pp. 22645–22656.

[77] Muñoz-Pineda, E., et al., "Symmetries and Relationships between Elements of the Mueller Matrix Spectra of the Cuticle of the Beetle *Cotinis mutabilis*," *Thin Solid Films*, Vol. 571, No. 3, 2014, pp. 660–665.

[78] Arwin, H., L. Fernandez del Río, and K. Järrendahl, "Comparison and analysis of Mueller-matrix Spectra from Exoskeletons of Blue, Green and Red *Cetonia aurata*," *Thin Solid Films*, Vol. 571, No. 3, 2014, pp. 739–743.

[79] Arwin, H., et al., "Sum Decomposition of Mueller-Matrix Images and Spectra of Beetle Cuticles," *Optics Express*, Vol. 23, No. 3, 2015, pp. 1951–1966.

[80] Fernandez del Río, L., H. Arwin, and K. Järrendahl, "Polarizing Properties and Structural Characteristics of the Cuticle of the Scarab Beetle *Chrysina gloriosa*," *Thin Solid Films*, Vol. 571, No. 3, 2014, pp. 410–415.

[81] Valyukh, S., H. Arwin, and K. Järrendahl, "Simulation of Light Scattering from Exoskeletons of Scarab Beetles," *Optics Express*, Vol. 24, No. 6, 2016, pp. 5794–5808.

[82] Ingram, A. L., et al., "Dual Gratings Interspersed on a Single Butterfly Scale," *Journal of the Royal Society Interface*, Vol. 5, 2008, pp. 1387–1390.

[83] Vigneron, J.-P., et al., "Reverse Color Sequence in the Diffraction of White Light by the Wing of the Male Butterfly *Pierella luna* (Nymphalidae: Satyrinae)," *Physical Review E*, Vol. 82, No. 2, 2010, 021903.

[84] Vigneron, J.-P., et al., "Natural Layer-by-Layer Photonic Structure in the Squamae of *Hoplia coerulea* (Coleoptera)," *Physical Review E*, Vol. 72, No. 6, 2005, 061904.

[85] Liu, F., et al., "Structural Color Change in Longhorn Beetles *Tmesisternus isabellae*," *Optics Express*, Vol. 17, No. 18, 2009, pp. 16183–16191.

[86] Mouchet, S. R., et al., "Photonic Scales of *Hoplia coerulea* Beetle: Any Colour you Like," *Materials Today: Proceedings*, Vol. 4, No. 4, 2017, pp. 4979–4986.

[87] Zi, Y., et al., "Coloration Strategies in Peacock Feathers," *Proceedings of the National Academy of Sciences of the United States of America*, Vol. 100, No. 22, 2003, pp. 12576–12578.

[88] Li, Y., et al., "Structural Origin of the Brown Color of Barbules in Male Peacock Tail Feathers," *Physical Review E*, Vol. 72, No. 010902(R), 2005, pp. 1–4.

[89] Feyer, P., B. D. Wilts, and D. G. Stavenga, "Reflections on Iridescent Neck and Breast Feathers of the Peacock, *Pavo cristatus*," *Interface Focus*, Vol. 9, No. 1, 2018, 20180043.

[90] Welch, V. L., *Structural Colouration in Jellyfish, Fish and Ctenophores*, PhD Thesis, Oxford University, UK, 2003.

[91] Welch, V. L., "Photonic Crystals in Biology," In *Structural Colors in Biological Systems - Principles and Applications*, pp. 53–71, S. Kinoshita and S. Yoshioda (eds.), Osaka: Osaka University Press, 2005.

[92] Welch, V. L., J.-P. Vigneron, and A. R. Parker, "The Cause of Colouration in the Ctenophore *Beroe cucumis*," *Current Biology*, Vol. 15, No. 24, 2005, pp. R985–R986.

[93] Welch, V. L., et al., "Optical Properties of the Iridescent Organ of the Comb-Jellyfish, *Beroe cucumis* (Ctenophora)," *Physical Review E*, Vol. 73, No. 4, 2006, 041916.

[94] Vigneron, J.-P., et al., "Structural Origin of the Colored Reflections from the Black-Billed Magpie Feathers," *Physical Review E*, Vol. 73, No. 021914, 2006, pp. 1–7.

[95] Parker, A. R., et al. "Opal Analogue Discovered in a Weevil," *Nature*, Vol. 426, 2003, pp. 786–787.

[96] Welch, V., et al., "Orange Reflection Form a Three-Dimensional Photonic Crystal in the Scales of the Weevil *Pachyrrhynchus congestus pavonius* (Curculionidae)," *Physical Review E*, Vol. 75, No. 4, 2007, 041919.

[97] Welch, V. L., and Vigneron, J.-P., "Beyond Butterflies—the Diversity of Biological Photonic Crystals," *Optical and Quantum Electronics*, Vol. 39, No. 4, 2007, pp. 295–303.

[98] Wilts, B. D., and V. Saranathan, "A Literal Elytral Rainbow: Tunable Structural Colors Using Single Diamond Biophotonic Crystals in *Pachyrrhynchus congestus* Weevils," *Small*, Vol. 14, No. 46, 2018, 1802328.

[99] Wilts, B.D., et al., "Brilliant Camouflage: Photonic Crystals in the Diamond Weevil, *Entimus imperialis*," *Proceedings of the Royal Society B*, Vol. 279, No. 1738, 2012, pp. 2524–2530.

[100] Wilts, B.D., et al., "Hemispherical Brillouin Zone Imaging of a Diamond-type Biological Photonic Crystal," *Journal of the Royal Society Interface*, Vol. 9, No., 2012, pp. 1609–1614.

[101] Galusha, J. W., et al., "Discovery of a Diamond-Based Photonic Crystal Structure in Beetle Scales," *Physical Review E*, Vol. 776, No. 5, 2008, 050904(R).

[102] Ebihara, R., et al., "Cuticle Network and Orientation Preference of Photonic Crystals in the Scales of the Weevil *Lamprocyphus augustus*," *Journal of the Royal Society Interface*, Vol. 15, No. 145, 2018, 20180360.

[103] Deparis, O., and J.-P. Vigneron, "Modeling the Photonic Response of Biological Nanostructures Using the Concept of Stratified Medium: The Case of a Natural Three-Dimensional Photonic Crystal," *Materials Science and Engineering B: Solid-State Materials for Advanced Technology*, Vol. 169, No. 1–3, 2010, pp. 12–15.

[104] Mouchet, S. R., et al., "Additive Photonic Colors in the Brazilian Diamond Weevil, *Entimus imperialis*," *Proceedings of SPIE*, Vol. 8480, 2012, 848003.

[105] Van Opdenbosch, D., et al., "Fabrication of Three-dimensional Photonic Crystals with Tunable Photonic Properties by Biotemplating," *Photonics and Nanostructures – Fundamentals and Applications*, Vol. 10, No. 4, 2012, pp. 516–522.

[106] Mouchet, S. R., et al., "Method for Modeling Additive color Effect in Photonic Polycrystals with Form Anisotropic Elements: The Case of *Entimus imperialis*," *Optics Express*, Vol. 21, No. 11, 2013, pp. 13228–13240.

[107] Wu, X., et al., "Extreme Optical Properties Tuned through Phase Subtitution in a Structurally Optimized Biological Photonic Polycrystal," *Advanced Functional Materials*, Vol. 23, No. 29, 2013, pp. 3615–3620.

[108] Michielsen, K., and D. G. Stavenga, "Gyroid Cuticular Structures in Butterfly Wing Scales: Biological Photonic Crystals," *Journal of the Royal Soicety Interface*, Vol. 5, 2008, pp. 85–94.

[109] Michielsen, K., H. De Raedt, and D. G. Stavenga, "Reflectivity of the Gyroid Biophotonic Crystals in the Ventral Wing Scales of the Green Hairstreak Butterfly, *Callophrys rubi*," *Journal of the Royal Society Interface*, Vol. 7, 2010, pp. 765–771.

[110] Saranathan, V., et al., "Structure, Function, and Self-Assembly of Single Network Gyroid (I4132) Photonic Crystals in Butterfly Wing Scales," *Proceedings of the National Academy of Sciences of the United States of America*, Vol. 107, 2010, pp. 11676–11681.

[111] Pouya, C., and P. Vukusic, "Electromagnetic Characterization of Millimetre-scale Replicas of the Gyroid Photonic Crystal Found in the Butterfly *Parides sesostris*," *Interface Focus*, Vol. 2, No. 5, 2012, pp. 645–650.

[112] Colomer, J.-F., et al., "Photonic Polycrystal in the Greenish-White Scales of the African Longhorn Beetle *Prosopocera lactator* (Cerambycidae)," *Physical Review E*, Vol. 85, No. 011907, 2012, pp. 1–9.

[113] Burresi, M., et al., "Bright-White Beetle Scales Optimize Multiple Scattering of Light," *Scientific Reports*, Vol. 4, 2014, 6075.

[114] Stavenga, D. G., M. A. Giraldo, and B. J. Hoenders, "Reflectance and Transmittance of Light Scattering Scales on the Wings of Pierid Butterflies," *Optics Express*, Vol. 14, No. 11, 2006, pp. 4880–4890.

[115] Stavenga, D. G., et al., "Butterfly Wing Colours: Scale Beads Make White Pierid Wings Brighter," *Proceedings of the Royal Society B*, Vol. 71, 2004, pp. 1577–1584.

[116] Morehouse, N. I., P. Vukusic, and R. Rutowski, "Pterin Pigment Granules are Responsible for Both Light Scattering and Wavelength Selective Absorption in the Wing Scales of Pierid Butterflies," *Proceedings of the Royal Society B*, Vol. 274, 2007, pp. 359–366.

CHAPTER 3

Color Additive Devices

In nature, many visual appearances arise from additive color mixing, including butterflies [1–3], beetles [4–8], fruits [9], fish [10], and cephalopods [11]. One example described in the previous chapter is the case of the African shield bug *Calidea panethiopica* [5]. These effects usually occur at multiple length-scales. They arise from the juxtaposition of different colors due to, for example, some surface texturing at the micrometric-length scale [3, 5] or some form of disorder, such as photonic polycrystals with grains corresponding to different structures or the same structure repeated at different orientations [1, 2, 6–8, 12]. In this chapter, after a general introduction to color mixing, we will describe a few biological systems giving rise to colors by additive mixing due to multiscale interferential effects.

3.1 Color Mixing

Additive color mixing consists of the addition of colored light beams [13–15]. For instance, it occurs when spotlights of different colors overlap at the same place and in a simultaneous manner on a screen or on the retina of the eye. Human perception of mixed colors can be described by Grassmann's laws, which were discovered in 1853 and are named after their discoverer, Hermann Grassmann [16]. They state that a color can be produced by an additive mix of three primary colors. In a color mixing system, three colors are considered as primaries if it is not possible to reproduce one of them by a combination of the other two. This number of three consists in an irreducible minimum. The color space defined by these colors is called the gamut. In the framework of additive color mixing, primaries are usually red, green, and blue. Mixing them in the correct proportions gives rise to white (Figure 3.1(a)). Mixing two of them produces magenta (blue and red), cyan (blue and green), or yellow (red and green). The lack of incident light naturally leads to black. Same colors can usually be produced from different spectral intensity distributions. These colors are called metamers. For instance, yellow incident light, with an intensity peak at 580 nm, would give rise to the same color as a mix of some red and green light without leading to any difference in perception to the human eye and brain. One of the main advantages of additive color mixing lies in the simplicity of the mixing rules that follow the laws of linear algebra. Additive color mixing is used in television, computer, and smartphone screens, where it arises from a set of tiny and nonoverlapping light sources, acting as pixel components called subpixels. These subpixels appear as one color in the distance due to blurring by the optics and spatial integration by the photoreceptors and nerves in the eye.

In addition, colors may be produced by subtractive color mixing [13–15]. This consists of the subtraction of colors from the white light spectrum. With such a color

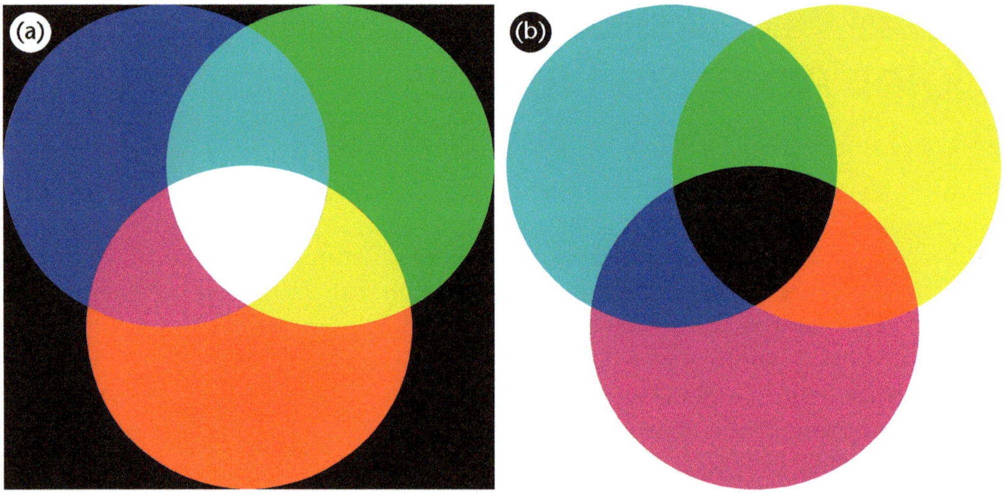

Figure 3.1 Color mixing occurs either by an additive process (a), as illustrated here with blue, green, and red colors or a subtractive process (b), as here with cyan, magenta, and yellow. (Adapted (a) and reproduced (b) from Terin, https://commons.wikimedia.org/wiki/File:CMY_ideal_version.svg, public domain.)

mixing, an object or substance containing pigments would, for instance, selectively absorb incident white light and scatter nonabsorbed light. Primaries defined for subtractive color mixing are magenta, cyan, and yellow. Most people have known since preschool that mixing two of these colors in pigmentary forms (e.g., watercolors or gouaches) gives rise to red (magenta and yellow), green (yellow and cyan), or blue (cyan and magenta)—namely, the primaries of additive color mixing. Black arises from mixing magenta, cyan, and yellow together in the correct proportions (Figure 3.1(b)). Subtractive color mixing is found in, for example, dyeing, color film photography, painting, and color printing.

However, this explanation of subtractive color mixing relies upon some simplification. This process is more complex than its additive counterpart due to the nature of pigments. A color produced by such a mixing process not only depends on the visual appearances of the primaries, but also on their exact spectral responses and relative intensities. While most colors can be produced by a computer screen, the pigment industries tend to broaden the range of available colors by developing new pigments instead of relying on three pigments. Ink cartridges for printers contain pigments that were designed in order to give rise to as large a gamut as possible. However, producing black by subtractive color mixing only occurs if incident light is totally absorbed at all wavelengths of the visible spectrum. This means that the absorbing pigments must correspond to highly saturated colors and, together, they must cover the whole visible spectrum. The saturation of a color relates to its intensity and purity: whether or not it is "polluted" by other colors. With very desaturated colors (i.e., pastel colors) that involves strong scattering by pigments at some wavelengths and no total absorption at any other wavelengths, producing true black is impossible. In addition to cost efficiency considerations, that is why many color printers require an extra black cartridge. For this reason, restoring ancient paintings usually proves to be quite tricky. Spectral absorption properties of pigments change due to aging. Selecting the right pigments in order to produce the metamer colors needed for some restorations is a real challenge.

3.2 The Colorful Stripes of *Argyrophorus argenteus* Butterfly

The male *Argyrophorus argenteus* butterflies have naturally evolved a structure that displays a broadband silver color [3] (Figure 3.2). The upper lamina of the scale structure has a symmetric sawtooth profile, whereas the basal lamina is approximately flat. This results in air gaps forming a corridor-like configuration displaying a variation in the gap thickness. This quasi-ordered structure gives rise to light interference. Due to the range of thicknesses occurring in the structure, multicolored bands are displayed by the butterfly scales under an optical microscope. Through an additive color mixing process, the juxtaposed bands exhibit the silver color that is observed macroscopically.

3.3 The Shiny Colors of *Pollia condensata* Fruits

One remarkable example of additive photonic colors due to a complex biological structure is encountered in *Pollia condensata* fruits [9]. This plant grows in the understor of African forests and produces spherical fruits that display a metallic-blue visual appearance when they reach maturity. As described in the previous chapter, the blue colors are due to both left- and right-handed Bouligand structures occurring within the cell walls of the fruit epicarp tissues (Figure 2.13 (b,c)). Each cell reflects either left- or right-handed polarized light. These helicoidal multilayers are made of cellulose microfibrils and cover a few layers of brown tannin pigments. These tannin-rich absorbing layers increase the saturation of the displayed colors. In addition to the change in polarization, the pitch of the Bouligand structure varies from one cell to another, giving rise to a variety of colors. These are mostly blue, with some green and red, and described as "pointillist" by the authors [9] (Figure 3.3). In a similar way to the painting style of an impressionist artist, the overall color perceived by a normal observer is blue.

3.4 Photonic Crystal Grains in Beetle Scales

The elytra and bodies of several beetles, including weevils and longhorn beetles, are known to be covered in scales comprising photonic polycrystals. These scales can be almost round or quite elongated. Typically, the photonic structures are periodic along the three dimensions and described as 3D photonic crystals, at least all over a region of limited size, called a grain. Within one single grain, also known as domain, the photonic crystal structure exhibits the same crystallographic orientation. The size of such a domain may reach up to 2000 μm^2. One species that is specifically relevant here is *Entimus imperialis*, the Brazilian diamond weevil. This insect is quite common in Brazil and is known to live on tree leaves. Its size can reach about 3 cm. For a long time it has been used for making jewelery by Amazonian populations, sometimes in combination with gold—hence its vernacular name of

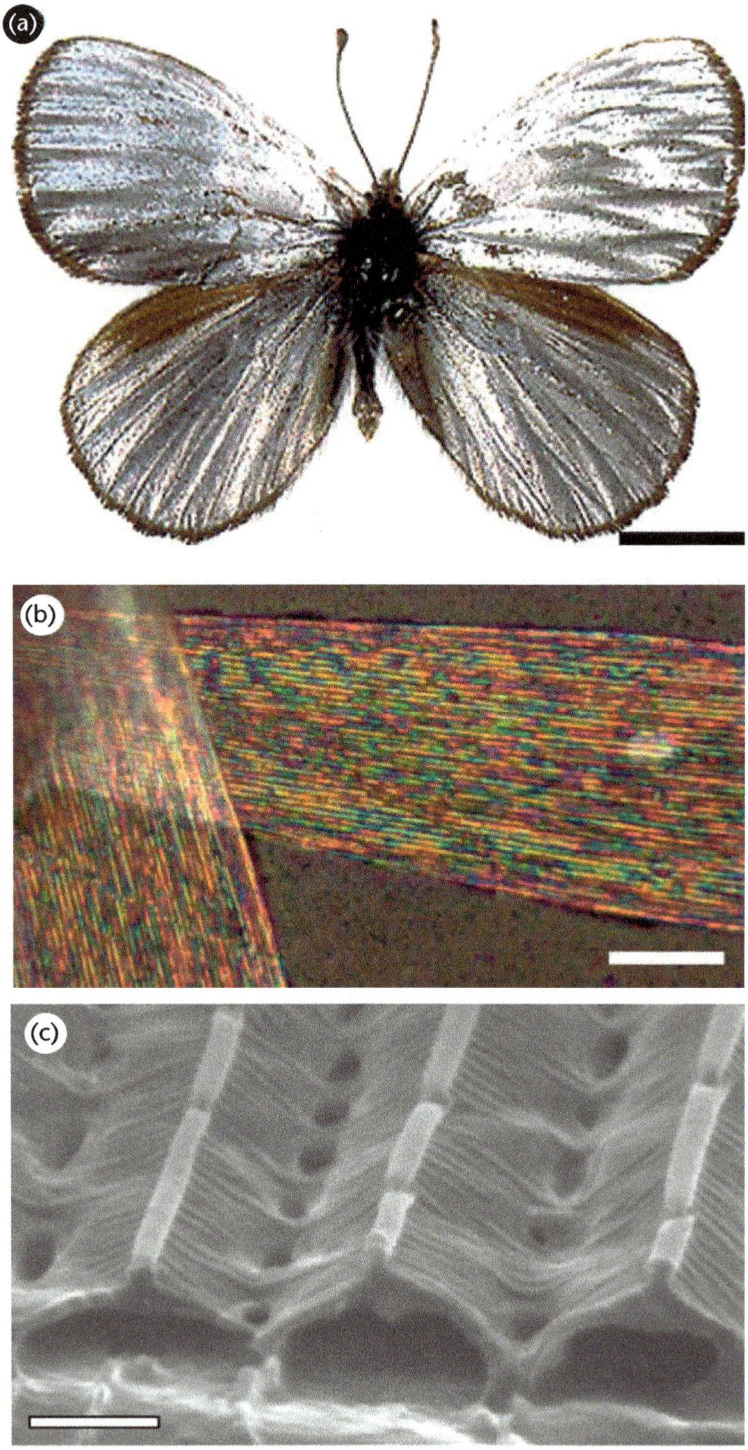

Figure 3.2 The silver color of *A. argenteus* butterfly wings (a) arises from an additive color mixing process of the colorful quasi-ordered bands (b). These bands are due to variations in the thickness of the air gap between the upper lamina and the basal lamina (c). The scale bars correspond to 0.5 cm (a), 15 μm (b), and 0.8 μm (c). (Reproduced from [3], with permission from the Royal Society.)

Figure 3.3 *Pollia condensata* fruits are a remarkable example of additive photonic colors. (a,b) The fruit's blue color appearance is due to left- and right-handed Bouligand structures occurring within the cell walls of the fruit epicarp tissues. Each cell reflects either left- or right-handed polarized light, as observed from optical micrographs of the same area of the fruit with (c) left- or (d) right-handed light, as well as (e) between crossed polarizers. (Reproduced from [9], with permission from PNAS.)

"diamond" weevil. In the far field, *E. imperialis* elytra display noniridescent matte green spots on its elytra. This visual appearance is reported to play a role in the beetle's camouflage [8]. These green spots arise from arrangements of scales gathered around pits covering the elytral surface. Each scale comprises a few grains, displaying bright colors ranging from orange to blue. Back in 1883, Dimmock noticed the palette of colors arising from this beetle's scales and hypothesized their interferential origin [17]. This observation was confirmed 45 years later by Mason [18]. Over the last decades, this insect has attracted much attention from the natural photonics community [6–8, 19–22]. The photonic structure occurring in its grains can be described by a diamond-type triply-periodic bicontinuous cubic minimal surface structure [8, 19], as mentioned in the previous chapter (Figures 2.16 and 2.17). A similar photonic structure was observed in the scales of *Pachyrrhynchus congestus pavonius* [23, 24], *Lamprocyphus augustus* [25, 26], *Eupholus cuvieri* [27], *Eupholus magnificus* [28] weevils, and *Prosopocera lactator* longhorn beetle [29]. It was recently shown to convert between linear polarization states [22]. In addition, the juxtaposition of grains of photonic crystals with various orientations on the dark background of the *Entimus imperialis* weevil's elytra gives rise to an additive color [6–8, 20]. On the one hand, each grain is iridescent and independently interacts with incident light leading to bright light reflection, the color of which ranges from orange to blue (Figure 3.4). On the other hand, the color of the spots perceived by the human naked eye results from a multiscale effect, involving a disordered set

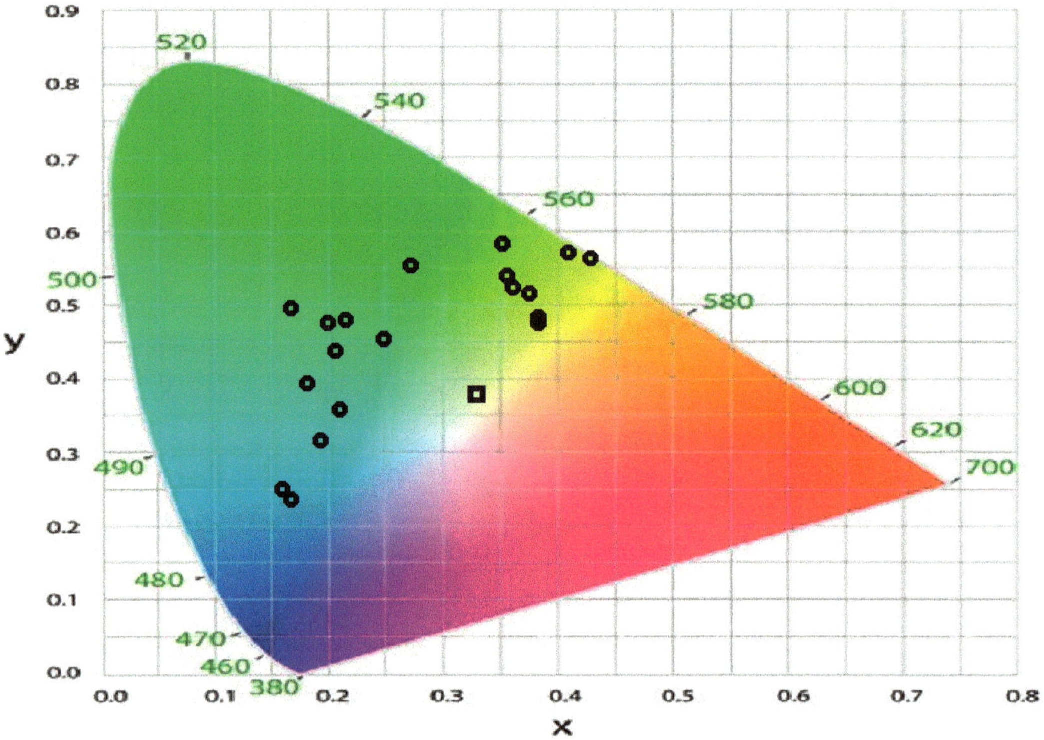

Figure 3.4 The noniridescent matte green spots displayed by *E. imperialis* elytra result from additive color mixing of the bright colorful polycrystals that are found in the scales covering this beetle's elytra. Two important features to the observed color are illustrated here. Firstly, each grain is iridescent and interacts independently with incident light, leading to bright light reflection, the color of which ranges from orange to blue. This was measured by microspectrophotometry and plotted in the CIE 1931 chromaticy diagram (circles), following the method described in Chapter 1. Secondly, the color of the spots perceived by the naked human eye results from a multiscale effect, involving a disordered set of very ordered grains with several crystallographic orientations. The color measured by spectrophotometry lies in the desaturated green area of the chromaticity diagram (square). (Reproduced from [30], with permission from the Electromagnetics Academy.)

of very order grains with several crystallographic orientations (Figure 3.4). This additive color mixing suppresses the iridescence.

References

[1] Vigneron, J.-P., et al., "Reflectance of Topologically Disordered Photonic-Crystal Films," *Proceedings of SPIE*, Vol. 5733, 2005, pp. 308–315.

[2] Kertész, K., et al., "Gleaming and Dull Surface Textures from Photonic-Crystal-Type Nanostructures in the Butterfly *Cyanophrys remus*," *Physical Review E*, Vol. 74, No. 2, 2006, 021922.

[3] Vukusic, P., R. Kelly, and I. Hooper, "A Biological Sub-micron Thickness Optical Broadband Reflector Characterized Using Both Light and Microwaves," *Journal of the Royal Society Interface*, Vol. 6, No. S2, 2009, pp. S193–S201.

[4] Schultz, T. D., and G. D. Bernard, "Pointillistic Mixing of Interference Colours in Cryptic Tiger Beetles," *Nature*, Vol. 377, No. 6202, 1989, pp. 72–73.

[5] Vigneron, J.-P., et al., "Spectral Sideband Produced by a Hemispherical Concave Multilayer on the African Shield-Bug *Calidea panaethiopica* (Scutelleridae)," *Physical Review E*, Vol. 79, No. 2, 2009, 021907.

[6] Deparis, O., and J.-P. Vigneron, "Modeling the Photonic Response of Biological Nanostructures Using the Concept of Stratified Medium: The Case of a Natural Three-Dimensional Photonic Crystal," *Materials Science and Engineering B*, Vol. 169, No. 1-3, 2010, pp. 12–15.

[7] Mouchet, S. R., et al., "Additive Photonic Colors in the Brazilian Diamond Weevil, *Entimus imperialis*," *Proceedings of SPIE*, Vol. 8480, 2012, 848003.

[8] Wilts, B. D., et al., "Brilliant Camouflage: Photonic Crystals in the Diamond Weevil, *Entimus imperialis*," *Proceedings of the Royal Society B*, Vol. 279, No. 1738, 2012, pp. 2524–2530.

[9] Vignolini, S., et al., "Pointillist Structural Color in *Pollia* Fruit," *Proceedings of the National Academy of Sciences of the United States of America*, Vol. 109, No.39, 2012, pp. 15712–15715.

[10] Denton, E. J., and J. A. C. Nicol, "A Survey of Reflectivity in Silvery Teleosts," *Journal of the Marine Biological Association of the United Kingdom*, Vol. 46, No. 3, 1966, pp. 685–722.

[11] Mäthger, L. M., et al., "Mechanisms and Behavioural Functions of Structural Coloration in Cephalopods," *Journal of the Royal Society Interface*, Vol. 6, No. S2, 2009, pp. S149–S163.

[12] Mouchet, S. R., et al., "Optical Costs and Benefits of Disorder in Biological Photonic Crystals," *Faraday Discussions*, Vol. 223, 2020, pp. 9–48.

[13] Judd, D. B., and G. Wyszecki, *Color in Business, Science and Industry*, New York: John Wiley & Sons, 1975.

[14] Chamberlin, G. J., and D. G. Chamberlin, *Colour: Its Measurement, Computation and Application*, London: Heyden International Topics in Science, 1980.

[15] Elias, M., and J. Lafait, *La Couleur: Lumière, Vision et Matériaux*, Paris: Éditions Belin, 2006.

[16] Grassmann, H., "Zur Theorie der Farbenmischung," *Annalen der Physik und Chemie*, Vol. 165, No. 5, 1853, pp. 69–84.

[17] Dimmock, G., "The Scales of Coleoptera," *Psyche*, Vol. 4, No. 109–110, 1883, pp. 43–47.

[18] Mason, C. W., "Structural Colors in Insects. II," *Journal of Physical Chemistry*, Vol. 31, No. 3, 1927, pp. 321–354.

[19] Wilts, B. D., et al., "Hemispherical Brillouin Zone Imaging of a Diamond-Type Biological Photonic Crystal," *Journal of the Royal Society Interface*, Vol. 9, No. 72, 2012, pp. 1609–1614.

[20] Mouchet, S. R., et al., "Method for Modeling Additive Color Effect in Photonic Polycrystals with Form Anisotropic Elements: The Case of *Entimus imperialis* Weevil," *Optics Express*, Vol. 21, No. 11, 2013, pp. 13228–13240.

[21] Wu, X., et al., "Extreme Optical Properties Tuned through Phase Subtitution in a Structurally Optimized Biological Photonic Polycrystal," *Advanced Functional Materials*, Vol. 23, No. 29, 2013, pp. 3615–3620.

[22] Wu, X., et al., "Polarization Conversion Effect in Biological and Synthetic Photonic Diamond Structures," *Advanced Optical Materials*, Vol. 6, No. 24, 2018, 1800635.

[23] Welch, V., et al., "Orange Reflection from a Three-Dimensional Photonic Crystal in the Scales of the Weevil *Pachyrrhynchus congestus pavonius* (Curculionidae)," *Physical Review E*, Vol. 75, No. 4, 2007, 041919.

[24] Wilts, B. D., and V. Saranathan, "A Literal Elytral Rainbow: Tunable Structural Colors Using Single Diamond Biophotonic Crystals in *Pachyrrhynchus congestus Weevils*," *Small*, Vol. 14, No. 46, 2018, 1802328.

[25] Galusha, J. W., et al., "Discovery of a Diamond-Based Photonic Crystal Structure in Beetle Scales," *Physical Review E*, Vol. 77, No. 5, 2008, 050904(R).

[26] Ebihara, R., et al., "Cuticle Network and Orientation Preference of Photonic Crystals in the Scales of the Weevil *Lamprocyphus augustus*," *Journal of the Royal Society Interface*, Vol. 15, No. 145, 2018, 20180360.

[27] Wilts, B., et al., "Shine and Hide: Biological Photonic Crystals on the Wings of Weevils," *MRS Proceedings*, Vol. 1504, 2013, pp. 1–6.

[28] Pouya, C., D. G. Stavenga, and P. Vukusic, "Discovery of Ordered and Quasi-Ordered Photonic Crystal Structures in the Scales of the Beetle *Eupholus magnificus*," *Optics Express*, Vol. 19, No. 12, 2011, pp. 11355–11364.

[29] Colomer, J.-F., et al., "Photonic Polycrystal in the Greenish-White Scales of the African Longhorn Beetle *Prosopocera lactator* (Cerambycidae)," *Physical Review E*, Vol. 85, No. 1, 2012, 011907.

[30] Mouchet, S. R., et al., "Layer Homogenization Method for Modeling Additive Color Effect in Natural Photonic Polycrystals – The Case of *Entimus imperialis* Weevil," *PIERS Proceedings*, Stockholm, Sweden, Aug. 12–15, 2013, pp. 43–47

CHAPTER 4

Transparent Devices

In insects, the evolution of antireflective structures on the corneal surface of the eye has long been a way to increase photon capture and therefore enhance visual efficiency. More precisely, these structures are found on the outer surface of the optical units making up the compound eye of dipterans and lepidopterans (the ommatidia) [1–3]. In a fossil of an extinct fly species dating back to the Ecocene period (45 Ma), Parker and coworkers discovered an antireflection device on well-preserved eyes embedded in Baltic amber matrix [4]. The structure consists of many ridge-type gratings of small areas, which form a cross-hatched pattern on the ommatidium. A close observation shows that the grating ridges are made of aligned protuberances that tend to disorganize on the flattened central region of each ommatidium. This structure is known today as a moth eye structure. More or less ordered patterns of protuberances, sometimes called nipple arrays, also cover the wings of some butterflies, moths, dragonflies, damselflies, and cicadas. They act as antireflection structures, which make them transparent. These antireflection devices often give rise to additional surface properties such as (super-)hydrophobicity [5–13] or antibacterial activity [12, 14–18]. This chapter provides a brief overview of transparency properties in insect wings and eyes, as well as their combination with other surface properties.

4.1 Enhanced Transparency in Insect Wings

In lepidopteran insects (i.e., butterflies and moths), antireflection structures can be found. They make wings more transparent to light by reducing reflectance below typically 2% over the whole visible range [7, 10, 11, 19–26]. In the hawkmoth *Cephonodes hylas*, a rather unusual case of insect wings without scales, Yoshida found a regular hexagonal array of protuberances on the wing surface (Figure 4.1) [19–21]. Parker and Vigneron found a similar type of protuberance array in another moth species, *Cacostatia ossa*, yet with a slightly larger spatial period and some disorder [24]. The glasswing buterfly *Greta oto* was an interesting case for studying the role of disorder as well [25, 26]. Similar structures were also found in odonatan insects, for instance on the transparent wings of the dragonfly *Aeshna cyanea* [22]. Amongst hemipteran insects, several species of cicadas possess transparent wings with antireflection structures [7, 8, 10, 11, 23]. Structures with very similar morphologies and optical behaviors have been found in many species and in both male and female insects [7, 23].

The overall explanation for the antireflection effect is the gradual refractive index matching between the incidence medium (air) and the wing membrane material (usually assumed to be chitin) that is provided by the dome-shaped

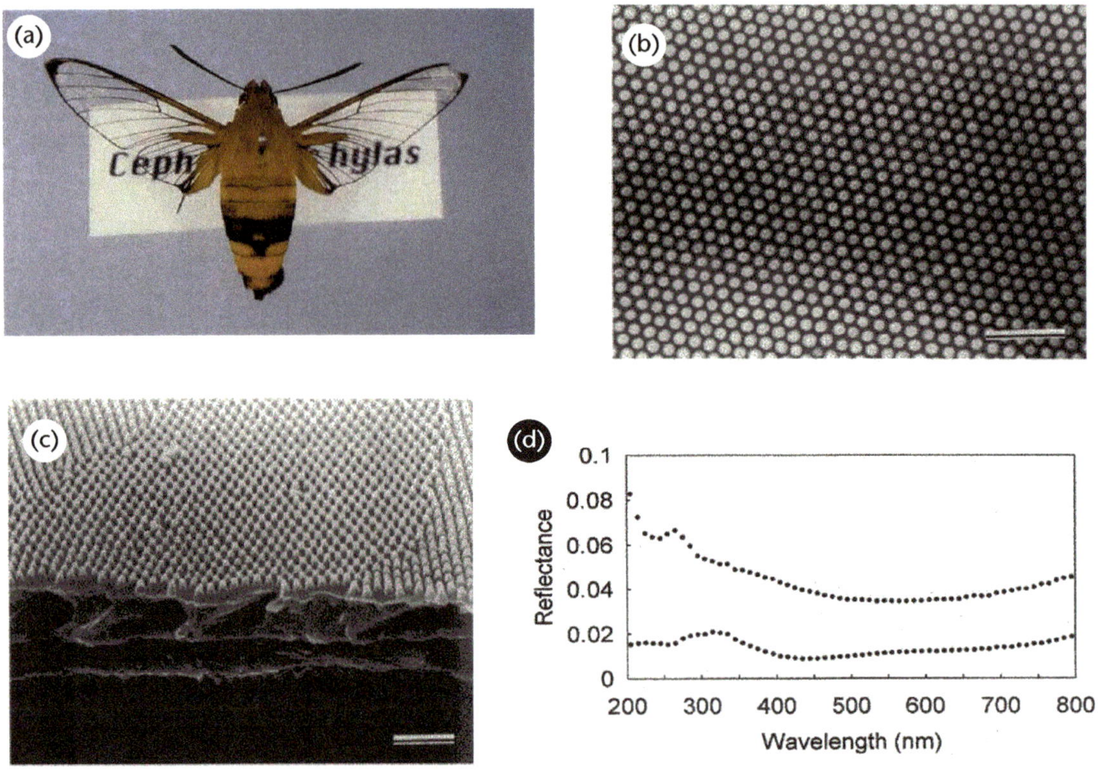

Figure 4.1 The enhanced transparency in the wings of the hawkmoth *Cephonodes hylas* (a) is due to the ordered array of nanosized protuberances (b,c) covering the wing surface observed here under SEM. The reflectance spectra (d) were measured on a wing of the moth with intact protuberance structures (lower trace) and on the wing after crushing these protuberances (upper trace). Scale bars: 1 μm (b,c). (Reproduced from [20], with permission from the Zoological Society of Japan.)

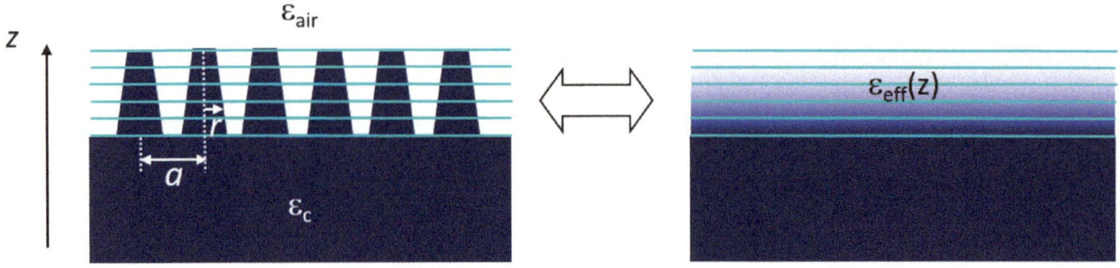

Figure 4.2 Effective medium treatment of a nipple array on top of a wing substrate (assumed to be chitin). The nipples (modeled here as truncated cones) are divided into thin layers, for which an effective permittivity ε_{eff} is calculated as a function of the layer altitude z. The explicit relationship between ε_{eff}, ε_c (chitin) and ε_{air} depends on the geometry and the type of effective medium approximation used (a: lattice constant, r: cone radius within each individual layer).

protuberances (Figure 4.2). When the average distance between protuberances (i.e., the spatial period when the protuberance array is perfectly ordered) is smaller than the shortest wavelength of the spectral range of interest, the structure can be regarded as a subwavelength 2D grating, for which all diffraction orders except the zero-order are evanescent. In this case, an effective medium approximation approach can be used to calculate the reflectance. The typical spacing between neighboring protuberances (namely, the period) of approximately 150–200 nm [19, 20, 22, 23] justifies this modeling approach in the visible range, at least at normal incidence. Using an effective medium modeling technique, Stoddart and coworkers, indeed, achieved excellent agreement with transmittance spectra measured on transparent cicada wings [23]. In such an effective medium modeling, the structure is divided into layers and the effective dielectric constant in each layer is calculated from a mixing formula, in which the geometry-dependent volume filling fractions of air and material enter as parameters. A sufficiently large number of layers has to be assumed in order to obtain a smooth, slowly varying refractive index profile along the direction normal to the wing surface. The mixing formula may vary according to the authors. Stoddart and coworkers used a simple proportion rule (see Chapter 2); that is, $\varepsilon_{\text{eff}}(z) = \varepsilon_{\text{air}} A_{\text{air}}(z) + \varepsilon_{\text{c}} A_{\text{c}}(z)$ where A_{air} and A_{c} are, respectively, the air and chitin volume filling fractions, which depend on the depth z across the structure (Figure 4.2).

The non-close packed, less-ordered nipple array structure found on the transparent wings of the moth *Cacostatia ossa* had an average period of approximately 380 nm [24] (Figure 4.3). This relatively large period was shown to be responsible for a small, yet nonnegligible, contribution of diffraction orders to the total (hemispherical) reflectance, as attested by 3D scattering calculations using the rigorous coupled wave analysis (RCWA) method. The nipple array structure, observed on both sides of the wing, was modeled by a periodic array of nontouching truncated cones, the choice of the array type (square or hexagonal) being not critical, given the presence of disorder in such structures. Comparison between RCWA and effective medium calculations demonstrated that the latter approach could not render accurately the reflectance behavior at short wavelengths, even at normal incidence. The mismatch was even worse when the angle of incidence was increased. In such a case, RCWA calculations, which take into account diffraction orders, are robust in describing both the spectral and angular broadband features of the antireflection. After removing the nipple array structure from the wing back side in the model, RCWA simulations predicted a small yet significant contribution of the back side corrugation to the antireflection.

The case of the glasswing butterfly *Greta oto* has brought important lessons for our understanding of nipple or pillar-like antireflection devices: disorder may improve performance. Indeed, in this species, randomness in pillar height distribution was shown to confer a high degree of omnidirectionality on the antireflection property [26]. The diffuse reflectance was as low as 2% (as compared with 8% for a glass plate) over the visible range, with a low haze. The dependence of the specular reflectance on the incidence angle was small, with a level maintained below 2.2% up to approximately 65°. The surface corrugation had the particularity that pillars were irregularly arranged and had random height and width distributions. Pillars had typical heights ranging from 160 nm to 200 nm, an aspect ratio of approximately 5; they had cylindrical shapes and stood on conical pedestals. The relatively small average distance between two adjacent pillars (approximately 120 ± 20 nm) allowed an effective medium modeling approach. With the standard deviation of height (σ_h) as a parameter, the refractive index profile became smoother as randomness increased, approaching the optimal quintic profile for a graded-index

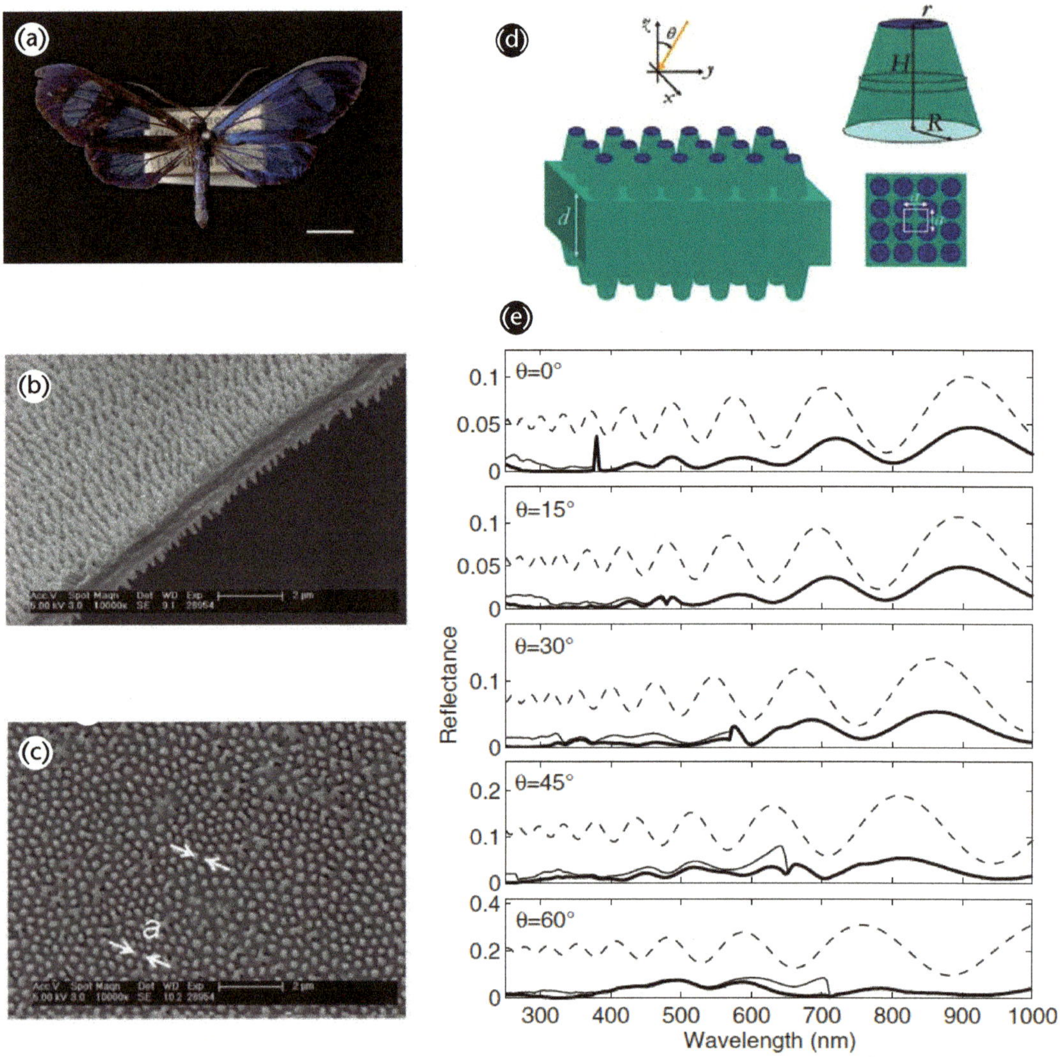

Figure 4.3 *Cacostatia ossa* moth (a) has wings with enhanced transparency due to a disordered nipple array structure covering its transparent wings (b,c). These nipples were modeled as truncated cones for RCWA reflectance calculations (d). The total and specular reflectance spectra (e) of the model (thin and thick solid lines, respectively) were calculated at various incidence angles for TE-polarized light. Specular reflectance spectra of a bare wing are shown for comparison. Scale bars: 0.5 cm (a) and 2 μm (b,c). (Reproduced from [24], with permission from the American Physical Society.)

antireflection coating when pedestals were included in the model. For a nonrandom structure ($\sigma_h = 0$), the omnidirectional aspect of the antireflection property was nearly lost.

4.2 Enhanced Transparency in Insect Eyes

Numerous, separate, and independent clusters of photoreceptors, called ommatidia, form the compound eyes of arthropods such as insects and crustaceans. The anatomy of the faceted surface of insect compound eyes was observed for the first time in 1965 by Berhnard, Miller, and Moller, thanks to the progress in electron microscopy at that time [1]. According to this early study, the corneal surface in numerous insect species appeared to be smooth in electron micrographs but, for certain species, it was entirely covered by conical protuberances, which they named corneal nipples without knowing, a priori, their function (actually, in a later study, they showed that corneal nipple arrays were widespread among insects [2]).

What makes our basic understanding of nipple arrays as broadband antireflection devices was already present in this seminal work. The regularity of protuberance shape and spacing, their conical shape, their approximately hexagonal display and close packing, the average height and center-to-center distance (estimated at approximately 200 nm) smaller than the visible light wavelengths—all these characteristics were recognized as the key ingredients for producing an efficient biological impedance transformer device, which is able to reduce reflection from the corneal front surface and thereby to increase transmission to the photoreceptor cells thanks to the gradual transition of refractive index from air ($n = 1.00$) to the lens material ($n \simeq 1.57$). It was also pointed out that the nipple array device reduces the reflection more efficiently over a broad range of wavelengths than does the quarter-wave plate antireflection coating, which is designed to cancel reflection at a single wavelength. The lower sensitivity of the nipple array antireflection property to the angle of incidence was explained along the same lines [1]. Interestingly, the experimental demonstration of the antireflection property was obtained from microwave reflection and transmission measurements (confirmed by calculations) on scaled models of the nipple array tens of centimeters in size. This is the same scaling method researchers have used 30 years later in order to demonstrate experimentally the existence of a forbidden energy gap in photonic crystals! Regarding the biological function of the nipple array antireflection device, the authors speculated it could serve as greater camouflage during the day or increase night vision, depending on whether the species concerned had diurnal or nocturnal activity, respectively. Considering the very small increase in transmittance (a few percent) wrought by the nipple array, the effect on light sensitivity is questionable. The other option, camouflage, is likely to be more plausible for moths that are inactive in the daytime, since glittering eyes would increase their vulnerability to predators.

Corneal nipple arrays in butterflies were investigated by Stavenga and coworkers in 19 species among the families Papilionidae, Pieridae, Lycaenidae, and Nymphalidae [3]. They classified nipples according to three types of shape (conical, paraboloidal, Gaussian) and performed reflectance calculations based on effective medium modeling, assuming a hexagonal lattice and varying the nipple height and width (Figure 4.4). They concluded that paraboloid nipples nearly touching each other, with a corresponding almost linear effective refractive index profile, led to the strongest antireflection effect. They found, however, that the precise shape of the nipples was not critical for reducing reflection, as antireflection was mainly determined by the height of the nipples, with their width playing only a secondary role. Noticing that the five papilionid species investigated had only shallow nipples (height smaller than 30 nm), they suggested an evolutionary scenario in order to explain the loss of nipple trait in papilionids [3].

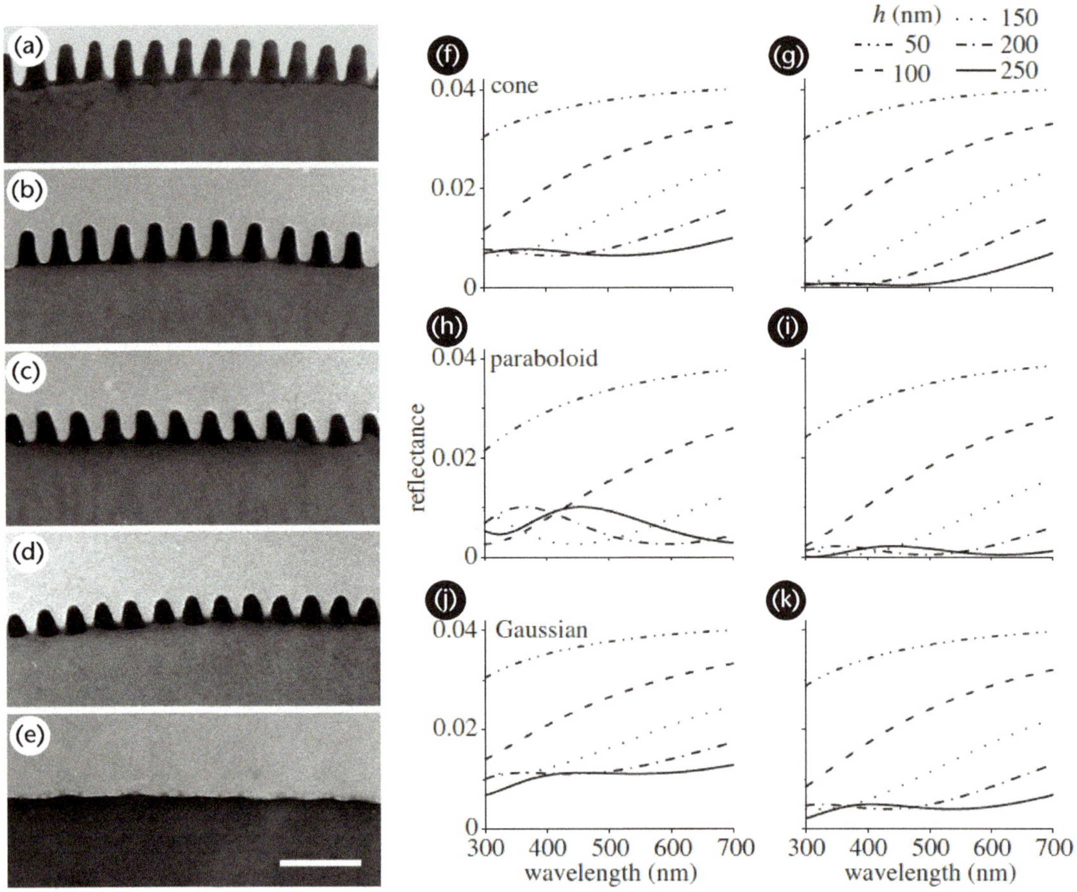

Figure 4.4 Cross-sectional TEM images of corneal nipple arrays in the nymphalid butterflies *Bicyclus anynana* and *Polygonia c-aureum* (a,b), the pierid *Pieris rapae* (c), the lycaenid *Pseudozizeeria maha* (d) and the papilionid *Papilio xuthus* (e). Calculated reflectance spectra at normal incidence for three types of nipple shape (cone (f,g), paraboloid (h,i), Gaussian(j,k)) with two values of width parameter ($p = 0.40$ (f,h,j) and $p = 0.53$ (g,i,k)) and five values of height parameter (h, see inset). Scale bar: 500 nm (a–e). (Reproduced from [3], with permission from the Royal Society.)

4.3 Transparency Combined with Other Properties

Nipple array structures on insect wings often have a function other than antireflection: hydrophobicity [5–13]. Both properties were studied in the case of cicada wings although this was done independently. Since the properties originate from the same structure, Dellieu and coworkers decided to investigate both with the aim of understanding the role of the nipple geometry in the interplay between optical and wetting properties [10, 11]. They selected the gray cicada *Cicada orni* (Figure 4.5), one of the most abundant cicada species in the south of France. The authors found that antireflection and superhydrophobicity properties were combined in a two-in-one device at different levels of the nanostructure's geometry. Displayed in an approximately hexagonal array (with a period $a = 173$ nm), the nipples had, on average, a truncated cone shape (base radius $R = 85$ nm) with a hemispherical top (radius $r = 40$ nm), and a cumulative height $H = 220$ nm. Transmittance

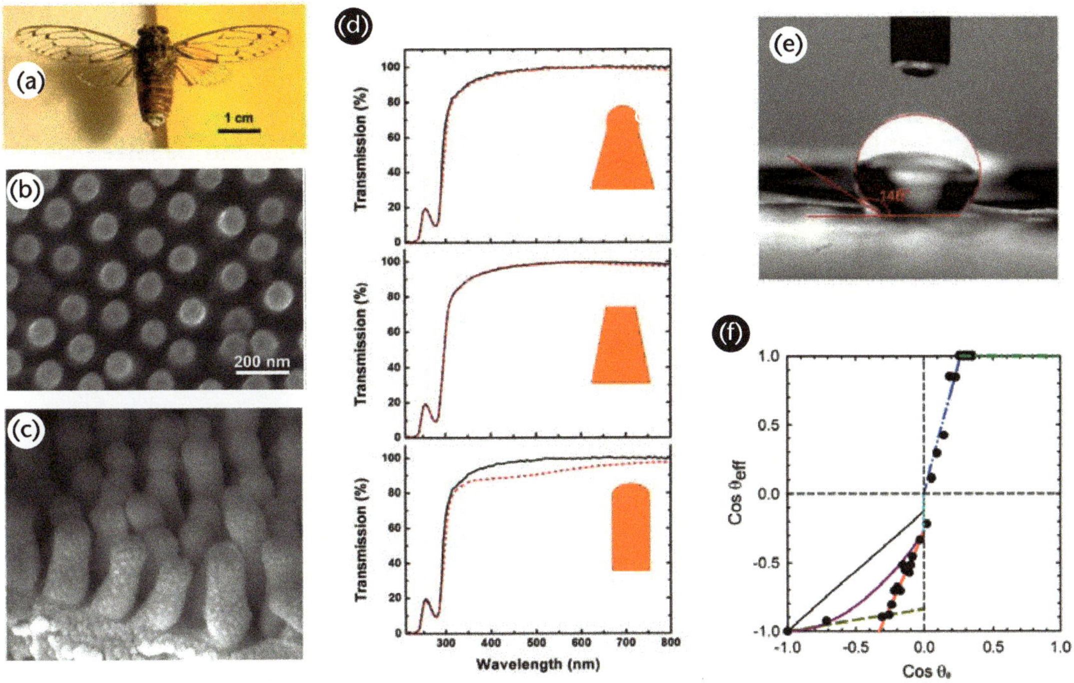

Figure 4.5 The gray cicada *Cicada orni* (a) exhibits a nipple array (b,c) on its wings. Transmittance spectra were measured and calculated using three nipple geometrical models (d). Water static contact angle measurements on the wing showed an effective value of 146°, close to superhydrophobicity ($> 150°$)(e). The measured effective static contact angle θ_{eff} was calculated as a function of the static contact angle θ_0 on flat chitin surface for liquids of different surface tensions (f). The purple curve and dashed dotted blue line are theoretical predictions from Cassie-Baxter model (hydrophobic regime: $\cos\theta_{eff} < 0$) and Wenzel model (hydrophilic regime: $\cos\theta_{eff} > 0$) applied to the hemisphere-topped truncated-cone nipple geometry. The black line is a Cassie-Baxter prediction obtained by removing the hemisphere from the nipple model. The horizontal green line corresponds to the hemiwicking regime. Red and gold lines are asymptotic experimental trends. (Reproduced from [10], with permission from AIP Publishing.)

spectra measurement, with and without calibrated index-matching liquid on the wing, confirmed the broadband antireflection with a transmittance improvement of approximately 6% [10]. Calculated transmittance spectra using three nipple shapes (truncated cone with hemispherical or flat top, cylindrical cone with hemispherical top) and effective medium approximation showed that the top hemisphere (upper level of the nanostructure) did not contribute to the transparency, whereas the truncated cone (lower level of the nanostructure) was fundamental for achieving antireflection (Figure 4.5). Dellieu and coworkers concluded that the lower level determined solely the antireflection property of the device. The upper level, by contrast, determined only the hydrophobicity property, as demonstrated by wetting experiments and analytical modeling. Using liquids, the surface tensions of which allowed testing of both hydrophobic and hydrophilic regimes, the effective static contact angle θ_{eff} (measured on the wing's corrugated surface) was studied as a function of the static contact angle θ_0 on a flat chitin surface (calculated by Young's equation). In the hydrophobic regime, the Cassie-Baxter model applied to the hemisphere-topped truncated cone nipple structure predicted a nonlinear relationship that agreed reasonably well with experimental data, at least when the effective angle approached 0° or 180°. The calculated contact angle for water was 151°, close to the measured value of 146°, implying that the wing was nearly superhydrophobic. On the other hand, when the hemisphere was removed from the model, the relationship between both angles became linear and departed significantly from the experimental data (calculated contact angle for water: 111°). From these results, the authors concluded that the hemispheric top shape was essential for achieving superhydrophobicity in the wing of the gray cicada, whereas the truncated cone bases had no significant role. In the hydrophilic regime, when the Wenzel model was applied to the nipple geometry, it predicted a linear dependence that agreed very well with the experimental data, until the capillarity-driven superhydrophilic regime (hemiwicking) was reached.

Hydrophobic wing surfaces offer a survival advantage to terrestrial insects, as they protect them against wetting by rainwater and contamination by particles, which both have a detrimental effect on flight and can impair antireflection. Sun and coworkers investigated 15 species of cicadas with distinct hydrophobicity properties (with contact angles ranging from $\theta \simeq 77°$ to $\theta \simeq 146°$) and classified their wing nanostructures into four categories, according to the pattern, diameter, spacing, and height of protrusions [6]. Wetting properties were found to depend upon slight species-related variations in nanostructure morphology as well as on the chemical composition of the wax layer covering the wing. Biological adaptation of the wing hydrophobicity to the environment may explain why larger cicadas have evolved more hydrophobic wings ($\theta > 135°$) than smaller cicadas ($\theta > 95°$) because the former species live in more open environments, while the latter species live lower in the canopy and can benefit from rainwater protection provided by the leaf cover.

The transparent wings of cicadas may also convey another interesting property for living organisms: antibacterial activity [12, 14–18]. The transparent and hydrophobic wings of the cicada *Psaltoda claripennis*, for instance, were shown to be deadly to Gram-negative rod-shaped *Pseudomonas aeruginosa* bacterial cells [14, 15]. The wing surface was not effective at repelling bacteria, as many cells were able to adhere to it. However, viability experiments showed that individual bacterial cells were killed within a few minutes of contact and that those cells were penetrated by the nanopillars covering the wing surface. Thus the nanopillars had a third function in addition to forming an antireflection device and conferring hydrophobicity (measured average contact angle of 159°). As the lethality to bacterial cells was not inhibited by coating the surface with gold, the bactericidal effect was attributed mainly to the wing's physical nanostructure rather than its chemical properties. The bactericidal property of *P. claripennis* was further tested against a range of bacterial species of various morphologies and cell wall types [14]. The wings were found consistently to kill Gram-negative cells, while Gram-positive cells remained resistant—probably because of their thicker layers of peptidoglycan (i.e., polymers forming a meshlike layer outside the plasma membrane of the majority of bacteria), making them more rigid. By contrast, the cell morphology did not appear to play any role.

Kelleher and coworkers compared the viability of *Pseudomonas fluorescens*, a Gram-negative and autofluorescent bacteria, on the wings of three cicada species

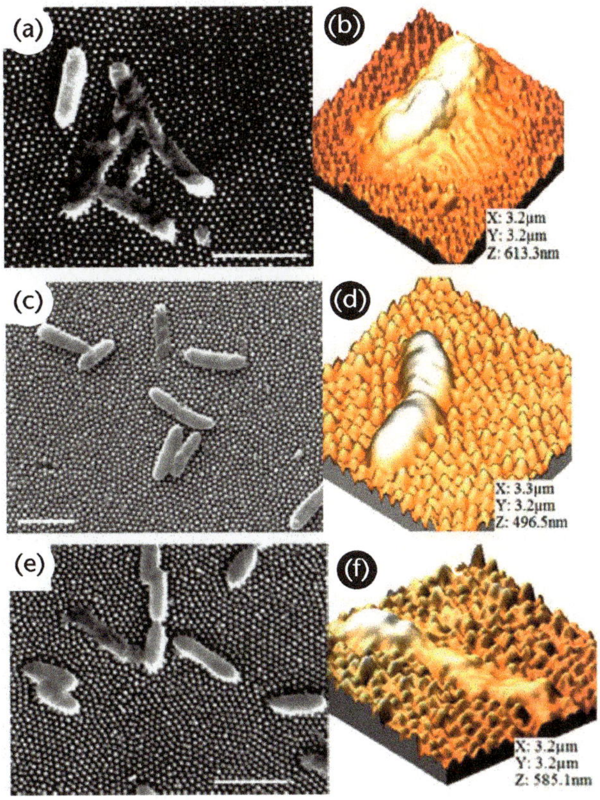

Figure 4.6 Antibacterial activity of the wings of the cicadas *Megapomponia intermedia* (ME), *Cryptotympana aguila* (CA), *Ayuthia spectabile* (AY). SEM images (a,c,e) and AFM images (b,d,f) of *Pseudomonas fluorescens* bacterial cells on ME (a,b), CA (c,d), and AY (e,f) wing surfaces. (Reproduced from [18], with permission from the American Chemical Society.)

(*Megapomponia intermedia*, *Cryptotympana aguila*, and *Ayuthia spectabile*) [18]. SEM and AFM images (Figure 4.6) showed the collapsing of bacterial cells and indentation by the nanopillars into the cells. Although there was no trend observed regarding wettability and bactericidal activity, they did find a correlation between the latter property and morphology of the wing nanostructure. Specifically, they noted that the higher the packing of nanopillars and the sharper their features, the higher the dead/live cell ratio. The average number of pillars with which a cell interacts (*M. intermedia*: 24, *C. aguila*: 19, *A. spectabile*: 10) correlated with the bactericidal activity (dead/live ratio: 0.222, 0.123 and 0.067, respectively).

References

[1] Bernhard, C. G., W. H. Miller, and A. R. Moller, "The Insect Corneal Nipple Array," *Acta Physiologica Scandinavica*, Vol. 63, No. 243, 1965, pp. 1–79

[2] Bernhard, C. G., G. Gemne, and J. Sällström, "Comparative Ultrastructure of Corneal Surface Topography in Insects with Aspects on Phylogenesis and Function," *Zeitschrift für Vergleichende Physiologie*, Vol. 67, 1970, pp. 1–25.

[3] Stavenga, D. G., et al., "Light on the Moth-Eye Corneal Nipple Array of Butterflies," *Proceedings of the Royal Society B*, Vol. 273, No. 1587, 2006, pp. 661–667.

[4] Parker, A. R., "Solar-Absorber Antireflector on the Eye of an Ecocene Fly (45 Ma)," *Proceedings of the Royal Society B*, Vol. 265, No. 1398, 1998, pp. 811–815.

[5] Zhang, G., et al., "Cicada Wings: A Stamp from Nature for Nanoimprint Lithography," *Small*, Vol. 2, No. 12, 2006, pp. 1440–1443.

[6] Sun, M., et al., "Wetting Properties on Nanostructured Surfaces of Cicada Wings," *Journal of Experimental Biology*, Vol. 212, No. 19, 2009, pp. 3148–3155.

[7] Sun, M., et al., "A study of the Anti-Reflection Efficiency of Natural Nano-arrays of Varying Sizes," *Bioinspiration & Biomimetics*, Vol. 6, No. 2, 2011, 026003.

[8] Sun, M., et al., "Influence of Cuticle Nanostructuring on the Wetting Behaviour/States on Cicada Wings," *PLoS ONE*, Vol. 7, No. 4, 2012, e35056.

[9] Wisdom, K. M., et al., "Self-Cleaning of Superhydrophobic Surfaces by Self-Propelled Jumping Condensate," *Proceedings of the National Academy of Sciences of the United States of America*, Vol. 110, No. 20, 2013, pp. 7992–7997.

[10] Dellieu, L., et al., "A Two-in-One Superhydrophobic and Anti-Reflective Nanodevice in the Grey Cicada *Cicada orni* (Hemiptera)," *Journal of Applied Physics*, Vol. 116, No. 2, 2014, 024701.

[11] Deparis, O., et al., "Nanostructured Surfaces: Bioinspiration for Transparency, Coloration and Wettability," *Materials Today: Proceedings*, Vol. 1, No. S, 2014, pp. 122–129.

[12] Román-Kustas, J., et al., "Molecular and Topographical Organization: Influence on Cicada Wing Wettability and Bactericidal Properties," *Advanced Materials Interfaces*, Vol. 7, No. 10, 2020, 2000112.

[13] Šigutoá, H., et al., "Wing Wettability Gradient in a Damselfly *Lestes sponsa* (Odonata: Lestidae) Reflects the Submergence Behaviour During Underwater Oviposition," *Royal Society Open Science*, Vol. 7, No. 12, 2020, 201258.

[14] Hasan, I., et al., "Selective Bacterial Activity of Nanopatterned Superhydrophobic Cicada *Psaltoda claripennis* Wing Surfaces," *Applied Microbiology and Biotechnology*, Vol. 97, No. 22, 2012, pp. 9257–9262.

[15] Ivanova, F. P., et al., "Natural Bactericidal Surfaces: Mechanical Rupture of *Pseudomonas aeruginosa* Cells by Cicada Wings," *Small*, Vol. 8, No. 16, 2012, pp. 2489–2494.

[16] Ivanova, F. P., et al., "Bactericidal Activity of Black Silicon," *Nature Communications*, Vol. 4, No. 1, 2013, p. 2838.

[17] Diu, T., et al., "Cicada-Inspired Cell-Instructive Nanopatterned Arrays," *Scientific Reports*, Vol. 4, 2014, 7122.

[18] Kelleher, S. M., et al., "Cicada Wing Surface Topography: An Investigation into the Bactericidal Properties of Nanostructural Features," *ACS Applied Materials and Interfaces*, Vol. 8, No. 24, 2016, pp. 14966–14974.

[19] Yoshida, A., et al., "Nanoprotuberance Array in the Transparent Wing of a Hawkmoth, *Cephonodes hylas*," *Zoological Science*, Vol. 13, No. 4, 1996, pp. 525–526.

[20] Yoshida, A., et al., "Antireflective Nanoprotuberance Array in the Transparent Wing of a Hawkmoth, *Cephonodes hylas*," *Zoological Science*, Vol. 14, No. 5, 1997, pp. 737–741.

[21] Yoshida, A., "Antireflection of the Butterfly and Moth Wings through Microstructure," *Forma*, Vol. 17, No. 2, 2002, pp. 75–89.

[22] Hooper, I. R., P. Vukusic, and R. J. Wootton, "Detailed Optical Study of the Transparent Wing Membranes of the Dragonfly *Aeshna cyanea*," *Optics Express*, Vol. 14, No. 11, 2006, pp. 4891–4897.

[23] Stoddart, P. R., et al., "Optical Properties of Chitin: Surface-Enhanced Raman Scattering Substrates Based on Antireflection," *Nanotechnology*, Vol. 17, No. 3, 2006, pp. 680–686.

[24] Deparis, O., et al., "Assessment of the Antireflection Property of Moth Wings by Three-Dimensional Transfer-Matrix Optical Simulations," *Physical Review E*, Vol. 79, No. 4, 2009, 041910.

[25] Stavenga, D. G., "Thin Film and Multilayer Optics Cause Structural Colors of Many Insects and Birds," *Materials Today: Proceedings*, Vol. 1, No. S, 2014, pp. 109–121.

[26] Siddique, R. H., G. Gomard, H. Hölscher, "The Role of Random Nanostructures for the Omnidirectional Anti-Reflection Properties of the Glasswing Butterfly," *Nature Communications*, Vol. 6, No. 1, 2015, 6909.

CHAPTER 5

Liquid-Induced Structural Color Changes

When observed in their natural environment, the coloration of living organisms appears to be mostly a static property, except in a few cases. The most popular example, and also probably the most studied, is the chameleon; for example *Chamaeleo chamaeleon*, also known as the common chameleon [1, 2]. Its fast, complex and reversible changes occur, for example, during fights between males, for courtship, or for camouflage. They are usually explained by the aggregation, spreading, or reorientation of pigments within cells, called chromatophores, which are located in the animals' integuments, or, alternatively, by changes in the size of these chromatophores [1–3]. These cells are termed xanthophores if they contain yellow pigments, erythrophores with red pigments, and iridophores if they contain photonic structures made of guanine crystals spaced by cytoplasm. Chameleons are thought to modify these cells through their hormonal or neural systems in order to adapt the light absorption and reflection from these different cells and enable them to generate a large variety of visual appearances. Many other organisms, including cephalopods and fish species, are known to control their color using chromatophores in a similar way [4–6]. As this chapter is about changes in structural colors, we will not discuss pigmentary color changes further. However, the mechanism underlying the color control by iridophores is relevant to structural color changes induced upon contact with liquids. Indeed, color changes in the skin of *Furcifer pardalis* (known as the panther chameleon) have been reported to arise from variations in the distances between guanine crystals forming hexagonal lattices within the dermal iridophores [3].

It is hard to conceive, a priori, of how the color produced by a structure within a natural organism could change (apart from a change in incident light conditions). In general, color-producing structures are not expected to be modified—neither by the organism bearing them itself, nor by the environment (unless, for instance, a compression or traction force is applied to the structure or slow processes occur such as bacterial load, abrasion, or wear). Accordingly, structural colors are generally expected to be relatively static and more stable than pigmentary colors, which may suffer from aging or other forms of degradation, leading to color fading such as bleaching. Photonic structures are often regarded as having neither innervation nor vascularization. However, entomologists and physicists have known for a long time that the change in color happening when a drop of ethanol is poured on a butterfly wing or a bird feather is conclusive evidence of the structural origin of the color [7–12].

In this chapter, we will present and discuss several examples of structural colors that change upon contact with liquids. We introduce the chapter with a brief comment on the fact that hydrophilicity may be an unexpected feature to encounter in insect cuticle, as will be discussed later in the case of the beetle *Hoplia coerulea*. The story of this beetle, the cerulean chafer beetle (synonymous with the *chasseur bleu* of Voet, 1806) [13], is told next because it has been the subject of a series of exciting studies by Jean-Pol Vigneron's group at the University of Namur. Vigneron started by investigating the physical origin of the beetle's iridescent blue scales and came to the accidental discovery of reversible blue to green color changes, induced by contact of the cuticle with liquid—a hygrochromic property that was further scrutinized by his PhD student, Marie Rassart. Hygrochromism was also studied in detail by Rassart, who explained why the elytra of *Dynastes hercules* change from khaki green to black when wet. Sébastien Mouchet renewed interest in the *chasseur bleu* by investigating the fluorescent properties of its scales during his PhD thesis. Based on his detailed study of the cuticle wettability, he eventually proposed the bioinspired concept of a photonic cell. The color changes in *H. coerulea* and *D. hercules* arise from a modification of the animal's environment (atmospheric relative humidity or direct contact with a liquid) and, for this reason, can be classified as *passive* color changes. However, the switchable golden mirror discovered by Vigneron in the tortoise beetle *Charidotella egregia* belongs to the class of *active* color changes. Upon a tactile stimulus, the tortoise beetle changes the color of its dorsal exoskeleton, from shiny metallic golden, to matte red, within a couple of minutes. This fully reversible phenomenon is obviously a source of bioinspiration for smart coating devices. The chapter ends by examining the case of a swelling multilayer found in *Tmesisternus isabellae*, before presenting a short review of other cases of structural color changes.

5.1 Unexpected Hydrophilicity

Wings of insects, including butterflies, moths, and cicadas are, in many reported cases, hydrophobic [14–16], as explained in Chapter 4. This property is critical, as water staying on a wing's surface would make it heavier and flight much less easy. Hydrophobicity can be provided by corrugations of the wing surface, which can take the form of nipple array covering the whole wing's surface. In some cases, these structures provide the insect with another property, such as structural color or transparency, which can have a biological function, too. As described in Chapter 4, the gray cicada *Cicada orni*, for instance, has nipple arrays on both sides of its wings, which provide it with the properties of both broadband antireflection and superhydrophobicity [16]. This is an inspiring example of a natural two-in-one nanostructure that confers transparency and self-cleaning to the wings.

In lepidopterans, it seems that hydrophobicity is a rather general property of the wings [14]. In coleopterans, however, there are several instances of hydrophilic cuticle surfaces, either on scales or elytra, as will be discussed hereafter. An early report of hydrophilic integuments in beetles that change color upon contact with water is the camouflage of bark bugs by integumentary wetting [17]: upon wetting under the rain, tree bark turns to darker hues than when it is dry. Some bark-inhabiting beetles were found to have a relative reflectance dropping immediately by as much as 62% upon contact with water [17]. This color change is reversible and enables the camouflage of the beetles. Such a change in appearance is similar to any absorbing material, including asphalt, a field not covered with vegetation, black mold, or sand [18, 19]. When wet, such a surface is overlaid with a thin water film. On one hand, diffuse reflection by the surface enables total internal reflection in the film, leading to a higher absorption by the surface. On the other hand, absorption is enhanced by the decrease of refractive index contrast due to the presence of the water film [18, 19]. Moreover, some species have evolved control of the hydrophobic and hydrophilic balance in quite a sophisticated way. The cuticle of a beetle from the *Stenocara* genus living in the Namib desert has developed a surface exhibiting hydrophilic bumps surrounded by hydrophobic areas coated by wax [20]. These alternating hydrophilic and hydrophobic areas specifically help with collecting water in low hygroscopic environments. In the light of the above examples, the wettability of the insects' cuticle appears to be a feature that is selected by evolution in both hydrophobic and hydrophilic cases.

5.2 The Story of the *Chasseur Bleu*

An emblematic case of structural color change induced by liquids is *Hoplia coerulea* (Scarabaeidae), a beetle living in the South of France and the North of Spain [9, 13, 21–24]. The male has a bright blue coloration that turns to emerald green after a few seconds of contact with a drop of water [24–28]. Obviously, one needs to be very lucky to observe this effect in the field! In fact, water-induced color changes were observed accidentally in our laboratory: a story that will be told below. Following this discovery, color changes induced by water, and then by other liquids, were studied in detail and the mechanism responsible for such changes was examined. This led us to the bioinspired concept of a photonic cell, which will be discussed later.

After the first observations were made on *H. coerulea*, we found both a sketch and a quote about that beetle in a book by J. E. Voet published in 1806. The manuscript, written in Latin, French, and Dutch, which would have been welcome in the Low Countries of that time, was entitled *Catalogus Systematicus Coleopterorum—Tomus I* [13]. In this impressive volume, the author gives a short description of the "*Venator coeruleus*/Le chasseur bleu" (item 73 in [13]; Figure 5.1(a,b)), as named by him and recognizable by its fine color sketch, which mentions the iridescent blue color ("sa couleur bleue a l'éclat d'une pierre précieuse"), but not any color change effect. Elsewhere in the book, we found a sketch and a quote of the Hercules beetle (*Dynastes hercules*), item 98 (called "Hercule" in French and *Hercules* in Latin; see Figure 5.1(c,d)). This was a species we also studied later: in this case the elytral color changes from khaki green to black upon contact with water. The case of *D. hercules* will also be discussed herein.

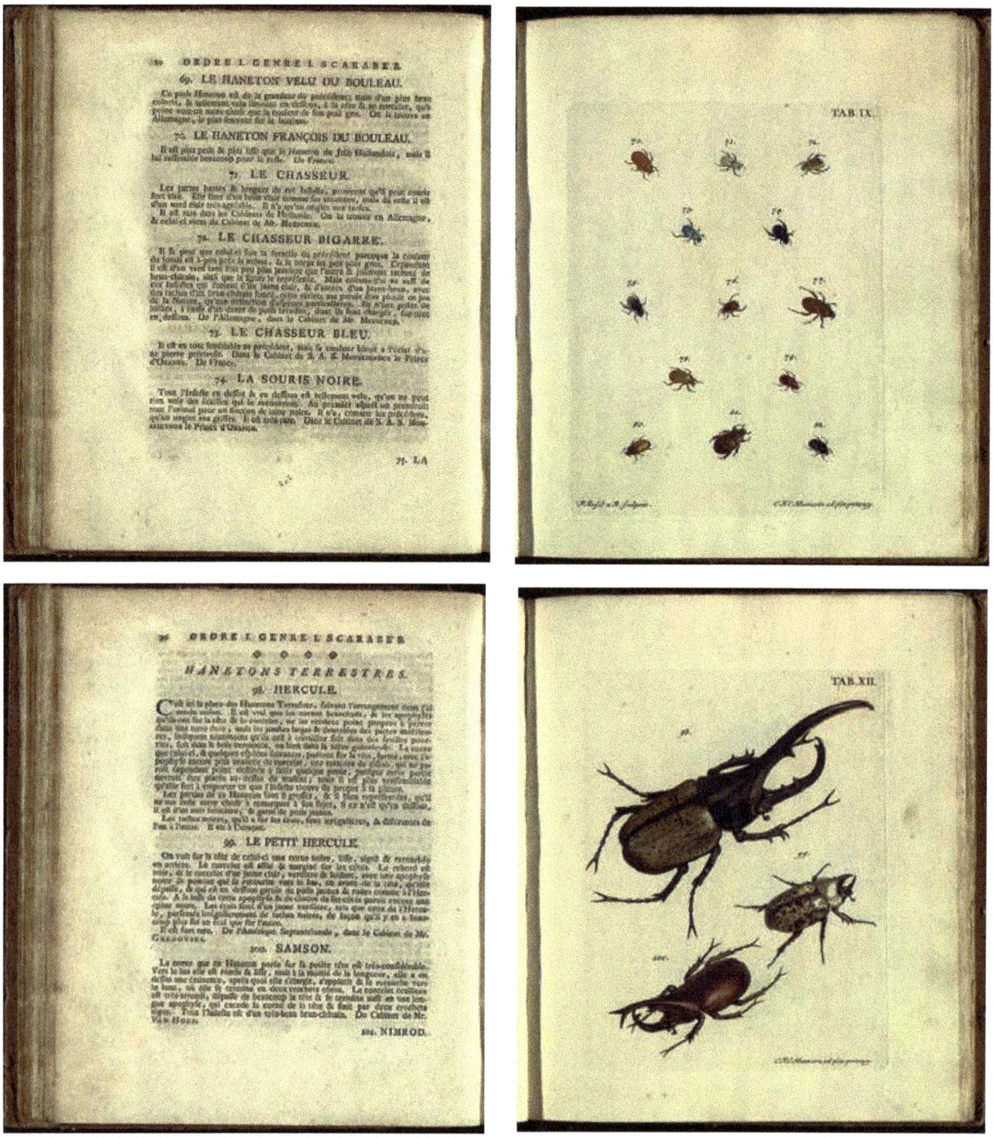

Figure 5.1 Description (in French) and sketch (item 73) of the male *Hoplia coerulea* beetle (pictures at top), description (in French) and sketch (item 98) of *Dynastes hercules* (pictures at bottom). (Reproduced from [13], with permission from the University Library Moretus Plantin.)

5.3 The Blue Scales of *Hoplia coerulea*

The male *H. coerulea* is a blue beetle from the family Scarabaeidae, also known as the cerulean chafer beetle [9, 22, 23]. It can often be encountered from May to August, for instance, in the green vegetation bordering rivers and streams in the South of France and the North of Spain. Whereas the male displays a spectacular iridescent blue color, which turns to violet at large angles, the female is not readily seen in its habitat. Instead, its appearance is indicated by the gathering of several blue iridescent males, since the latter are more readily visible.

Figure 5.2 A male and female *Hoplia coerulea* beetle (a) on the tip of the same tall blade of grass in the village of Meyras (Ardèche, France). The male (above) exhibits a striking blue coloration, whereas the female (below), which is rare to observe, has a dark appearance, with a faint blue shade. Dr Louis Dellieu (b) collecting specimens from the tops of high grass in the Cévennes, France, in summer 2014. (Reproduced from [25], with permission from the American Physical Society (a). Photograph (b) was taken by S. R. Mouchet, June 2014.)

A photograph, taken by Jean-Pol Vigneron at the end of a sunny afternoon, shows both a blue male and a slightly larger, darker female (Figure 5.2(a)). In [25], Vigneron wrote, "Many male blue beetles were at that moment distributed at the upper end of bushes growing in a rather humid place on the bank of a stream. The small blue spots were visible from several tens of metres under a bright sun. The adult males adopt a strange position holding the stem with the two pairs of forelegs and lifting the large rear legs, waiting seemingly for a female inspection. Eventually, the dark heavy female appears, flies for about one or two minutes over the exposed males and suddenly dives straight on the stem occupied by one of them" [25]. He suggested, "Most likely, the blue coloration plays a role in the choice made by the female, as the motion toward the chosen male starts several metres away from its sitting place, from where the colored visual signal is best perceived" [25], an idea he commented upon critically in the next sentence: "It is however difficult, without biological field experiments, to know whether the driving force of the evolution toward a conspicuous sky-blue coloration for the male is the result of a better visibility from a distance or of a complex sexual color selection by the female, as was shown to happen with many other animals" [25].

The blue-violet iridescence of the male *H. coerulea* is explained by light interference within a macroporous photonic structure that is mostly made of chitin and enclosed within a thin envelope, which is found inside the circular shaped scales (squamae) covering the beetle elytra and thorax (Figure 5.3) [23,24].

In an early study, the photonic structure was described as a periodic multilayer stack (period $a = 160$ nm) composed of dense chitin layers (refractive index $n = 1.56$ [8, 11, 29, 30]) that are regularly separated by chitin rods [23]. The rods are more or less parallel and equidistant, with a size smaller than the

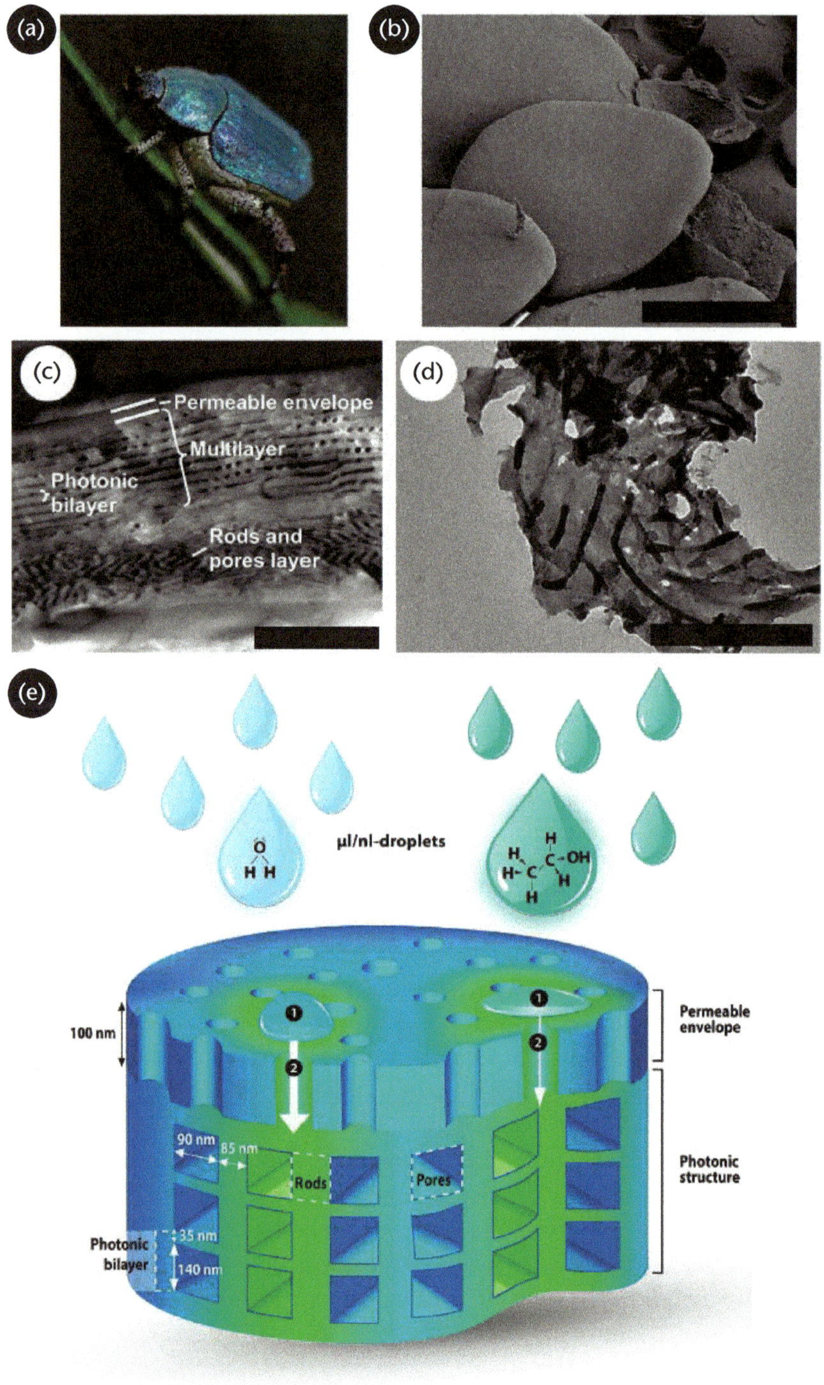

Figure 5.3 The blue scales of the male *Hoplia coerulea* beetle and its internal photonic structure. Photograph (a) of the beetle in its habitat, in the Cévennes, France, June 2014. Scanning electron micrograph (b) of the beetle's scales. The porous multilayer, viewed here in a scanning electron micrograph (c) of a scale cross section, is enclosed by a 100-nm thick envelope and is at the origin of the beetle's blue-violet color. Transmission electron micrograph (d) of the mixed air-cuticle layer. Schematic illustration (e) of color change mechanisms. Despite the larger contact angle formed by water droplets (1) on the scales, the permeability of the envelope (2) is higher to water than to ethanol. Scale bars: 40 μm (b), 2 μm (c) and 1 μm (d). (Reproduced from [28], License CC-BY-4.0.)

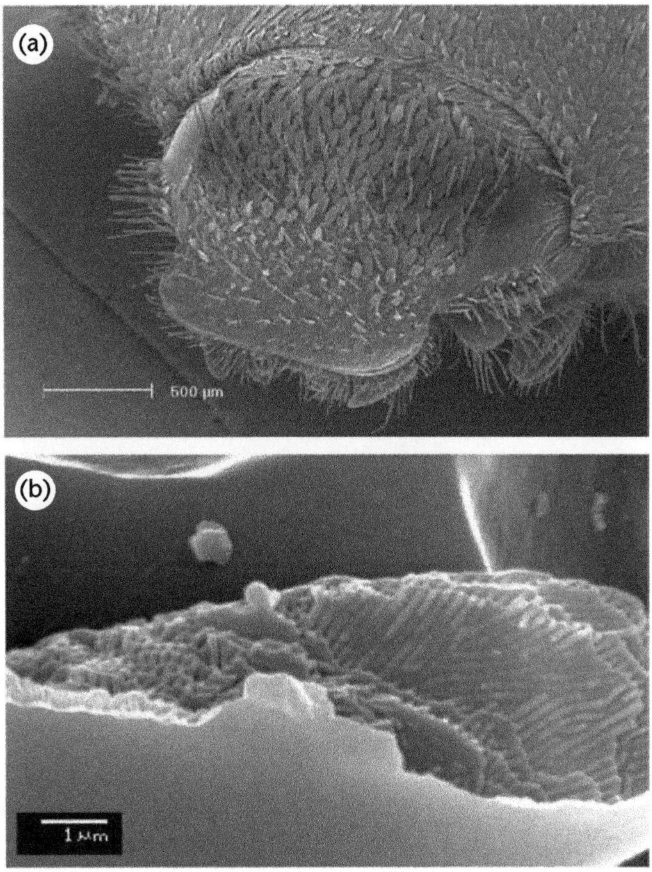

Figure 5.4 The cuticle (a) of *Hoplia coerulea* beetle is covered by scales (diameter: 50 μm, thickness: 3.5 μm). Scanning electron micrograph (b) of a broken scale, unveiling the scale lumen. The structure consists of a periodic multilayer made of dense layers spaced by rods. Within a layer of rods, these rods are locally parallel with a period of about 170 nm. (Reproduced from [23], with permission from the American Physical Society.)

wavelengths of interest (visible range). This local order is lost at distances of about 5 to 10 spacing rods (Figure 5.4). For the purpose of optical modeling, the rod spacing layers were treated as an effective medium having a refractive index of about $n_{eff} = 1.34$ [23]. A simple formula derived from the Bragg equation predicted a dominant (peak) reflected wavelength of $\lambda = 448$ nm at normal incidence, while simulations of the reflectance spectrum confirmed details of the blue iridescence observed experimentally [23].

Color change from blue to green was first observed on a dead *H. coerulea* specimen that had been inadvertently stored in a wet glass vial [25]. After this serendipitous observation, the phenomenon was examined in more detail. As soon as it was brought into contact with water, the insect's color turned green. When subsequently exposed to dry air for about 10 minutes, it recovered its natural blue color (Figure 5.5). It was also noted that blue coloration reappeared in patches, rather than through a gradual change of color [25]—an interesting fact that was later scrutinized more closely [28]. The reversible color change induced by exposure to water vapor, called *hygrochromism* for this reason, was explained by the infiltration

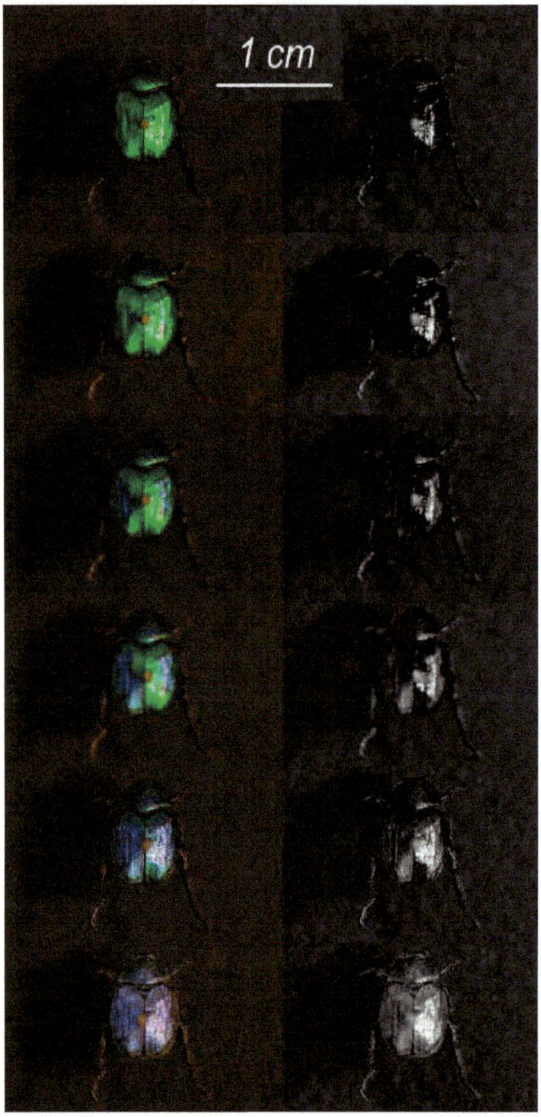

Figure 5.5 Time series of the drying process of a *H. coerulea* specimen (from top to bottom) previously brought into contact with water. The left column comprises unprocessed pictures, whereas the right column corresponds to images processed so that the green color appears dark and the blue color, lighter. The beetle's color turns to blue in patches. (Reproduced from [25], with permission from the American Physical Society.)

of liquid water into the empty spaces formed by the rods within the scale's photonic structure. This infiltration increases the effective refractive index of the rod layers and induces the shift of the Bragg wavelength toward a longer wavelength [25]. The mechanism of water infiltration was, however, not understood at that point.

In a later study, exposing the elytra to droplets of various liquids and through in situ optical microscopy observations, as well as with contact angle measurements and scale's composition chemical analysis, Mouchet and coworkers elucidated the liquid infiltration mechanism and its dynamics [24, 26–28]. They proposed a conceptual explanation, which is illustrated in Figure 5.3(d). A first counterintuitive observation was that the liquid-induced color change was much faster with water than with ethanol, in spite of the fact that water had a contact angle of $76 \pm 15°$, whereas the droplet of ethanol was not stable on the surface and directly flattened after deposition [28]. A second puzzling observation was that the scale's internal photonic structure was enclosed by a thin envelope, which prevented direct contact with the liquid. The presence of sodium, potassium, and calcium salts in the scale's chitin material (evidenced by X-ray photon spectroscopy analysis) gave a crucial clue. Indeed, as salts are very much soluble in water but only slightly soluble in alcohol, they most likely foster the capillary flow of water into micropores (i.e., pores with a diameter smaller than 2 nm) that were assumed to be present in the chitin envelope. The presence of such pores was inferred as neither macropores (pores with a diameter larger than 50 nm) nor mesopores (pores with a diameter ranging from 2 nm to 50 nm) could be detected despite the known permeability of chitin to water [31–33]. Therefore, the envelope's chemical composition was explained as mediating the penetration of liquid through these micropores. For instance, water molecules penetrated much more easily through the envelope because of the higher permeability of chitin to water than to ethanol. Moreover, the molecule size appeared to influence liquid penetration through the micropores, with the envelope's microporous structure acting as a sieve and allowing smaller molecules to penetrate more easily. In addition, the molecules' dipole moment had a sizeable effect on the permeability of different alcohols. These considerations were able to explain the dynamics of the color changes observed when using liquids whose molecules varied in size and polarity (Figure 5.3) [28].

In the system described above, the scale's envelope acts as a permeable barrier that controls exchanges between the enclosed photonic structure and the outer environment. By analogy with a biological cell, in which the cell membrane plays a similar mediating role, the term *photonic cell* was coined in order to describe the hygrochomic device found in the male *H. coerulea* [24,28].

5.4 The Khaki Green Elytra of *Dynastes hercules*

The khaki green color of the male Hercules beetle (*D. hercules*, Scarabaeidae) is another early case of hygrochromic behavior investigated by Jean-Pol Vigneron and his team [34]. This insect is part of the subfamilly Dynastinae, also known as rhinoceros beetles. This vernacular name relates to the horns exhibited by the males of a large number of species within this family. *D. hercules* subspecies are spread across Central and South America, as well as some Caribbean islands. This beetle is one of the biggest beetles on Earth and produces an impressively large amount of noise when flying. In his *Catalogus Systematicus Coleopterorum*, Voet dedicated most of the descriptions of two hercules beetles to their large horns (Figure 5.1). He finished his account of the first one by "Les parties de ce Haneton sont si grosses, et si bien représentées, qu'il ne me reste autre chose à remarquer à son sujet, si ce n'est qu'en dessous, il est d'un noir brunâtre et garni de poils jaunes" ("The parts of this chafer beetle are so big, and so well represented, I have nothing else to comment on it, apart from the fact that under, it is brownish black and adorned

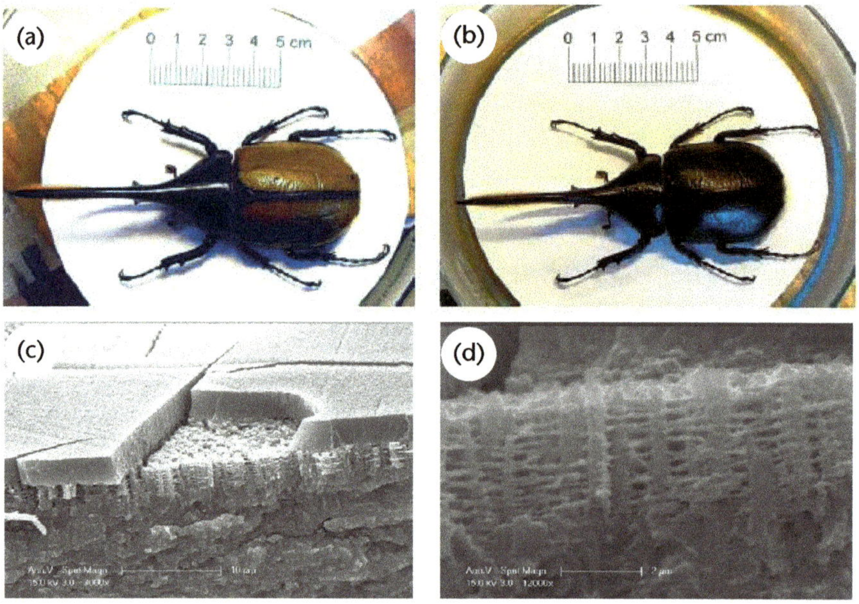

Figure 5.6 The khaki green color of *Dynastes hercules* (a) turns black upon contact with water (b). Water penetrates into the porous structure through cracks in the external wax layer (c). The photonic structure (d) was modeled by a 3D photonic crystal. (Reproduced from [34] with permission from IOP Publishing.)

with yellow hairs") [13]. As pointed out by Voet in the description of *Hercules minor*, black spots appear randomly scattered on the beetle's elytra. The khaki coloration in the dry state was explained by coherent scattering and light diffraction occurring in a porous structure located in the cuticle of the beetle under a 3-μm-thick homogeneous layer [34–36]. The photonic structure comprises columns made of cuticle materials, the direction of which is normal to the elytral surface; and chitinous filaments connecting the columns. This spongy structure was modeled by a 3D photonic crystal [34] (Figure 5.6). The lack of observable iridescence (i.e., the matte appearance of the khaki color), was explained by the existence of a strong disorder in the photonically active structure and by the presence of a thick diffusive layer covering it [34].

The appearance of this insect is known for turning reversibly to black upon contact with water or in an atmosphere with a relative humidity higher than 80% [34,35]. The liquid was described as penetrating the photonic structure via cracks in the 3-μm-thick external layer, filling the pores of the structure. As a consequence, the refractive index contrast is reduced and the photonic structure becomes almost transparent. The black color observed arises from an underlying layer containing absorbing pigments, such as melanin [34,35]. In the case of a similar rhinoceros beetle, *Dynastes tityus*, which also exhibits such color changes [37,38], Jiyu Sun and her team showed that these color variations are joined with changes in mechanical properties. The elytra have a reduced stiffness and a higher dampening coefficient upon contact with humidity.

These color changes were mainly passively tested: they were experimentally induced. However, this beetle was described as being able to regulate actively the

Figure 5.7 Following a tactile stimulus, *Charidotella egregia* adult beetle's golden color turns red within a few minutes (a–c). The photonic structure at the origin of the golden coloration is a chirped multilayer presenting some defects (d,e). (Reproduced from [40] with permission from the American Physical Society.)

water contained in its elytra and, hence, modify the color displayed. This active control of its visual appearance is unfortunately not perfectly understood [35,39].

Once again, the biological function associated with this color change is not obvious. Hinton and Jarman explained that it could be related to camouflage [35]. The beetle would be black at night, when the relative humidity is high and khaki green in the day, when the relative humidity is low. Another possible role could be thermoregulation [39]. A dark appearance would help heating at sunrise. However, as pointed out by Rassart and coworkers, both hypotheses are questionable, as only the male displays such color changes, the female being mostly dark on the elytra despite having an activity similar to the male [34]. In addition, displaying either a khaki green or black appearance would not make a big difference from a camouflage point of view, as this beetle is active in the night in low-light environments [34].

5.5 The Switchable Golden Armor of *Charidotella egregia*

The tortoise beetle *C. egregia* (Chrysomelidae) changes the appearance of its elytra reversibly, from golden to red, following a tactile stimulus (Figure 5.7(a–c)) [40]. Such a feature is not uncommon among tortoise beetles [41–45]. For instance, *Aspidomorpha tecta* has also been reported to change its color from golden to red [46]. The appeal of these beetles' color changes lies in the fact that they are active (i.e., entirely controlled by the beetles' metabolism). In the case of *C. egregia*, the process takes place within less than a minute. This occurs in a similar manner to that observed in *Charidotella ambita* [47] described in Chapter 2: a chirped multilayer (i.e., a stack of layers with slowly varying thickness and refractive index across depth) is at the origin of the golden coloration of *C. egregia*'s body (Figure 5.7(d,e)). This multilayer, which is found here in the exocuticle, consists of laterally inhomogeneous porous layers and is filled by a physiological liquid, probably hemolymph. The resulting golden coloration is a trait found only in the living animal, since dead specimens, as seen in museum collections, are red-brown when dry. Optically, the layers filled with liquid act as effective layers with a refractive index contrast between them that is low, yet sufficient to produce a chirped multilayer. Incident light interacting with this structure mainly reflects specularly over a broad spectral range, extending throughout the visible, except the blue region, hence producing a golden appearance. Under stress, the beetle is thought to drain the physiological liquid out of the multilayer mirror [40]. The structure consequently loses its coherence and becomes translucent, unveiling an underlying layer that has a diffused red color, due to pigmentation. In addition to the hue, the light scattering behavior is affected, too, changing from a metallic aspect to a matte coloration. The pores contribute to further scattering of incident light, since the empty pores act as randomly dispersed scatterers in an otherwise transparent chitinous material.

As the color change only occurs when the beetle is alive and *C. egregia* only feeds from the leaves of its host plant, *Ipomoea lindenii* [48], Jean-Pol Vigneron and his team had to perform some of the experiments in the beetle's natural environment. They carried out in situ measurements on samples that were collected in the forest surrounding Panama City in the Republic of Panama: an approach that is far from usual for researchers in photonics. Despite this fieldwork, it was not possible to make enough observations to ascertain the ecological function of this color change [40]. Barrows reported the rapid color change of the closely related *Charidotella sexpunctata*, previously known as *Metriona bicolor*, from metallic gold to brownish orange with black spots during copulation [44]. He speculated that *C. sexpunctata* could be a Batesian mimic of the immaculate ladybird beetle, *Cycloneda munda*, reasoning that such mimicry could reduce bird predation during the sexual act, which lasts between 15 and 583 minutes [44]. Furthermore, the golden colors of *C. sexpunctata* and *C. ambita* were reported to act as intraspecific sexual signals as they develop when the beetles' sexual maturity has been reached [44,47]. A sexual behavior similar to the one of *C. sexpunctata* has, however, not been reported for *C. egregia*, nor has *C. egregia*'s color been related, so far, to sexual communication.

5.6 The Swelling Multilayer of *Tmesisternus isabellae*

A porous multilayer structure similar to the one observed in the scales of *H. coerulea* is located in the lumen of the elongated and flat scales of the longhorn beetle *T. isabellae* (Cerambycidae) (Figure 5.8) [49]. It consists of periodically stacked homogeneous and inhomogeneous layers; the former are made only of cuticle material and the latter of cuticle material and pores. Thanks to this structure, the scales exhibit a metallic golden color. Upon contact with water, they turn passively

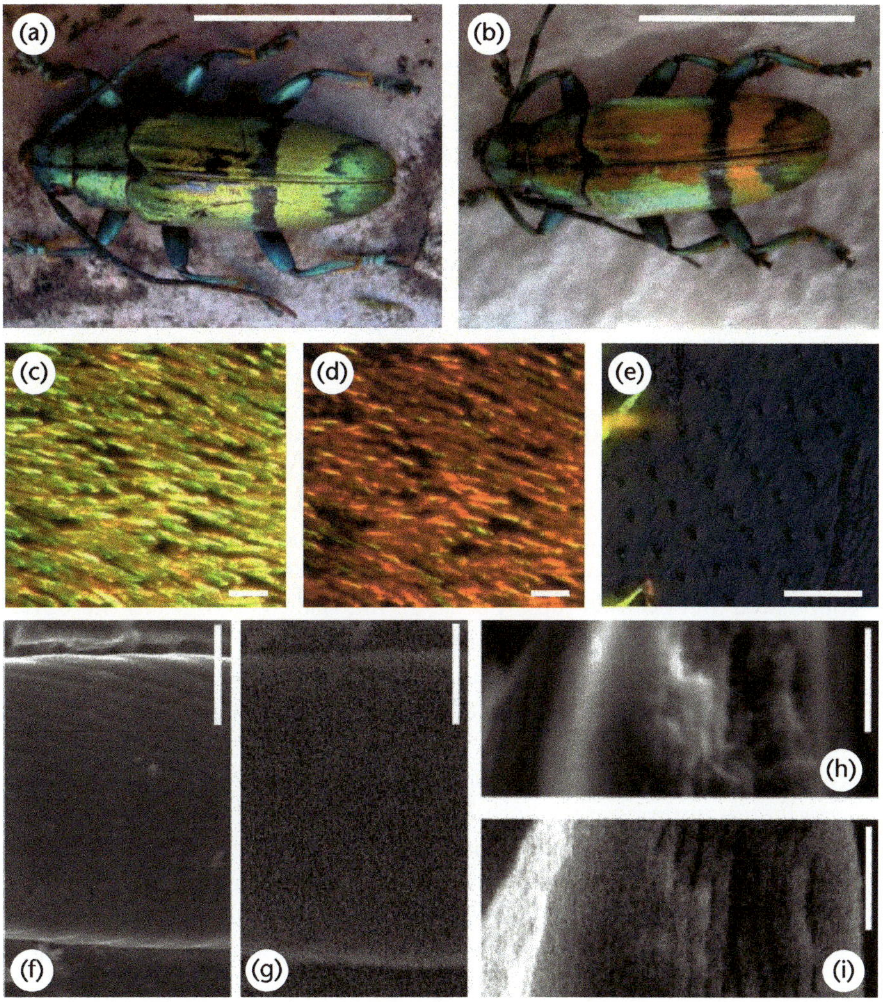

Figure 5.8 Upon contact with water, the dry-state golden color of *Tmesisternus isabellae* (a) turns to red-orange (b). This color is due to scales covering the insect's elytra (c). The scales' appearance changes to red-orange in the wet state (d). The underlying exocuticle exhibits a black coloration, most likely due to melanin (e). Using environmental scanning electron microscopy, the scales can be observed in the dry (f,h) and wet (g,i) states in top views (f,g) and cross sections (h,i). Scale bars: (a,b) 100 mm; (c,d) 100 μm; (e) 50 μm; (f,g) 5 μm; and (h,i) 1 μm. (Reproduced from [49], with permission from the Optical Society.)

to red (Figure 5.8). However, this color change was described as arising from another phenomenon related to passive color changes. In addition to the filling of air pores, the homogeneous layers were reported to swell when wet [49]. Following water infiltration into the scales, the water molecules disrupt hydrogen bonds between biopolymer chains, giving rise to the swelling of the scales' material and increasing the structure's period. Using environmental scanning electron microscope (ESEM) observations, Liu and coworkers assessed that the multilayer period swells from 175 nm in the dry state to 190 nm in the wet state (Figure 5.8) [49]. The homogeneous layers were found to swell from 105 nm to 120 nm, leaving the inhomogeneous layers almost unchanged. Measuring the swelling of the layers is, however, not an easy task from the ESEM observations (Figure 5.8) [49].

A similar longhorn beetle, *Tmesisternus rafaelae*, also known as *Sulawesiella rafaelae*, displays a metallic greenish yellow color arising from a porous photonic structure [50, 51] akin to that in *T. isabellae*; again, the beetle turns to a reddish coloration upon contact with water. This change was attributed by Rassart to the filling of the pores with water [50], as in the cases of *H. coerulea* and *D. hercules*. Although the absence of swelling is not demonstrated in this study on *T. rafaelae*, it is worth noting that the pores filling with water could, on its own, explain the color changes of both *T. isabellae* and *T. rafaelae*.

5.7 A Short Review of Other Cases

A well-known case of liquid-induced color change is the wings of *Morpho* butterflies. Despite their well-known hydrophobicity (see Chapter 4) [14], they turn from shiny blue to green upon contact with alcohol [11,12,52,53]. As described in Chapter 1, these wings have hierarchical Christmas tree structures, whose branches are decorated with nanometer-size protrusions forming open cavities or pores. The deposited alcohol fills these pores, giving rise to changes in the optical response of the structure. Depositing a droplet of a refractive index-matching liquid on butterflies' wings is often performed in order to remove the optical effects of the photonic structures. This allows, for example, the study of light absorption by embedded pigments [7–11,54,55]. A similar effect to that seen in *Morpho* species was reported in the color change from green to red of the structurally colored wings of *Trogonoptera brookiana* upon contact with ether [56]. In another study, Wang and coworkers compared the change in coloration of *Morpho rhetenor* (Nymphalidae), known as the Rhetenor blue morpho, with that observed in *Papilio blumei* (Papilionidae), better known as the green swallowtail [12]. The latter has a concave porous multilayer structure on its scales (see Chapter 2). The authors showed that the reflectance peak changes were more striking, in terms of intensity and position, in the case of *P. blumei*, due to the larger porous volume of this butterfly's structure. It seems unlikely that there is a biological role for these alcohol-induced color changes, as these butterflies are most probably never in contact with liquid alcohol in nature. Interestingly, Wilts and coworkers investigated the water-induced color change of the wings of the green forester moth *Adscita statices* from green in the dry state to red in the wet state [57]. The photonic structure occurring in the scales of this moth is a porous multilayer similar the one of the butterfly *P. blumei* [58] and giving rise to an additive color effect (as described in Chapter 3). Although the authors were not able to explain whether the water uptake was the result of the specific morphology of *A. statices* or the chemical composition of the scale materials, they provided a discussion of the biological role of this change [57]. The moth's red color would be beneficial for camouflage in the habitat where this insect lives, namely among the brown and red stems of meadow plants, including the common sorrel. The underlying uptake of water, however, leads to an increase in the mass of the moth's wings, which is disadvantageous for flying. The green visual appearance is advantageous in green vegetation from a camouflage point of view and is an aposematism signal to predators, as this color is related to poison in the moth's family (Zygaenidae). In addition, this green color would be a sexual signal. The moth's brilliant color, combined with a possible polarization effect, would, thus, play a role in intraspecific communication. Exploring this phenomenon in an artificial way, Ahmed and coworkers reported liquid-induced color change of the scales covering the wings of *Morpho peleides* [59]. Specifically, the research team used an interpenetrating polymer network and embedded these scales with some polyvinyl alcohol. This allowed them to measure some color changes induced by the swelling and shrinking of the interpenetrating polymer network upon contact with different liquids (water, oil, ethanol, and acetone).

Water-induced swelling of photonic structures is not confined to insects; in fact, color changes similar to those of *T. isabellae* have been observed in the feathers of a few bird species. These behaviors are reported to be fast and reversible. As described in Chapter 2, many iridescent colors in bird feathers are known to be due to thin film interference in a single keratin layer comprising the external cortex of the barbules [60,61]. *Tachycineta bicolor* (Hirundinidae), commonly known as the tree swallow, exhibits green-blue feathers. As detailed in Chapter 6, these feathers turn green-yellow in high relative humidity due to the swelling of their keratin layers [60]. When they dry, the structure shrinks back and the original color is displayed. An additional mechanism was discovered in the case of the feathers of *Zenaida macroura* (Columbidae), also known as the mourning dove. This bird displays a violet pink appearance. Upon contact with water, the barbules swell and rotate at the point of attachment to the barbs due to mechanical constrains [61]. The authors noted that repeated rounds of wetting and drying of the feathers produced both a change in their hue and an increase in their brightness. An overall increase in reflectance of 50% was measured and Shawkey's team correlated their spectroscopic and electron microscopic data with the appearance changes visible to the naked eye.

In plants, an example of water-induced color change lies in the integuments of the fruit of the tree *Margaritaria nobilis* studied by Mathias Kolle and Silvia Vignolini [62,63], and also known as the bastard hogberry. This tree is native to Central and South America, West Indies, and Mexico. As described in Chapter 2, the endocarp tissues of the fruit comprise a Bouligand structure. When the fruit is hydrated, the endocarp perfectly adheres to the seed (Figure 5.9(a)) and the Bouligand structure gives rise to a structurally colored appearance: light transmitted through the structure is absorbed by the dark coat of the seed [63]. When dry, the seed is separated from the endocarp by an air gap (Figure 5.9(b)). The additional interface between the endocarp and air partially backscatters transmitted light regardless of the wavelength. This leads to a significant desaturation of the color [63].

Finally, a very different color change mechanism was unveiled by scrutinizing the immediate change in appearance of the egg sacs of some Asian salamander species—*Hynobius* sp. [64]. These sacs are envelopes, in which eggs are embedded when laid by the female salamanders. Under water, they exhibit a brilliant iridescence from blue to yellow, whereas when they are removed from water, they lose their iridescence and appear translucent [64]. Observation by electron microscopy showed that the iridescent color is due to quasi-periodic diffraction grating arising from corrugations with a period of approximately 190 nm located within the wall of the egg sac. When the sac is removed from water, the refractive index of the incidence media is decreased, which shifts the optical response by 90 nm into the UV range [64].

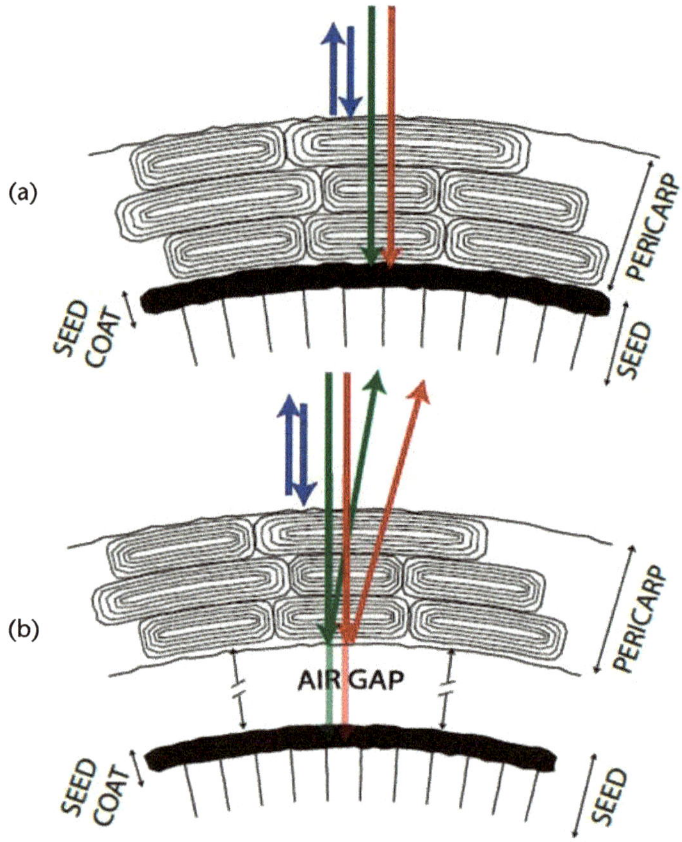

Figure 5.9 When wet, the endocarp of the fruit of the tree *M. nobilis* perfectly adheres to the seed (a). The endocarp is the most inner layer of the pericarp. In the case of this tree's fruits, a Bouligand structure is embedded within the endocarp. Light transmitted through the endocarp is absorbed by a the seed coat containing dark pigments. When dry, an air gap forms between the seed coat and the endocarp (b). The interface between the endocarp and air backscatters light regardless of the incident wavelength, which leads to a significant desaturation of the fruit's color. (Reproduced from [63], License CC-BY-4.0.)

References

[1] de Grijs, P., "Notes on the Faculty of Changing colour in Reptiles," *Annals and Magazine of Natural History*, Vol. 3, No. 17, 1899, pp. 396–402.

[2] Best, A. E., "The Discovery of the Mechanism of Colour-Changes in the Chameleon," *Annals of Science*, Vol. 24, No. 2, 1968, pp. 147–167.

[3] Teyssier, J., et al., "Photonic Crystals Cause Active Colour Change in Chameleons," *Nature Communications*, Vol. 6, No. 1, 2015, 6368.

[4] Mäthger, L. M., et al., "Rapid Colour Changes in Multilayer Reflecting Stripes in the Paradise Whiptail, *Pentapodus paradiseus*," *Journal of Experimental Biology*, Vol. 206, No. 20, 2003, pp. 3607–3613.

[5] Mäthger, L. M., et al., "Mechanisms and Behavioural Functions of Structural Coloration in Cephalopods," *Journal of The Royal Society Interface*, Vol. 6, No. S2, 2009, pp. S149–S163.

[6] Yoshioka, S., et al., "Mechanism of Variable Structural Colour in the Neon Tetra: Quantitative Evaluation of the Venetian Blind Model," *Journal of the Royal Society Interface*, Vol. 8, No. 54, 2011, pp. 56–66.

[7] Mason, C. W., "Structural colors in feathers. I," *The Journal of Physical Chemistry*, Vol. 27, No. 3, 1923, pp. 201–251.

[8] Mason, C. W., "Structural colors in insects. I," *The Journal of Physical Chemistry*, Vol. 30, No. 3, 1925, pp. 383–395.

[9] Mason, C. W., "Structural colors in insects. II," *The Journal of Physical Chemistry*, Vol. 31, No. 3, 1927, pp. 321–354.

[10] Mason, C. W., "Structural colors in insects. II," *The Journal of Physical Chemistry*, Vol. 31, No. 12, 1927, pp. 1856–1872.

[11] Vukusic, P., et al., "Quantified interference and diffraction in single *Morpho* butterfly scales," *Proceedings of the Royal Society B*, Vol. 266, No. 1427, 1999, pp. 1403–1411.

[12] Wang, W., et al., "Demonstration of higher colour response with ambient refractive index in *Papilio blumei* as compared to *Morpho rhetenor*," *Scientific Reports*, Vol. 4, 2014, 5591.

[13] Voet, J. E., *Catalogus Systematicus Coleopterorum*, Tomus I, La Haye: Backhuysen, G. (ed.), 1806.

[14] Berthier, S., *Photonique des Morphos*, Paris: Springer, 2010.

[15] Wisdom, K. M., et al., "Self-Cleaning of Superhydrophobic Surfaces by Self-Propelled Jumping Condensate," *Proceedings of the National Academy of Sciences of the United States of America*, Vol. 110, No. 20, 2013, pp. 7992–7997.

[16] Dellieu, L., et al., "A Two-in-One Superhydrophobic and Anti-Reflective Nanodevice in the Grey Cicada *Cicada orni* (Hemiptera)," *Journal of Applied Physics*, Vol. 116, No. 2, 2014, 024701.

[17] Silberglied, R., and A. Aiello, "Camouflage by Integumentary Wetting in Bark Bugs," *Science*, Vol. 207, No. 4432, 1980, pp. 773–775.

[18] Ångström, A., "The Albedo of Various Surfaces of Ground," *Geografiska Annaler*, Vol. 7, 1925, pp. 323–342.

[19] Lekner, J., and M. C. Dorf, "Why Some Things Are Darker When Wet," *Applied Optics*, Vol. 27, No. 7, 1988, pp. 1278–1280.

[20] Parker, A. R., and C. R. Lawrence, "Water Capture by a Desert Beetle," *Nature*, Vol. 414, 2001, pp. 33–34.

[21] Varrentrapp, and Wenner, *Deutsche Encyclopädie oder Allgemeines Real-Wörterbuch aller Künste und Wissenschaften von einer Gesellschaft Gelehrten*, Volume 18, Frankfurt am Mayn: Varrentrapp Sohn und Wenner (eds), 1781.

[22] Berthier, S., *Iridescences, les couleurs physiques des insectes*, Paris: Springer, 2003.

[23] Vigneron, J.-P., et al., "Natural Layer-by-Layer Photonic Structure in the Squamae of *Hoplia coerulea* (Coleoptera)," *Physical Review E*, Vol. 72, No 6, 2005, 061904.

[24] Mouchet, S. R, et al., "Photonic Scales of *Hoplia coerulea* Beetle: Any Colour You Like," *Materials Today: Proceedings*, Vol. 4, No. 4A, 2017, pp. 4979–4986.

[25] Rassart, M., et al., "Scale Coloration Change Following Water Absorption in the Beetle *Hoplia coerulea* (Coleoptera)," *Physical Review E*, Vol. 80, No. 3, 2009, 031910.

[26] Mouchet, S. R, et al., "*Hoplia coerulea*, a Porous Natural Photonic Structure as Template of Optical Vapour Sensor," *Proceedings of SPIE*, Vol. 9127, 2014, 91270U.

[27] Deparis, O., et al., "Nanostructured Surfaces: Bioinspiration for Transparency, Coloration and Wettability," *Materials Today: Proceedings*, Vol. 1, No. S, 2014, pp. 122–129.

[28] Mouchet, S. R, et al., "Liquid-Induced Colour Change in a Beetle: The Concept of a Photonic Cell," *Scientific Reports*, Vol. 6, 2016, 19322.

[29] Sollas, I. B. J., "On the Identification of Chitin by Its Physical Constants," *Proceedings of the Royal Society B*, Vol. 79, No. 534, 1907, pp. 474–481.

[30] Yoshioka, S. and S. Kinoshita,"Direct Determination of the Refractive Index of Natural Multilayer Systems," *Physical Review E*, Vol. 83, No. 5, 2011, 051917.

[31] Yonge, C. M., "On the Nature and Permeability of Chitin II—The Permeability of the Uncalcified Chitin Lining the Foregut of *homarus*," *Proceedings of the Royal Society B*, Vol. 120, No. 816, 1936, pp. 15–41.

[32] Hannan, J. V., and D. H. Evans, "Water Permeability in Some Euryhaline Decapods and *Limulus polyphemus*," *Comparative Biochemistry and Physiology A*, Vol. 44, No. 4, 1973, pp. 1199–1213.

[33] Dunson, W. A., "Permeability of the Integument of the Horseshoe Crab, *Limulus polyphemus*, to Water, Sodium, and Bromide," *Journal of Experimental Zoology*, Vol. 230, No. 3, 1984, pp. 495–499.

[34] Rassart, M., et al.,"Diffractive Hygrochromic Effect in the Cuticle of the Hercules Beetle *Dynastes hercules*," *New Journal of Physics*, Vol. 10, 2008, 033014.

[35] Hinton, H. E., and G. M. Jarman, "Physiological Colour Change in the Hercules Beetle," *Nature*, Vol. 238, 1972, pp. 160–161.

[36] Roux-Pertus, C., et al., "Multiscale Characterization of the Hierarchical Structure of *Dynastes hercules* Elytra," *Micron*, Vol. 101, 2017, pp. 16–24.

[37] Prokop, M. E., "Longevity and Color Change in the Rhinoceros Beetle, *Dynastes tityus* l. (Coleoptera: Scarabaeidae)," *The Coleopterists Bulletin*, Vol. 23, No. 1, 1969, pp. 20–22.

[38] Sun, J., et al., "Investigating the Nanomechanical Properties and Reversible Color Change Properties of the Beetle *Dynastes tityus*," *Journal of Materials Science*, Vol. 52, No. 11, 2017, pp. 6150–6160.

[39] Hinton, H. E., and G. M. Jarman, "Physiological Colour Change in the Elytra of the Hercules Beetle, *Dynastes hercules*," *Journal of Insect Physiology*, Vol. 19, No. 3, 1973, pp. 533–549.

[40] Vigneron, J. P., et al., "Switchable Reflector in the Panamanian Tortoise Beetle *Charidotella egregia* (Chrysomelidae: Cassidinae)," *Physical Review E*, Vol. 76, No. 3, 2007, 031907.

[41] Mason, C. W., "Transient Color Changes in the Tortoise Beetles (Coleop.: Chrysomelidae)," *Entomological News*, Vol. 40, No. 2, 1929, pp. 52–56.

[42] Hinton, H. E., "Some Recent Work on the Colours of Insects and Their Likely Significance," *Proceedings and Transactions of the British Entomological and Natural History Society*, Vol. 6, 1973, pp. 43–54.

[43] Neville, A. C., "Metallic Gold and Silver Colours in Some Insect Cuticles," *Journal of Insect Physiology*, Vol. 23, No. 10, 1977, pp. 1267–1274.

[44] Barrows, E. M., "Life Cycles, Mating, and Color Change in Tortoise Beetles (Coleoptera:Chrysomelidae:Cassidinae)," *The Coleopterists Bulletin*, Vol. 33, No. 1, 1979, pp. 9–16.

[45] Jolivet, P., "Physiological Colour Changes in Tortoise Beetles." In *Novel Aspects of the Biology of Chrysomelidae*, pp. 331–335, P. Jolivet, M. L. Cox and E. Petitpierre (eds.), Dordrecht: Kluwer Academic, 1994.

[46] Hinton, H. E., "Colour Changes." In *Environmental Physiology of Animals*, pp. 389–412, J. Bligh, J. L. Cloudsley-Thompson and A. G. Macdonald (eds.), London: Blackwell Scientific, 1976.

[47] Pasteels, J. M., et al., "Structural and Physical Evidence for an Endocuticular Gold Reflector in the Tortoise Beetle, *Charidotella ambita*," *Arthropod Structure & Development*, Vol. 45, No. 6, 2016, pp. 509–518.

[48] Windsor, D. M., E. G. Riley, and H. P. Stockwell, "Biology and Systematics of Panamanian Tortoise Beetles (Coleoptera: Cassidinae)." In *Insect of Panama and Mesoamerica, Selected Studies*, pp. 372–391, D. Quintero and A. Aiello (eds.), Oxford, UK: Oxford University Press, 1992.

[49] Liu, F., et al., "Structural Color Change in Longhorn Beetles *Tmesisternus isabellae*," *Optics Express*, Vol. 17, No. 18, 2009, pp. 16183–16191.

[50] Rassart, M., *Etude de matériaux hygrochromes d'origine naturelle et de structures colorantes multi-échelles*, PhD thesis, University of Namur, Belgium, 2010.

[51] Bermúdez-Ureña, E., et al., "Structural Diversity with Varying Disorder Enables the Multicolored Display in the Longhorn Beetle *Sulawesiella rafaelae*," *iScience*, Vol. 23, No. 7, 2020, 101339.

[52] Yang X., et al., "Using Hierarchy Architecture of *Morpho* Butterfly Scales for Chemical Sensing: Experiment and Modeling," *Sensors and Actuators A*, Vol. 167, No. 2, 2011, pp. 367–373.

[53] Jiang, X., et al., "Investigation on Color Variation of Morpho Butterfly Wings Hierarchical Structure Based on PCA," *Science China Technological Sciences*, Vol. 55, No. 1, 2012, pp. 16–21.

[54] Vukusic, P., and I. Hooper, "Directionally Controlled Fluorescence Emission in Butterflies," *Science*, Vol. 310, No. 5751, 2005, p. 1151.

[55] Trzeciak, T. M., et al., "Variable Multilayer Reflection Together with Long-Pass Filtering Pigment Determines the Wing Coloration of Papilionid Butterflies of the *Nireus* Group," *Optics Express*, Vol. 20, No. 8, 2012, pp. 8877–8890.

[56] Han, Z., et al., "Excellent Color Sensitivity of Butterfly Wing Scales to Liquid Mediums," *Journal of Bionic Engineering*, Vol. 13, No. 3, 2016, pp. 355–363.

[57] Wilts, B. D., K. Mothander, and A. Kelber, "Humidity-Dependent Colour Change in the Green Forester Moth, *Adscita statices*," *Biology Letters*, Vol. 15, No. 9, 2019, 20190516.

[58] Büchi, R., "Pointillistischer Farbeffekt beim Grünwidderchen, *Adscita statices* im Vergleich zur Goldwespe *Stilbum cyanurum*," *Mikroskopie*, Vol. 3, 2016, pp. 134–139.

[59] Ahmed, R., et al., "*Morpho* Butterfly-Inspired Optical Diffraction, Diffusion, and Bio-Chemical Sensing," *RSC Advances*, Vol. 8, No. 48, 2018, pp. 27111–27118.

[60] Eliason, C. M., and M. D. Shawkey, "Rapid, Reversible Response of Iridescent Feather Color to Ambient Humidity," *Optics Express*, Vol. 18, No. 20, 2010, pp. 21284–21292.

[61] Shawkey, M. D., et al., "Structural Color Change Following Hydration and Dehydration of Iridescent Mourning Dove (*Zenaida macroura*) Feathers," *Zoology*, Vol. 114, No. 2, 2011, pp. 59–68.

[62] Kolle, M., et al., "Bio-Inspired Band-Gap Tunable Elastic Optical Multilayer Fibers," *Advanced Materials*, Vol. 25, No. 15, 2013, pp. 2239–2245.

[63] Vignolini, S., et al., "Structural Colour from Helicoidal Cell-Wall Architecture in Fruits of *Margaritaria nobilis*," *Journal of the Royal Society Interface*, Vol. 13, No. 124, 2016, 20160645.

[64] Zabuga, A. V., et al., "Translucent in Air and Iridescent in Water: Structural Analysis of a Salamander Egg Sac," *Soft Matter*, Vol. 16, No. 7, 2020, pp. 1714–1721.

CHAPTER 6

Other Forms of Structural Color Changes

In addition to direct contact with liquids, as presented in the previous chapter, both *passive* and *active* changes in structural colors occurring in nature may be induced by other phenomena, such as physicochemical, mechanical, or physiological factors. For instance, in 2014, Umbers and coworkers reviewed more than 120 arthropod species exhibiting changes in their appearances (due to pigments and/or photonic structures) following environmental and endogenous stimuli [1]. The animals studied make these changes in order to match the color of the background for camouflage purposes, following a change in the animal excitation or reproduction state in response to temperature or humidity changes or upon contact with fluids. These changes are dynamic and reversible. Usually, the underlying biological functions of active changes are better understood than those related to passive changes, as the latter may simply be by-product of other functions of the organisms. In this chapter, we review this variety of color changes induced upon contact with vapors and gases, by modifications in the surrounding temperature and pH, under the action of mechanical forces or by the nervous or endocrine systems of the organisms.

6.1 The Passive and Active Color Changes Induced by Vapors and Gases

In 2007, Potyrailo and coworkers demonstrated that change in the surrounding atmosphere's composition leads to modification in the spectral response of *Morpho sulkowskyi* wings (Figure 6.1) [2, 3]. These passive color variations were induced by vapors of water, methanol, ethanol, and dichloroethylene isomers. Molecules of these vapors penetrate open pores of the *Morpho* butterfly's photonic structures and adsorb on their surfaces. Interestingly, water vapor was found to induce such color changes despite the well-known hydrophobicity of these structures [4]. The detected changes were highly specific to the vapors, so much so that vapors could be individually identified on the basis of the measured spectral response. Despite the similar refractive indices and polarities of water, methanol, and ethanol, these vapors were discriminated at relative pressures as low as 2%. Moreover, three closely related dichloroethylene isomers; namely, 1,l-DCE, trans-1,2-DCE, and cis-1,2-DCE, were differentiated. Finally, the response of *M. sulkowskyi* wings was also shown to be highly sensitive to very low concentrations down to 1 or 2 ppm. Similar vapor-induced optical variations were reported later on with the wings of *Morpho didius* by other authors [5–7].

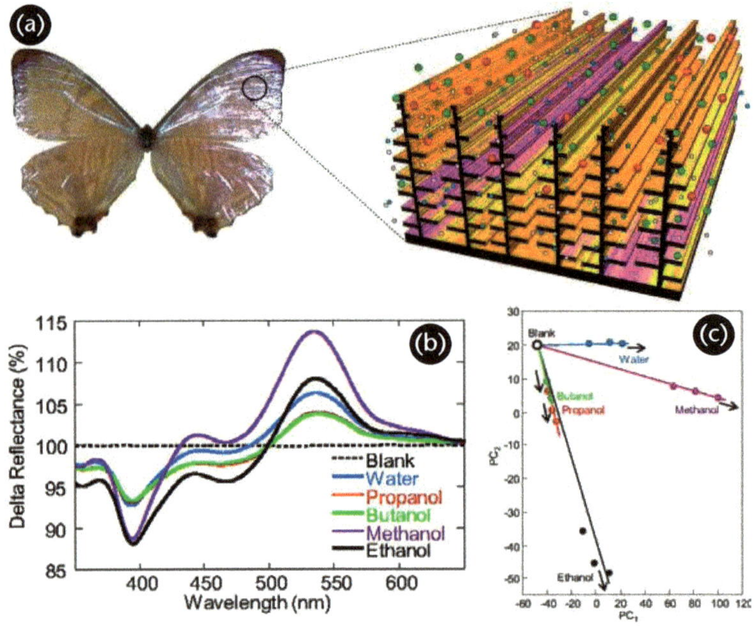

Figure 6.1 The spectral response of *M. sulkowskyi* wings varies upon contact with different vapors at concentrations as low as 1 or 2 ppm (a). This color change may not be perceptible by human eyes, but it is measurable by spectrophotometry (b). Principal component analysis (PCA) is often used to assess the selectivity of these natural photonic structures (c). (Reproduced from [3], with permission from SPIE (a) and from [8], with permission from PNAS (b,c).)

Very soon after Potyrailo and coworkers' 2007 article, Biró and his group studied the spectral response of the photonic structures occurring in 20 lepidopterans to six tested vapors; namely, water, ethanol, acetone, chloroform, pentane, and benzene, as well as one gas, ammonia [9]. The structures were varied and included porous multilayers, photonic crystals, and so-called "Christmas trees." Their responses, in terms of spectral reflectance, were found to be selective, sensitive, reproducible, and fast for the tested fluids. They were also found to be a function of the concentrations. For instance, in the cases of three lycaenid butterflies (*Polyommatus icarus*, *Polyommatus bellargus*, and *Polyommatus coridon*), the optical change is linear with respect to the partial pressure [10–12]. At concentrations of acetone, acetic acid, ethanol, 2-propanol, chloroform, toluene, and water lower than 50%, the color change is minute, with a peak shift of less than 10 nm. Such color changes are most likely not perceptible by humans but would be by butterflies such as *P. icarus*, according to Biró's group [13]. Among all the tested vapors, butterflies are usually only in contact with water vapors. However, so far, any underlying biological functions of these color changes remain unknown.

The dynamics of vapor- and gas-induced color changes in these natural photonic structures were explained by their openness to the surrounding environment. This allows the structure to interact quickly and directly with vapors and gases, which have access to the pores of the structures and can modify the material by triggering its swelling or physisorption on the inner surface of the accessible pores [14]. The physisorption can be followed by capillary condensation at higher partial pressures (above 97.7% of relative humidity for pores with a radius of 50 nm, giving rise to

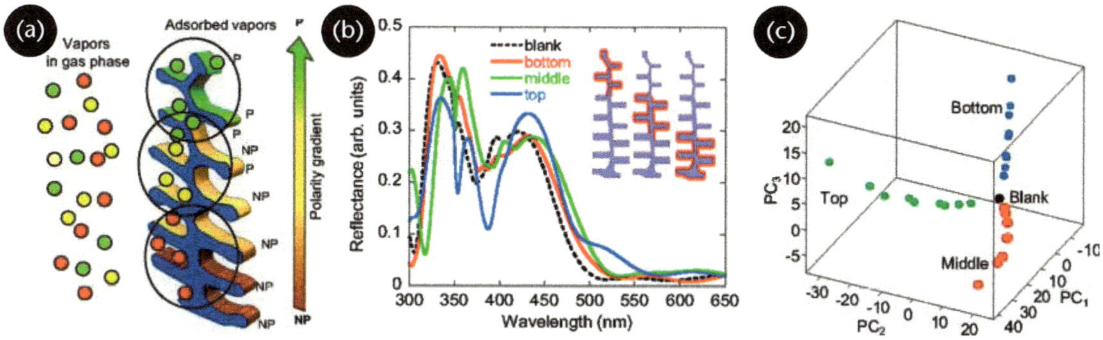

Figure 6.2 A polarity gradient was found along the axes of the so-called Christmas trees. Depending on the polarity of vapor molecules, they adsorbed at a different height (a). This gave rise to a variation of the spectral reflectance (b) that can be highlighted in terms of PCA (c). (Reproduced from [8], with permission from PNAS.)

a hemispherical meniscus) [14,15]. Experimentally, it was demonstrated that, after coating wings of *P. icarus* with 5 nm of Al_2O_3, both sensitivity and selectivity were reduced [13,16]. This obviously indicated that other factors related to the surface chemistry of the wing played a role in the response to vapors (i.e., factors other than the morphology and the refractive indices). In 2013, the high selectivity of *M. sulkowskyi* photonic structures was demonstrated by Potyrailo and coworkers to be due to a polarity gradient along the vertical axis of the structure ridges: the structure is polar at the top and apolar at the basis (Figure 6.2) [8]. Depending on their polarity, vapor molecules preferentially physisorb at a specific altitude of the ridges. In 2019, Biró's group performed parametric simulations of the reflectance spectra from a structural model developed from electron micrographs of *Albulina metallica* wings (Figure 6.3) [17]. The model took into account the (partial) filling of the pores with liquefied vapor and the swelling of the structure. They fitted simulations to the measured optical response of this butterfly's wings upon contact with ethanol vapor of various concentrations [17]. This analysis demonstrated that the color change is mainly due to the partial filling of the pore, with a small contribution from the structure's swelling. The former was found to depend on the shapes and sizes of the pores. Currently, research focuses, among other things, on the vapor selectivity of these natural photonic structures to binary and tertiary mixtures, often including volatile organic compounds [19,20,22] and including toxic agents (Figure 6.4) [20,21] and on the stability of their responses with time [22].

Unlike photonic structures in wings of butterflies where physisorption is dominant, the swelling of the photonic structure was found to play a major role in vapor-induced color changes of the feathers of tree swallow *Tachycineta bicolor* [23]. The barbules' visual appearance is due to thin-film interference in one layer of keratin covering disordered stacks of melanosomes (Figure 6.5(a,b)). Usually, bird plumage is considered unable to give rise to reversible color changes and only capable of undergoing irreversible and slow changes (e.g., due to abrasion, wear, or bacterial load). Furthermore, since this integument is not innervated nor vascularized, bird metabolism cannot actively control any color change. However, in the presence of water vapor, the tree swallow's color changes from blue-green to yellow-green (Figure 6.5(c)). Testing models assuming structural swelling,

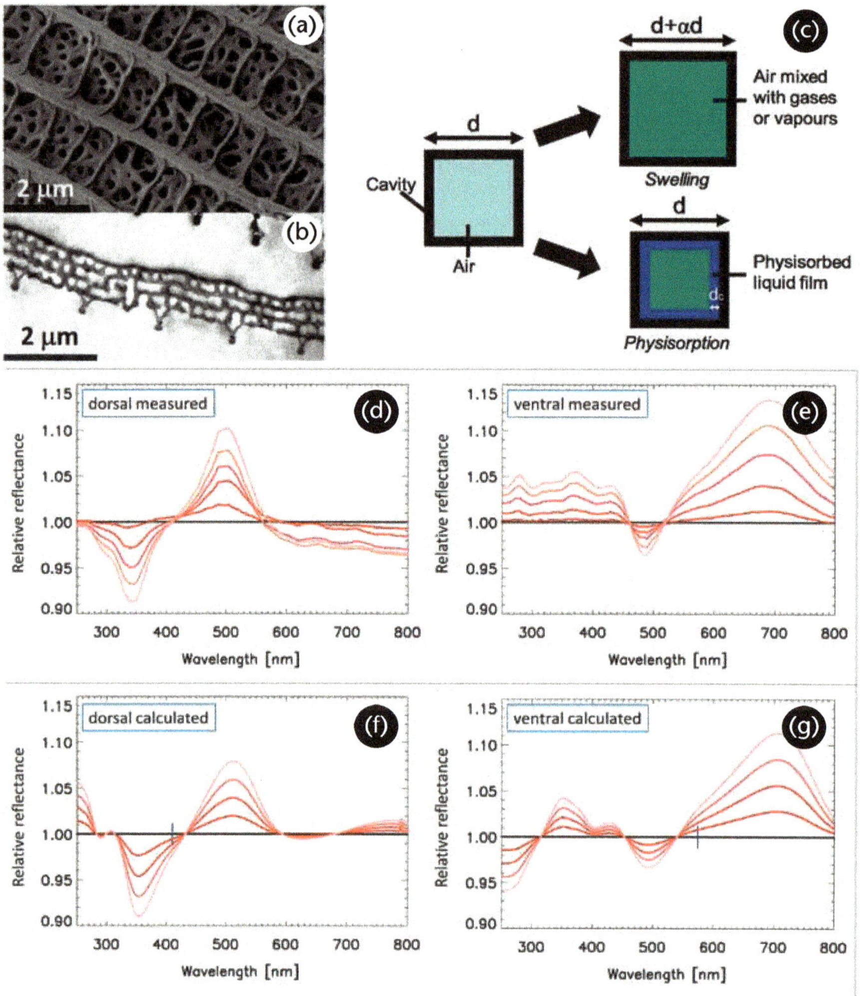

Figure 6.3 A typical case study of vapor- and gas-induced color changes investigated by Biró's group is the porous multilayer found in the scales covering the wings of lycaenid butterflies, such as the ventral wings of *A. metallica* observed here with SEM (a) and TEM (b) [17]. The two main physicochemical phenomena (c) taken into account in their modeling were the swelling of the photonic structure and the physisorption on the surface of the multilayer's pores [14,17]. d and d' are the dimensions of the structure model occurring on the butterfly wings; α is the swelling coefficient of the structure and d_c corresponds to the thickness of the physisorbed film on the pore surface. Fitting numerical predictions to measurements of the spectral differential reflectance ($\Delta R = \frac{R_c}{R_{0\%}}$, where R_c is the spectral reflectance at a given concentration of vapor or gas, and $R_{0\%}$ is the reference reflectance spectra—typically measured upon contact with nitrogen) from both ventral and dorsal wings at different ethanol vapor concentrations (10%, 20%, 30%, 40%, and 50% (d,e) and 12.5%, 25%, 37.5%, and 50% (f,g), respectively) showed that the color change is mainly due to the partial filling of the pores, with a small contribution from the structure's swelling. (Reproduced from [17] (a,b,d-g), License CC-BY-4.0 and from [18].)

surface adsorption of the vapor and modification in the keratin's refractive index (Figure 6.5(d–f)) led to the conclusion that the ambient humidity causes the swelling of the keratin layer from 148 to 158 nm [23]. Upon water infiltration in the keratin layer, the chemical network of keratin molecules loses its structure. Hydrogen bonds between neighboring keratin chains break, leading to the layer swelling.

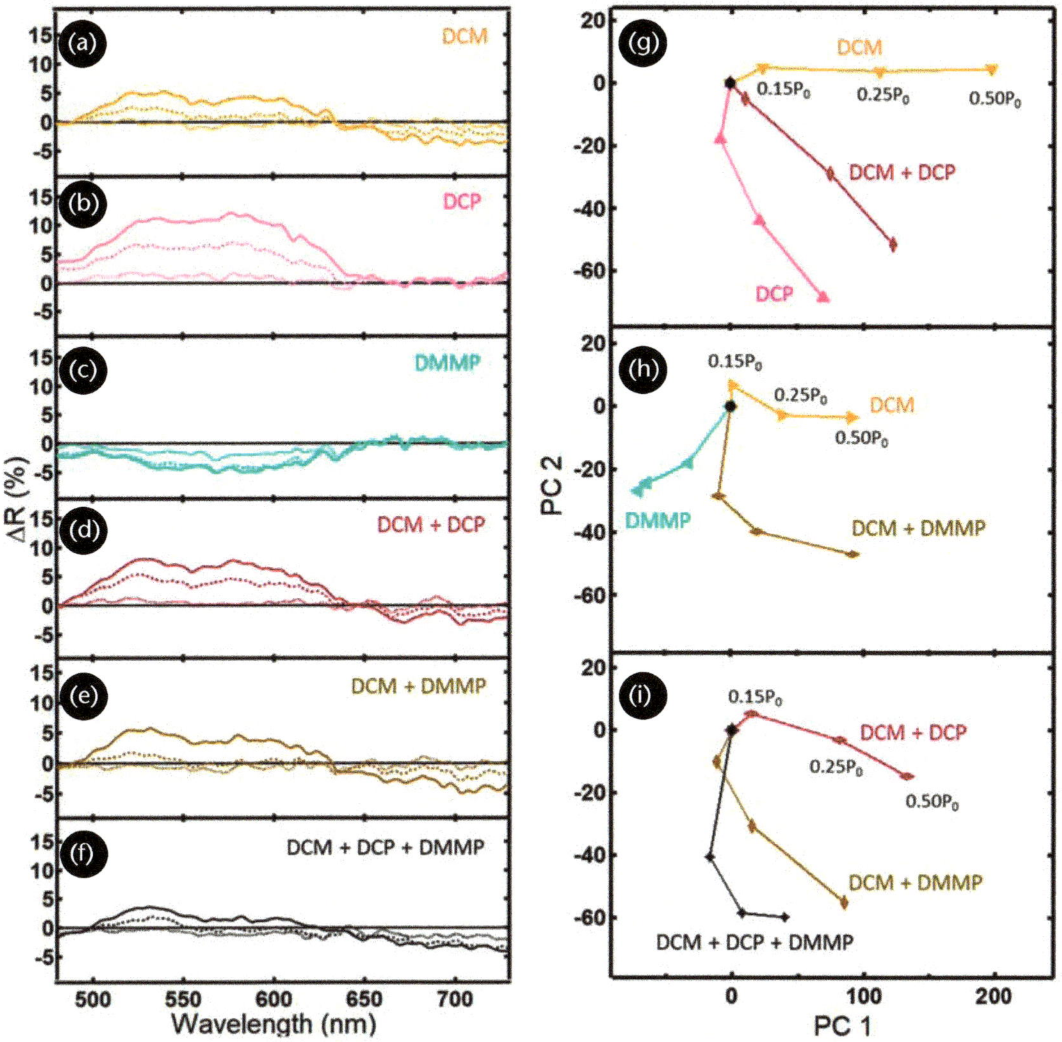

Figure 6.4 Spectral differential reflectance ΔR (a–f) measured from the wings of *M. didius* after 5 min. of contact with dichloromethane (DCM) (a), 1,5-dichloropentane (DCP, also known as mustard gas) (b), dimethyl methylphosphonate (DMMP) (c), which is a nerve agent simulant, mixtures of DCM and DCP (d), DCM and DMMP (e), as well as DCM, DCP, and DMMP (f) at three partial pressures; namely 15% (dotted lines), 25% (dashed lines), and 50% (solid lines). The PCA treatment of these responses highlights the selectivity of the natural photonic structures occurring in the wings of *M. didius*. (Reproduced from [20], License CC-BY-4.0.)

Upon drying, hydrogen bonds form again, the layer shrinks, and the original color is again exhibited by the feathers.

In the previous studies of gas- and vapor-induced color changes, the structures are in direct contact with the surrounding environment. Yet, some color changes have been found in beetle photonic structures that are not open to the environment [15,24–26]. One example is the scales covering the elytra and body of the male cerulean chafer beetle *Hoplia coerulea* described in the previous chapter [27,28]. Within these scales, a periodic multilayer comprising macropores (i.e., pores larger than 50 nm) is encased in a 100-nm envelope of chitin. The spectral reflectance

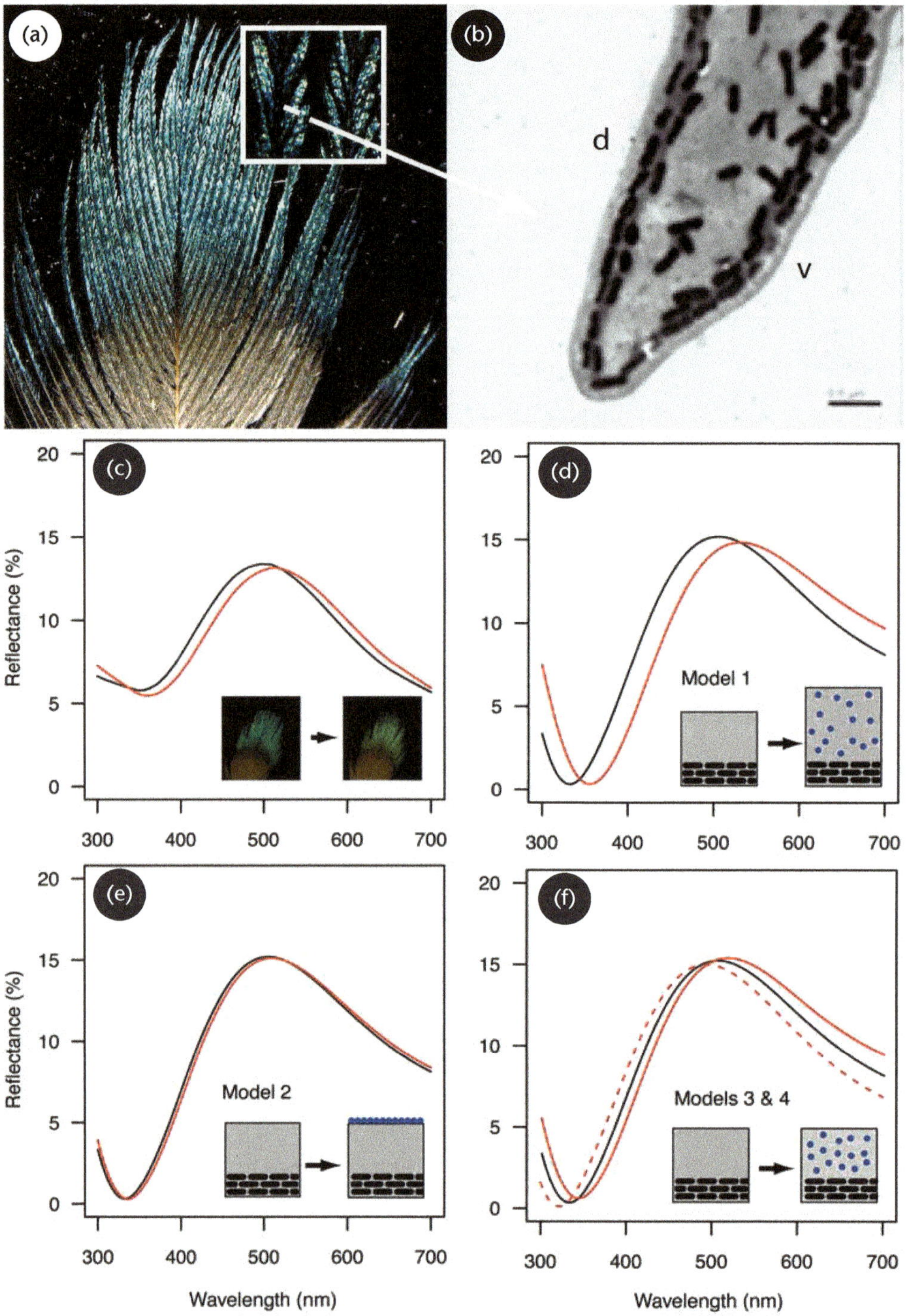

Figure 6.5 The feather barbules of *T. bicolor* exhibit a brilliant blue-green iridescence (a) due to thin-film interference within a keratin cortex encasing a layer of disordered melanosomes (b). Upon contact with water vapor, the color was shown to change to yellow-green (c) when the relative humidity increases from 20% (black line) to 80% (red line) [23]. Performing simulations of the optical response due to the swelling of the structure, surface adsorption of the vapor and modification in the keratin refractive index (d–f) concluded that the structure swells upon contact with humidity, increasing the keratin layer from 148 to 158 nm [23]. Scale bar: 500 nm (b). (Reproduced from [23], with permission from the Optical Society.)

measured from its elytra was found to exhibit a redshift in its peak position and an increase in its peak intensity upon contact with water and ethanol vapor (Figure 6.6) [24, 25]. In addition, environmental spectral ellipsometry [29] was employed to measure the change in optical properties of these structures upon contact with water, 2-propanol, and toluene vapors [28]. Spectral measurements of ellipsometric parameters cos Δ and tan Ψ (where Δ corresponds to the difference in phase modifications between both p- and s-polarizations and tan Ψ is the module of the ratio between Fresnel coefficients in p- and s-polarizations) showed a redshift of 10 nm and 4 nm, respectively, when the relative pressures of water and toluene vapors increased from 0% to 100%. Both reflectometry and ellipsometry measurements demonstrated a lower sensitivity with ethanol and 2-propanol vapors, respectively [15, 24, 25]. These results were in agreement with observations performed with liquids (see Chapter 5) [24, 30, 31]. Comparison between experimental results and numerical predictions that took several physicochemical phenomena into account (Figure 6.7) showed that the color change mostly originates from the physisorption of a film on the surface of the scales, followed by the infiltration of the liquefied vapor through the envelope of the scales and the filling of micropores (i.e., pores with diameters smaller than 2 nm) [15,24,25].

If all the previous cases are examples of *passive* color changes (namely, those not mediated by the metabolism of the organisms), the integuments of the desert tenebrionid beetle *Cryptoglossa verrucosa* are an interesting case of reversible humidity-induced color changes controlled by the insect's metabolism [32]. This beetle, which comes from the Sonoran Desert in the United States and Mexico, displays a bluish-white visual appearance in dry atmospheres and a black one in high-humidity environments, with intermediate color states. On the surface of its cuticle, many small tubercles can be observed (density assessed as 1200 per square millimeter). The tip of each tubercle is open, with some small amount of amorphous materials squeezing out, when the beetle is in a humid atmosphere. Upon contact with low relative humidity, some secretion exudes from the tubercles and forms a mesh of disordered filaments (with a diameter of about 140 nm) covering all the surface of the cuticle and radiating from the tubercles [32]. When the surrounding relative humidity returns to high values, the filaments dissolve into the wax covering the beetle's epicuticle. The secreted material appears to be mainly composed of lipids. The reversibility of this color change is limited after the death of the insects: dead black beetles do not return to their bluish-white visual appearance.

6.2 When Heat and Cold Induce Structural Color Changes

The wings of several structurally colored butterflies including three *Polymmatus* species [10, 33], *Callophrys rubi* [34], *Cyanophrys remus* [34], and *Morpho aega* [33, 34] were found to change color upon cooling from ambient temperature down to -10°C (Figure 6.8). Color changes induced by modification of temperature are often referred to as "thermochromism." At low temperatures, vapor-induced modification of the optical appearance of these wings was reported to be even more significant than at room temperature [10, 34]. The reason for this color change was described as being capillary condensation of vapor within the pores of the

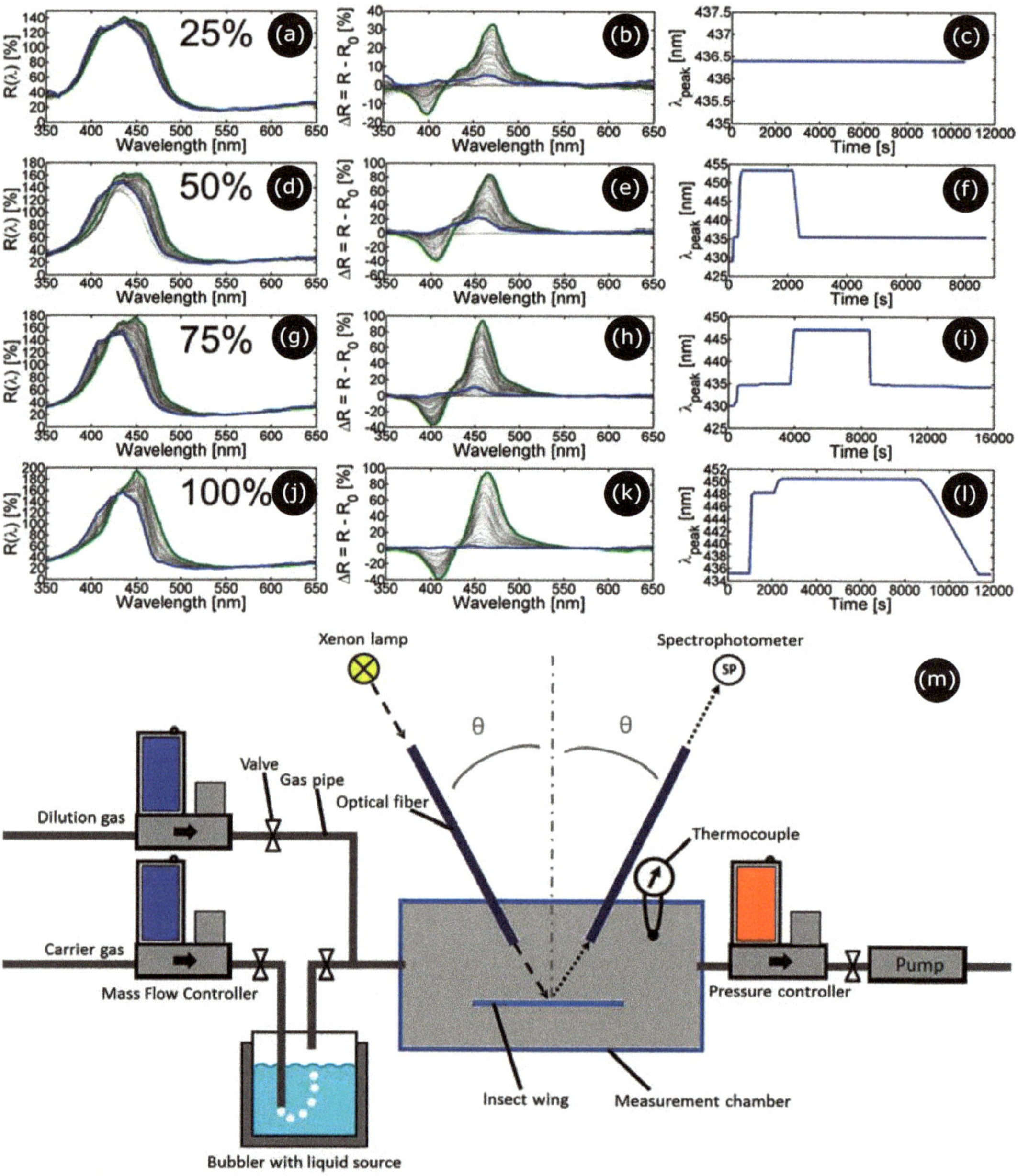

Figure 6.6 The peak intensity of spectral reflectance $R(\lambda)$ is shown here being measured from the elytra of *H. coerulea* at incidence and detection angles of 15°. Upon contact with water vapor at various concentrations, this peak intensity increases (a,d,g,j). Green curves: after exposure to water vapor. Blue curves: after complete drying. The differential reflectance $\Delta R = R - R_0$, where R_0 was measured in the dry state, highlights these differences (b,e,h,k). In addition to the increase in intensity, when the vapor concentration is higher than 50%, the peak also redshifts with respect to the peak position in the dry state, as can be seen from the variation in peak wavelength λ_{peak} as a function of exposure time (c,f,i,l). Typically, such optical measurements are carried out with optical fibers in a sealed chamber, the air composition of which is controlled by mass flow controllers (m). (Reproduced from [25], with permission from the Optical Society.)

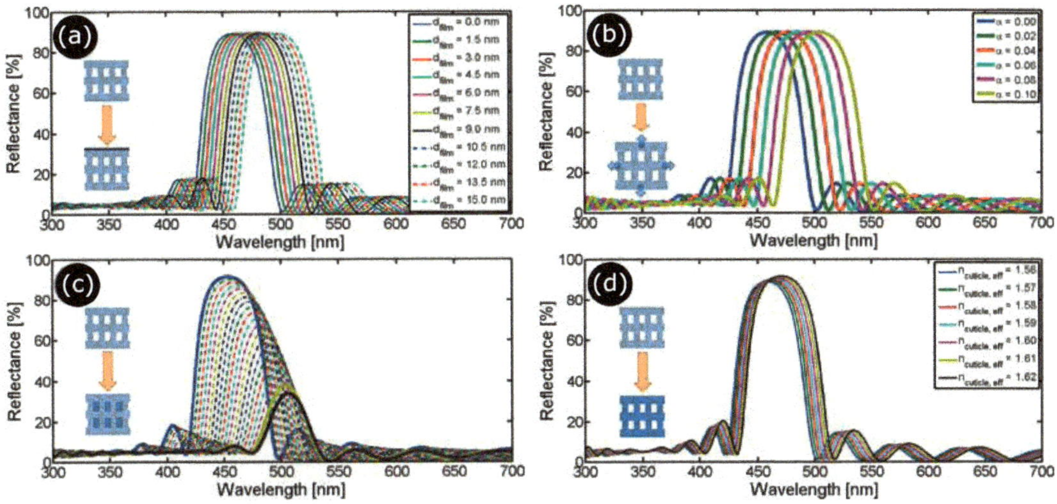

Figure 6.7 Numerical simulations were performed taking into account the possible physicochemical phenomena that were putatively at the origin of the measured changes in spectral reflectance of the elytra of *H. coerulea* [25]: adsorption of a liquid film with a thickness ranging from 0 nm to 15 nm (a); swelling of the structure with a swelling factor α ranging from 0 to 0.1 (b); filling of the structure's macropores with water (c); and filling of the structure's micropores with water (d). These predictions demonstrated that the measured reflectance changes are most likely due to the physisorption of a film on the scale, followed by the filling of the micropores. (Reproduced from [25], with permission from the Optical Society.)

photonic structures as a function of the pore dimensions, giving rise to the increase of the refractive index within the pores [10,33,34]. Based on this effect, Wang and coworkers demonstrated a reversible thermochromic response from the bright blue areas of the dorsal forewings and hindwings of *Papilio ulysses* [35]. They filled the photonic structures of these wings with ethanol that (partly) vaporizes or liquefies, leading to modifications in the refractive index of the pore content as a function of the temperature change. To ensure reversibility, as well as to avoid the ethanol completely evaporating from the structure, the butterfly wings were sealed between two glass slides.

By functionalizing scales of *M. sulkowskyi* with single-walled carbon nanotubes (SWNT), thermally induced changes in the optical response of these scales (both spectral reflectance and diffraction pattern) were measured by Pris and coworkers in 2012 [36]. The SWNT functionalization helped to increase the absorption of midwave infrared radiation (from approximately 3 to 8 μm) by the butterfly's photonic structures. Thanks to detailed optical simulations, the authors showed that the measured change was due to thermal expansion (i.e., increased spacing between ridges and lamellae with an increased temperature) and, to a much lesser extent, to a thermo-optic effect (namely, a decrease of refractive index due to a rise in temperature) [36].

6.3 pH-Driven Color Changes

The acidity or basicity of the surrounding environment of lepidopterans' wings was also shown to influence their optical response. Once again, these properties

Figure 6.8 Biró's group highlighted the thermochromic behavior of certain structurally colored butterfly wings by reducing the temperature of specific lycaenid butterflies in a refrigerator. Three butterflies, namely *C. rubi* (a), *C. remus* (b), and *P. bellargus* (c), exhibit such color change (top: room temperature; bottom: after cooling). Their wings are known for their structural colors. However, *Lycaena virgaureae* (d), which is pigmentary colored, does not display any color change upon cooling (left: room temperature; right: after cooling). (Reproduced from [33], License CC-BY-4.0.)

were not demonstrated by bare, natural photonic structures but by the addition of polymers to those lepidopteran wings [37–40]. As described in Chapter 5, the Christmas-tree structures from the Madagascan sunset moth *Chrysiridia rhipheus* [37] and various *Morpho* butterflies [38, 40] as well as the gyroid structures of the green hairstreak butterfly *Callophrys rubi* [39] were embedded with interpenetrating polymer network (IPN). In these instances, the IPN was pH-sensitive and was typically made of some polyvinyl alcohol (sometimes with some chitosan) or some polymer polymethylacrylic acid. In the former case, the chitin comprising the scales copolymerizes with polyvinyl alcohol by forming intermolecular hydrogen bonds [37]. For *C. rhipheus*, the reflectance peak was measured as redshifting by about 27 nm upon contact with acids (pH ranging from 7.07 to 1.00) and as blueshifting by approximately 260 nm with an increase of pH from 7.07 to 13.74 [37]. In the case of *Morpho peleides*, the reflectance peak wavelength was found to shift from 440 nm to 460 nm when the pH increased from 5.0 to 9.0 [40]. These pH-induced reversible changes are due to swelling and shrinking of the embedding hydrogel as a function of pH. For instance, an IPN may accept protons (a phenomenon called *protonation*) at low pH, whereas it releases these protons (*deprotonation*) at high pH values [37–39]. In an acidic environment, the uptake of protons weakens hydrogen bonding inside the IPN and fosters hydrogen bonding with surrounding water molecules, which leads to an increase in water content and, therefore, an increase in the effective refractive index that modifies the optical response. In addition, the electrostatic repulsion force between cationized polymer chains gives rise to a volume expansion, leading to a redshift of the reflectance spectrum.

6.4 Color Changes Induced by Mechanical Forces

So far, only a few examples of color changes induced by mechanical forces have been described in the literature. This phenomenon is sometimes referred to as "mechanochromism" and was demonstrated in the case of the scales covering the elytra of the weevil *Entimus imperialis* [41]. As described in Chapters 2 and 3, the matte green color of the spots on this creature's elytra are due to domains of different orientations of a diamond-type structure that can be generated by a triply-periodic bicontinuous cubic minimal surface [42–46]. These different orientations give rise to colors ranging from violet to orange. Researchers measured a blueshift in the scales' reflectance peak for both yellow and green domains by applying compression forces [41]. This blueshift, which may be up to 100 nm, was explained by a reduction of the photonic crystal period. At low compression rate, the deformation and related color change are usually reversible. Motivated by these results, the authors performed optical simulations with such diamond structures as well as with the related gyroid and primitive structures (Chapter 2). These predictions showed not only blueshift of the reflectance peak but also redshift, depending on the crystallographic directions and the direction of compression [41]. This predicted redshift would be due to an increase in the structure's effective refractive index following compression. The mechanochromic effect was experimentally tested on a gyroid photonic crystal structure inspired by the structures occurring on the wings of *Parides sesostris* butterfly [47]. This structure was fabricated by 3D printing in a rubber-like polymer. It had a lattice constant equal to 9 mm and a filling fraction of 40%, making it active in the microwave regime (i.e., 0.3–300 GHz). By applying various compression factors up to 22.2%, the researchers were able to measure that overlapping transmission minima spectrally separate into resolvable bands—mostly due to resonances in different crystallographic directions [47]. Due to the compression, the volume of the structure was reduced, thus the material filling fraction was increased. This led to a narrowing of the transmission minima and to a slight redshift in frequency. This behavior was numerically confirmed.

6.5 Nervous and Endocrine Controls of Photonic Colors

Several *active* changes of photonic colors mediated by neurons or hormones were described in Chapter 5, including the change in distances in the hexagonal lattices in the skin of the panther chameleon *Furcifer pardalis* [48] and the dehydration of the chirped multilayer of the tortoise beetle *Charidotella egregia* [49]. In this section, we will describe color changes in the iridophores occurring in the skin of some fish and cephalopod species.

The paradise whiptail *Pentapodus paradiseus* exhibits colorful bands on both its head and body (Figure 6.9) [50]. The appearance of these bands is due to a dense layer of iridophores within the skin. These iridophores comprise photonic

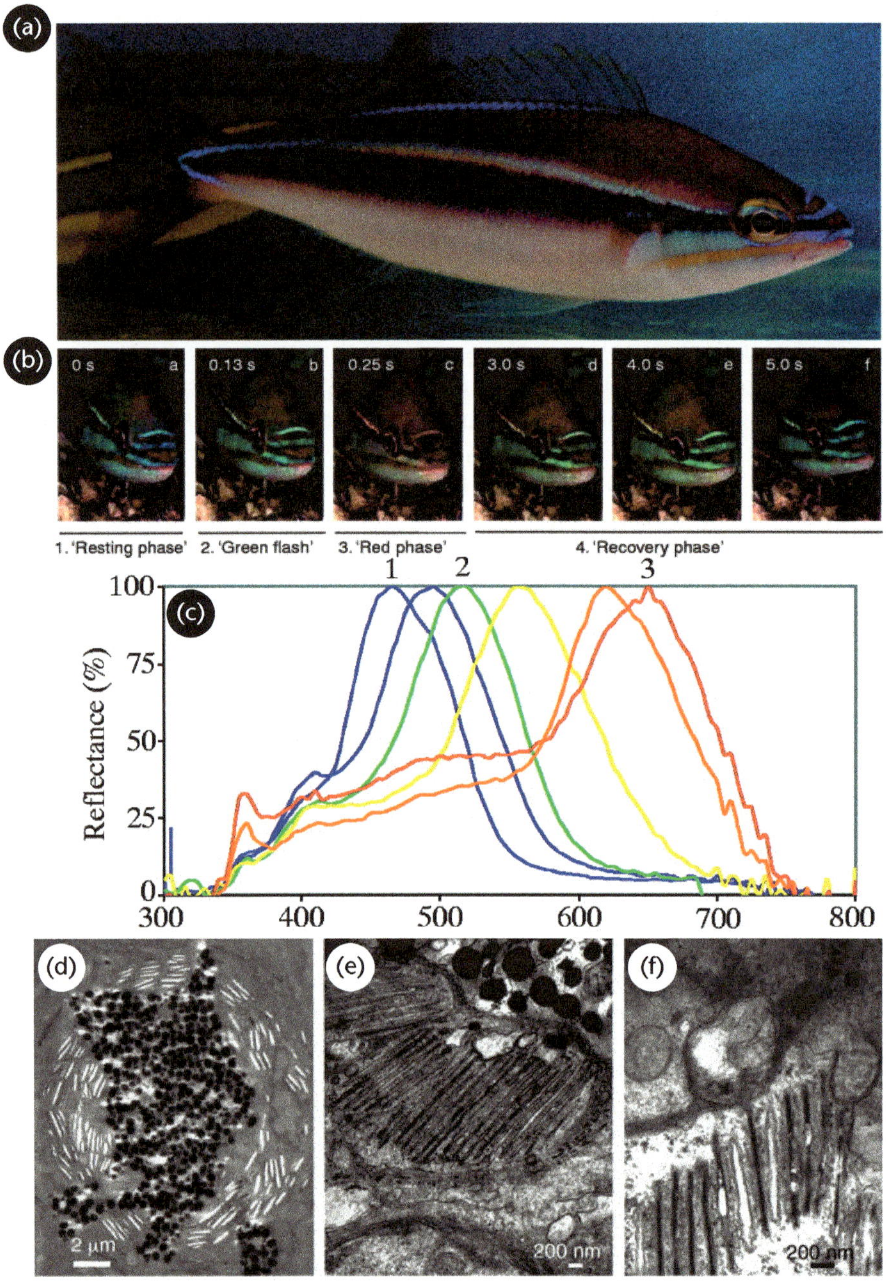

Figure 6.9 The paradise whiptail *P. paradiseus* exhibits iridescent bands on its head and body (a). These bands change reversibly from blue to red within 0.25 s, as observed in these photographs (b) and with spectral measurements (c). The integuments of these iridescent bands comprise both chromatophores and iridophores. The former are observed as black discs in (d). The latter are observed as white lines in (d) and black lines in (e,f). In fact, the white lines are tiny tears in the ultrathin TEM section marking the precise locations from which the platelets have inevitably fallen out of the cross section during preparation of the sample for TEM observation, leaving only their imprint behind. (Reproduced from [50], with permission from the Company of Biologists Ltd.)

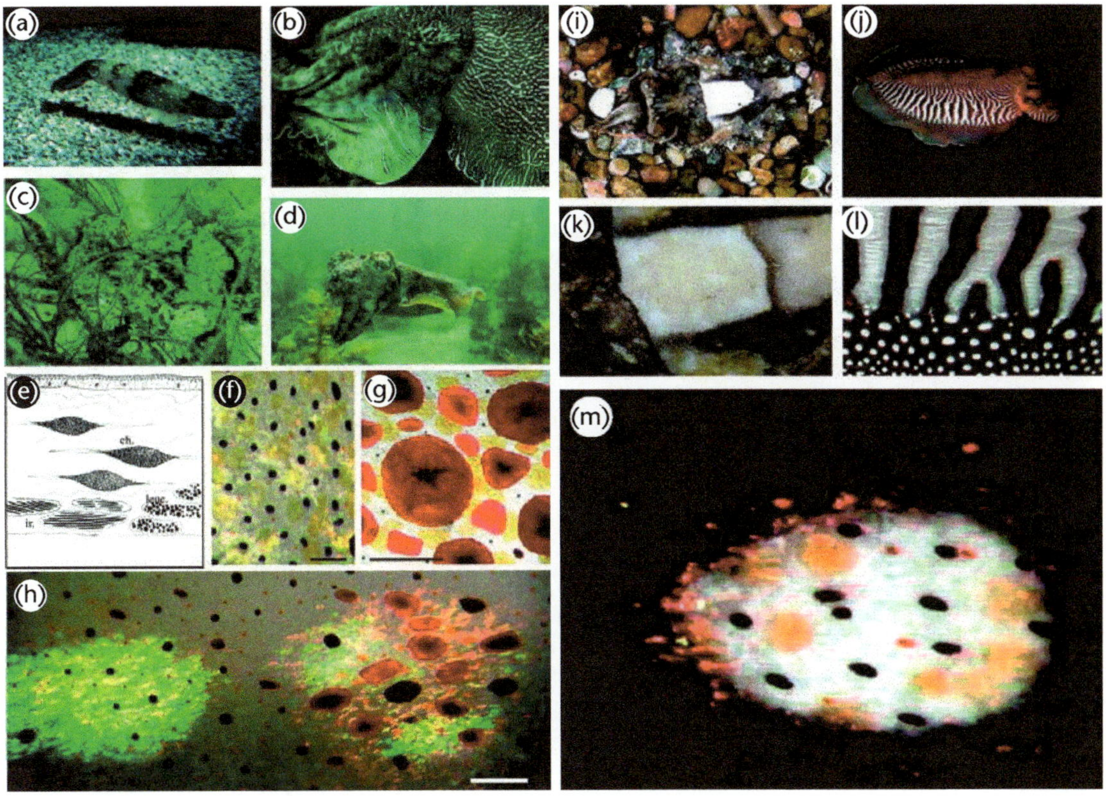

Figure 6.10 Cephalopods, including squids (like *Loligo pealeii* (a)), cuttlefishes (such as *Sepia apama* (b,d)), and octopuses, are able to change their visual appearances—for instance, for camouflage purposes (c). These changes are due to adaptive chromatophores (ch.), including iridophores (ir.) and leucophores (leuc.), which are located in their skins (e). Leucophores contain spherical proteins (called leucosomes) that scatter incident light incoherently. They appear white under incident white light. For instance, the expansion or the retraction of yellow, orange, red, and brown chromatophores in the skin of the cuttlefish *Sepia officinalis* (f) and *L. pealeii* (g) allows these animals to control their color. Chromatophores can be combined with iridophores (h) and leucophores (f–m). Scale bars: 1 mm (f–h). (Reproduced from [52], with permission from the Royal Society.)

multilayers made of guanine platelets that are separated by cytoplasm. The color reflected by these bands changes from blue to red within 0.25 s, following the increase in the guanine platelet interdistance of 70 nm [50]. The fish metabolism controls this color change via its nervous system, which releases two hormones—noradrenalin (giving rise to redshift of the reflectance peak) and adenosine (associated with a reflectance peak blueshift)—within the iridophores, inducing the change in the distance between two adjacent platelets. This mechanism is sometimes referred to as the accordion model.

Cephalopods, including cuttlefishes, squids, and octopuses, are well-known for being able to change their visual appearances (both color and pattern) for camouflage or intraspecific communication [51–54]. Among various chromatophores, iridophores contribute to these changes in appearance (Figure 6.10). They comprise some platelets made of reflectin (a protein associated with a high refractive index, and forming multilayers). Change in the protein's state modifies the refractive index and the thickness of the platelets [51–54].

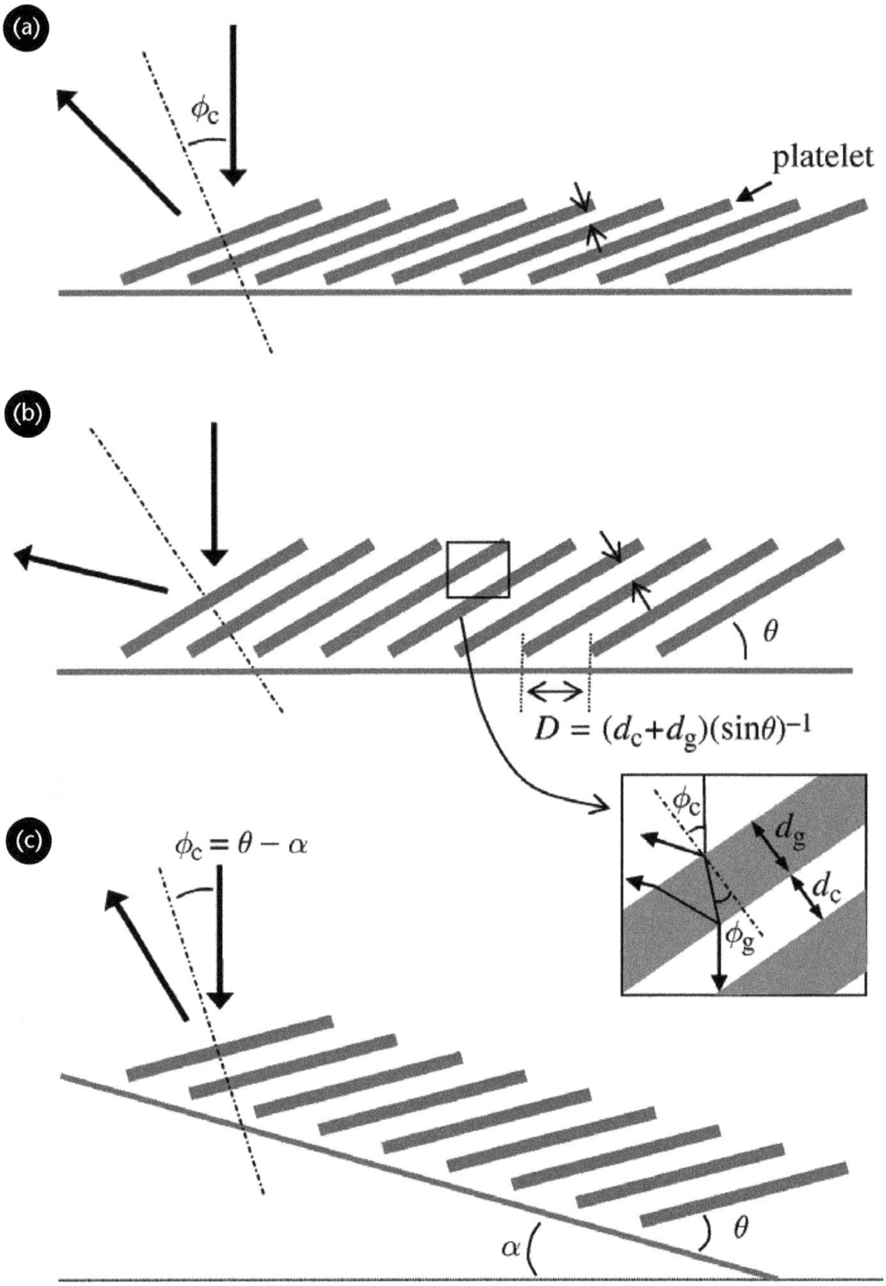

Figure 6.11 The Venetian blind model explains the color change observed in the neon tetra *Paracheirodon innesi* [58–60]. The angle θ between the platelets and the basal plane of the iridophore is modified under the action of the sympathetic nervous system of the fish, like the Venitian blinds of a window (a,b). The distance d_c between two neighboring platelets is modified. In addition, the direction of the reflected light changes, too. Often, the basal plane of the iridophore may also form an angle α with respect to the direction of observation (c). (Reproduced from [59], with permission from the Royal Society.)

A different mechanism was observed in the neon tetra *Paracheirodon innesi*. The lateral iridescent stripes exhibited by this fish are known for displaying a brilliant green color in the daytime and changing to violet at night [55–59]. This iridescent color originates from iridophores containing two or three stacks of guanine platelets [55–57]. For a long time, the underlying mechanism was an unsolved question for the natural photonic community. Although, at first, the color change was tentatively explained by the accordion model (i.e., the increasing in the spacing of the platelets [56]), it was later demonstrated to be due to the tilt of the platelets according to the Venetian blind model (Figure 6.11) [58, 59]. The sympathetic nervous system of the fish was found to control the optical response of these iridophores [57]; following the release of noradrenalin, filaments joining the platelets modify the tilt angle of the iridophores, like the Venitian blinds of a window [58–60]. The contribution of the increase in platelet interdistance was found to be negligible [59, 60].

References

[1] Umbers, K. D. L., et al., "Reversible Colour Change in Arthropoda," *Biology Reviews*, Vol. 89, No. 4, 2014, pp. 820–848.

[2] Potyrailo, R. A., et al., "*Morpho* Butterfly Wing Scales Demonstrate Highly Selective Vapour Response," *Nature Photonics*, Vol. 1, No. 2, 2007, pp. 123–128.

[3] Potyrailo, R. A., "Bio-Inspired Device Offers New Model for Vapor Sensing," *SPIE News*, published online: April 5th, 2011.

[4] Berthier, S., *Photonique des Morphos*, Paris: Springer, 2010.

[5] Yang, X., et al., "Using Hierarchy Architecture of *Morpho* Butterfly Scales for Chemical Sensing: Experiment and Modeling," *Sensors and Actuators A*, Vol. 167, No. 2, 2011, pp. 367–373.

[6] Gao, Y., et al., "Sensitivity Analysis of a Bioinspired Refractive Index Based Gas Sensor," *Journal of Bionic Engineering*, Vol. 8, No. 3, 2011, pp. 323–334.

[7] Jiang, T., et al., "Gas Sensing Using Hierarchical Micro/Nanostructures of *Morpho* Butterfly Scales," *Sensors and Actuators A*, Vol. 213, 2014, pp. 63–69.

[8] Potyrailo, R. A., et al., "Discovery of the Surface Polarity Gradient on Iridescent *Morpho* Butterfly Scales Reveals a Mechanism of Their Selective Vapor Response," *Proceedings of the National Academy of Science of the United States of America*, Vol. 110, No. 39, 2013, pp. 15567–15572.

[9] Biró, L. P., et al., "Photonic Nanoarchitectures Occurring in Butterfly Scales as Selective Gas/Vapor Sensors," *Proceedings of SPIE*, Vol. 7057, 2008, 705706.

[10] Kertész, K., et al., "Color Change of Blue Butterfly Wing Scales in an Air-Vapor Ambient," *Applied Surface Science*, Vol. 281, 2013, pp. 49–53.

[11] Piszter, G., et al., "Pretreated Butterfly Wings for Tuning the Selective Vapor Sensing," *Sensors*, Vol. 16, No. 9, 2016, 1446.

[12] Kertész, K., et al., "Optical Vapor Sensing on Single Wing Scales and on Whole Wings of the *Albulina metallica* Butterfly," *Sensors*, Vol. 18, No. 12, 2018, 4282.

[13] Piszter, G., et al., "Substance Specific Chemical Sensing with Pristine and Modified Photonic Nanoarchitectures Occurring in Blue Butterfly Wing Scales," *Optics Express*, Vol. 22, No. 19, 2014, pp. 22649–22660.

[14] Mouchet, S. R., O. Deparis, and J.-P. Vigneron, "Unexplained High Sensitivity of the Reflectance of Porous Natural Photonic Structures to the Presence of Gases and Vapours in the Atmosphere," *Proceedings of SPIE*, Vol. 8424, 2012, 842425.

[15] Mouchet, S. R., et al., "Assessment of Environmental Spectral Ellipsometry for Characterising Fluid-Induced Colour Changes in Natural Photonic Structures," *Materials Today: Proceedings*, Vol. 4, No. 4A, 2017, pp. 4987–1997.

[16] Piszter, G., et al., "Vapor Sensing of Pristine and ALD Modified Butterfly Wings," *Materials Today: Proceedings*, Vol. 1, No. S, 2014, pp. 216–220.

[17] Márk, G. I., et al., "Modeling the Reflectance Changes Induced by Vapor Condensation in Lycaenid Butterfly Wing Scales Colored by Photonic Nanoarchitectures," *Nanomaterials*, Vol. 9, No. 5, 2019, p. 759.

[18] Mouchet, S. R., "Mécanismes fondamentaux de la coloration structurale: addition et induction de couleurs," PhD thesis, University of Namur, Belgium, 2015.

[19] Piszter, G. et al., "Optical Detection of Vapor Mixtures Using Structurally Colored Butterfly and Moth Wings," *Sensors*, Vol. 19, No. 14, 2019, p. 3058.

[20] Kittle, J., et al., "Vapor Selectivity of a Natural Photonic Crystal to Binary and Tertiary Mixtures Containing Chemical Warfare Agent Simulants," *Sensors*, Vol. 20, No. 1, 2020, p. 157.

[21] Kittle, J., et al., "Sensing Chemical Warfare Agent Simulants via Photonic Crystals of the *Morpho didius* Butterfly," *ACS Omega*, Vol. 2, No. 11, 2017, pp. 8301–8307.

[22] Piszter, G., et al., "Stability and Selective Vapor Sensing of Structurally Colored Lepidopteran Wings Under Humid Conditions," *Sensors*, Vol. 20, No. 11, 2020, p. 3258.

[23] Eliason, C. M., and M. D. Shawkey, "Rapid, Reversible Response of Iridescent Feather Color to Ambient Humidity," *Optics Express*, Vol. 18, No. 20, 2010, pp. 21284–21292.

[24] Mouchet, S. R., et al., "*Hoplia coerulea*, a Porous Natural Photonic Structure as Template of Optical Vapour Sensor," *Proceedings of SPIE*, Vol. 9127, 2014, 91270U.

[25] Mouchet, S. R., et al., "Vapor Sensing with a Natural Photonic Cell," *Optics Express*, Vol. 24, No. 11, 2016, pp. 12267–12280.

[26] Sun, J., et al., "Investigating the Nanomechanical Properties and Reversible Color Change Properties of the Beetle *Dynastes tityus*," *Journal of Materials Science*, Vol. 52, No. 11, 2017, pp. 6150–6160.

[27] Vigneron, J.-P., et al., "Natural Layer-by-Layer Photonic Structure in the Squamae of *Hoplia coerulea* (Coleoptera)," *Physical Review E*, Vol. 72, No. 6, 2005, 061904.

[28] Mouchet, S. R., et. al., "Photonic Scales of *Hoplia coerulea* Beetle: Any Colour You Like," *Materials Today: Proceedings*, Vol. 4, No. 4A, 2017, pp. 4979–4986.

[29] Baklanov, M. R., et al., "Determination of Pore Size Distribution in Thin Films by Ellipsometric Porosimetry," *Journal of Vacuum Science & Technology B*, Vol. 18, No. 3, 2000, pp. 1385–1391.

[30] Mouchet, S. R., et al., "Liquid-Induced Colour Change in a Beetle: The Concept of a Photonic Cell," *Scientific Reports*, Vol. 6, 2016, 19322.

[31] Deparis, O., et al., "Nanostructured Surfaces: Bioinspiration for Transparency, Coloration and Wettability," *Materials Today: Proceedings*, Vol. 1, No. S, 2014, pp. 122–129.

[32] Hadley, N. F., "Wax Secretion and Color Phases of the Desert Tenebrionid Beetle *Cryptoglossa verrucosa* (LeConte)," *Science*, Vol. 203, No. 4378, 1979, pp. 367–369.

[33] Tamáska, I., et al., "Color Changes upon Cooling of Lepidoptera Scales Containing Photonic Nanoarchitectures, and a Method for Identifying the Changes," *Journal Insect Science*, Vol. 13, No. 87, 2013, pp. 1–15.

[34] Kertész, K., et al., "Temperature and Saturation Dependence in the Vapor Sensing of Butterfly Wing Scales," *Materials Science and Engineering C*, Vol. 39, 2014, pp. 221–226.

[35] Wang, W., et al., "Reversible Thermochromic Response Based on Photonic Crystal Structure in Butterfly Wing," *Nanophotonics*, Vol. 7, No. 1, 2018, pp. 217–227.

[36] Pris, A. D., et al., "Towards High-Speed Imaging of Infrared Photons with Bio-Inspired Nanoarchitectures," *Nature Photonics*, Vol. 6, No. 3, 2012, pp. 195–200.

[37] Zang, X., et al., "Moth Wing Scales as Optical pH Sensors," *Sensors and Actuators B*, Vol. 166-167, 2012, pp. 824–828.

[38] Yang, Q., et al., "Bioinspired Fabrication of Hierarchically Structured, pH-Tunable Photonic Crystals with Unique Transition," *ACS Nano*, Vol. 7, No. 6, 2013, pp. 4911–4918.

[39] Xue, R., et al., "Angle-Independent pH-Sensitive Composites with Natural Gyroid Structure," *Scientific Reports*, Vol. 7, 2017, 42207.

[40] Ahmed, R., et al., "*Morpho* Butterfly-Inspired Optical Diffraction, Diffusion, and Bio-Chemical Sensing," *RSC Advances*, Vol. 8, No. 48, 2018, pp. 27111–27118.

[41] Wu, X., et al., "From Insect Scales to Sensor Design: Modelling the Mechanochromic Properties of Bicontinuous Cubic Structures," *Bioinspiration & Biomimetics*, Vol. 11, No. 4, 2016, 045001.

[42] Deparis, O., and J.-P. Vigneron, "Modeling the Photonic Response of Biological Nanostructures Using the Concept of Stratified Medium: The Case of a Natural Three-Dimensional Photonic Crystal," *Materials Science and Engineering B*, Vol. 169, No. 1-3, 2010, pp. 12–15.

[43] Mouchet, S. R., et al., "Additive Photonic Colors in the Brazilian Diamond Weevil, *Entimus imperialis*," *Proceedings of SPIE*, Vol. 8480, 2012, 848003.

[44] Wilts, B. D., et al., "Brilliant Camouflage: Photonic Crystals in the Diamond Weevil, *Entimus imperialis*," *Proceedings of the Royal Society B*, Vol. 279, No. 1738, 2012, pp. 2524–2530.

[45] Wilts, B. D., et al., "Hemispherical Brillouin Zone Imaging of a Diamond-Type Biological Photonic Crystal," *Journal of the Royal Society Interface*, Vol. 9, No. 72, 2012, pp. 1609–1614.

[46] Mouchet, S. R., et al., "Method for Modeling Additive Color Effect in Photonic Polycrystals with Form Anisotropic Elements: The Case of *Entimus imperialis*," *Optics Express*, Vol. 21, No. 11, 2013, pp. 13228–13240.

[47] Pouya, C., et al., "Characterization of a Mechanically Tunable Gyroid Photonic Crystal Inspired by the Butterfly *Parides sesostris*," *Advanced Optical Materials*, Vol. 4, No. 1, 2016, pp. 99–105.

[48] Teyssier, J., et al., "Photonic Crystals Cause Active Colour Change in Chameleons," *Nature Communications*, Vol. 6, No. 1, 2015, 6368.

[49] Pasteels, J. M., et al., "Structural and Physical Evidence for an Endocuticular Gold Reflector in the Tortoise Beetle, *Charidotella ambita*," *Arthropod Structure & Development*, Vol. 45, No. 6, 2016, pp. 509–518.

[50] Mäthger, L. M., et al., "Rapid Colour Changes in Multilayer Reflecting Stripes in the Paradise Whiptail, *Pentapodus paradiseus*," *Journal of Experimental Biology*, Vol. 206, No. 20, 2003, pp. 3607–3613.

[51] Mäthger, L. M., and R. T. Hanlon, "Malleable Skin Coloration in Cephalopods: Selective Reflectance, Transmission and Absorbance of Light by Chromatophores and Iridophores," *Cell Tissue Research*, Vol. 329, No. 1, 2007, pp. 179–186.

[52] Mäthger, L. M., et al., "Mechanisms and Behavioural Functions of Structural Coloration in Cephalopods," *Journal of the Royal Society Interface*, Vol. 6, No. S2, 2009, pp. S149–S163.

[53] Izumi, M., et al., "Changes in Reflectin Protein Phosphorylation are Associated with Dynamic Iridescence in Squid," *Journal of the Royal Society Interface*, Vol. 7, No. 44, 2010, pp. 549–560.

[54] Ghoshal, A., et al., "Optical Parameters of the Tunable Bragg Reflectors in Squid," *Journal of the Royal Society Interface*, Vol. 10, No. 85, 2013, 20130386.

[55] Lythgoe, J. N., and J. Shand, "Changes in Spectral Reflexions from the Iridophores of the Neon Tetra," *Journal of Physiology*, Vol. 325, No. 1, 1982, pp. 23–34.

[56] Clothier, J., and J. N. Lythgoe, "Light-Induced Colour Changes by the Iridophores of the Neon Tetra, *Paracheirodon innesi*," *Journal of Cell Science*, Vol. 88, No. 5, 1987, pp. 663–668.

[57] Nagaishi, H., and N. Oshima, "Neural Control of Motile Activity of Light-Sensitive Iridophores in the Neon Tetra," *Pigment Cell Research*, Vol. 2, No. 6, 1989, pp. 485–492.

[58] Nagaishi, H., N. Oshima, and R. Fujii, "Light-Reflecting Properties of the Iridophores of the Neon Tetra, *Paracheirodon innesi*," *Comparative Biochemistry and Physiology*, Vol. 95, No. 3, 1990, pp. 337–341.

[59] Yoshioka, S., et al., "Mechanism of Variable Structural Colour in the Neon Tetra: Quantitative Evaluation of the Venetian Blind Model," *Journal of the Royal Society Interface*, Vol. 8, No. 54, 2011, pp. 56–66.

[60] Gur, D., et al., "The Mechanism of Color Change in the Neon Tetra Fish: A Light-Induced Tunable Photonic Crystal Array," *Angewandte Chemie International Edition*, Vol. 54, No. 42, 2015, pp. 12426–12430.

CHAPTER 7

Solar Energy Harvesting Devices

Harvesting solar energy is important for many natural organisms, either as a vital or simply a profitable function. Photosynthesis, as a vital function for plants, requires harvesting the visible part of the solar spectrum. Body thermoregulation, a profitable function for things such as, butterflies, requires a balance being struck between solar absorption and thermal emission involving both the visible and the near-infrared spectral regions. The UV part of the solar spectrum must not be forgotten, since excessive exposure of cellular tissues to UV radiation can cause irreversible DNA damage. UV protection mechanisms are therefore needed. Now, as throughout history, humans need to protect themselves against sunburn, but modern people also rely increasingly on solar energy in order to live in a sustainable world (Chapter 12). Nature can help humanity with both purposes, as will be highlighted in this chapter.

7.1 Light Harvesting in Plants

In nature, photonic structures may contribute to the enhancement of light absorption. Of all the energy harvesting mechanisms occurring in natural organisms, photosynthesis in plants is undeniably one of the most important processes.

7.1.1 Enhanced Light Absorption within Plant Chloroplasts

The structural colors exhibited by some plants, including blue iridescent leaves, have been known for a long time (Chapter 1) [1]. More recently, the multilayers observed in the blue iridescent epidermal chloroplasts of plant leaves including the shade-dwelling *Begonia* and *Selaginella erythropus* have been found to enhance light absorption [2–4]. Chloroplasts are plant organelles in which photosynthesis takes place through the absorption of sunlight by chlorophyll pigments; the trapped energy from sunlight is converted into chemical energy and stored as ATP and NADPH. More specifically, within the chloroplasts, the early, light-dependent stage of photosynthesis occurs in the thylakoid tissues. Two specific types of chloroplasts are iridoplasts and bizonoplasts [5,6]. The former are epidermal chloroplasts found only in certain specific plants. They contain photonic structures in the form of periodic stacks of absorbing thylakoid tissues. To date, bizonoplasts have only been found in one plant species, *Selaginella erythropus* [7,8]. In terms of ultrastructure, they resemble a combination of the typical arrangement of thylakoid tissue that is found in ordinary, nonphotonic (and irregular) chloroplasts of various plants and the highly ordered structure of iridoplasts. Iridoplasts and bizonoplasts have been observed in the cases of *Begonia* and *S. erythropus* leaves (Figure 7.1), respectively [5,7,8]. Thanks to increased light absorption in the green range (i.e., the incident

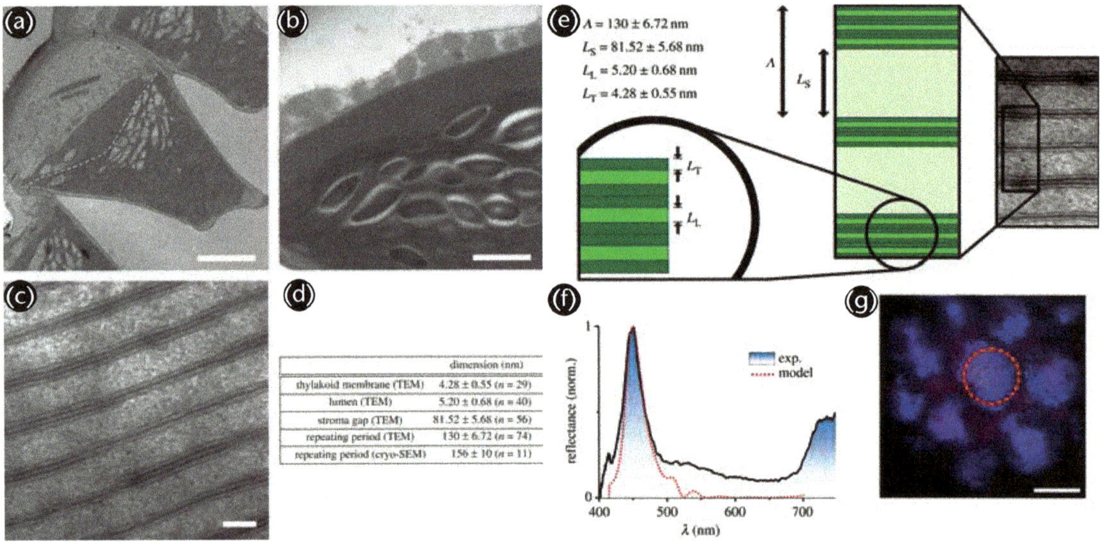

	dimension (nm)
thylakoid membrane (TEM)	4.28 ± 0.55 (n = 29)
lumen (TEM)	5.20 ± 0.68 (n = 40)
stroma gap (TEM)	81.52 ± 5.68 (n = 56)
repeating period (TEM)	130 ± 6.72 (n = 74)
repeating period (cryo-SEM)	156 ± 10 (n = 11)

Figure 7.1 The bizonoplasts (a) occurring in the epidermis of *S. erythropus* leaves comprise an upper periodical stack of thylakoid tissue (b,c) and lower irregular thylakoid tissue. Both tissues are separated by a white dashed line in (a). The characteristic dimensions of the structures were measured using the electron microscopic observations and are given in (d). The structure can be modeled as a multilayer (e) with a period Λ comprising thylakoid membranes (with a thickness L_{T}) interspersed with lumen layers (with a thickness L_{L}) and stromal layers (with a thickness L_{s}). This multilayer gives rise to light reflection in the blue range, as measured by spectrophotometry (f) and observed by optical microscopy (g). Scale bars: 5 μm (a), 0.5 μm (b), 100 nm (c), and 10 μm (g). (Reproduced from [3], with permission from the Royal Society.)

light environment of these shade-dwelling plants) and enhanced quantum yield in low-light conditions, the photosynthesis in iridoplasts and bizonoplasts is increased with respect to other chloroplasts [2–4]. This enhanced absorption is considered to be related to slow light effects at the red edge of the photonic bandgap of the photonic structures. It was demonstrated very recently that dynamic change of the iridoplasts' photonic structures could be induced by modification of the incident light conditions [4]. With an increase in the light intensity, the spectral reflectance peak and the light absorption enhancement are redshifted. This shift is understood to result from a photodamage protection mechanism [4].

7.1.2 Light Trapping at the Surface of Plant Integuments

The micro- and nanostructures occurring on the surface of plant petals and leaves (Figure 7.2) also enhance light harvesting. This absorption enhancement relies on antireflective properties and light trapping [9–12]. The former arises from subwavelength structuration, with conical-shaped epidermal cells giving rise to a gradual increase in the effective refractive index when light impinges on the plant integuments. This is similar to the profiles of the structures occurring on moth eyes and cicada wings (see Chapter 4). The latter involves redirection of light, increasing the path length in the plant integuments and, thereby, fostering light absorption. In addition, nanowrinkles were found to decrease further reflection losses [11,12]. In 2019, it was demonstrated in petals of *Bidens ferulifolia*, *Rudbeckia fulgida*, and *Erodium manescavii* flowers that this conical morphology could vary, depending on

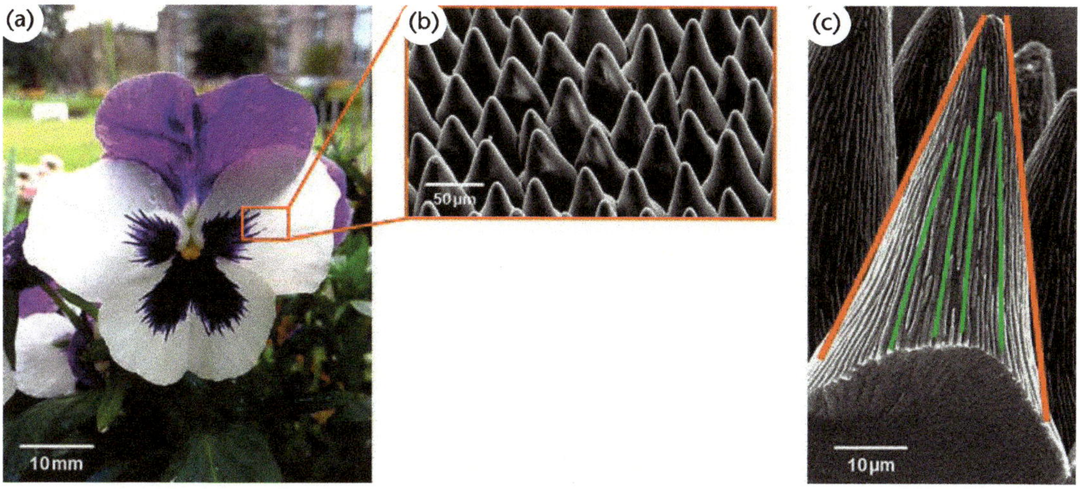

Figure 7.2 Pansy *Viola* sp. petals (a) are covered by conical microstructures overlaid by nanowrinkles (b,c). Both conical microstructures and nanowrinkles contribute to enhance light capturing. (Reproduced from [12], with permission from the American Chemical Society.)

the plant's strategy, to absorb or scatter UV light [13]. Each petal of these flowers exhibits distinct UV-absorbing and UV-scattering areas. The cones are taller in the former than in the latter. This enhanced light-capture plays a significant role for photosynthesis or strengthening of plants' color saturation, specifically in low-light environments. Furthermore, the microstructures give rise to superhydrophobicity and self-cleaning properties of the integuments [14, 15].

Similarly, a less primitive natural solar energy harvesting device is found in diatoms' frustules (i.e., the hard porous silica wall of diatom cells). Among diatoms, the valve of the diatom *Coscinodiscus* sp. exhibits a radial symmetry with highly organized multilevel pores [16]. Their hierarchical micro-/nano-scaled structure is made of three layers (called cribellum, cribrum, and the internal plate). All three layers are thin slabs comprising hexagonal arrays of disk holes. From one layer to another, both the size and the distance between the holes vary. This structure has also been demonstrated to be optimized for photosynthesis, which makes it an efficient solar light trapping device [16].

7.2 Structurally Enhanced Blackness

Black is the absence of colors. Consider a medium containing pigments that absorb in the visible range (e.g., melanin): a light wave propagating in such a medium would have an exponentially decaying intensity (following the Beer-Lambert law), which eventually tends to zero at all wavelengths, as the path length becomes sufficiently long. Does it mean that *perfect absorption* (i.e., equal to 100%) is possible? Of course not, simply because the light wave has to enter the medium and, therefore, Fresnel reflection at the medium surface sets an upper limit to the absorption. For a flat interface, it is easy to calculate the Fresnel loss from the complex refractive index of the medium, taking into account the spectral dependencies of its real and imaginary parts. In solar cell technology, it has been known for more than 50 years that structuring the silicon surface helps increase the optical absorption, and hence enhances the cell efficiency. Depending on the size of the surface corrugation, the absorption enhancement is explained by either geometrical optics arguments (i.e., multiple ray deflections) or wave optics arguments; namely, coupling to quasi-guided modes. Interestingly, nature has used these tricks for a long time. If you dip the black wing of certain butterfly species into an index matching fluid, you will observe a decrease in the absorption: this is a simple piece of experimental evidence for a structural enhancement of blackness.

In 2004, Vukusic and coworkers gave the first evidence that intricate nanostructures in the scales of the male *Papilio ulysses* butterfly enhanced the blackness of its wing regions [17]. As a qualitative explanation, they suggested the incident light is scattered by the scales' structure toward the ridges and the scale interior, such that the increased optical path length results in enhanced absorption by the pigments, which are assumed to be spread, diffusely, throughout the scale. To quantify the effect, they immersed black scales into index-matching fluid and found a substantial reduction in the measured absorption, from approximately 95% to about 55% and from circa 90% to around 70% in, respectively, matte black scales and lustrous black scales (Figure 7.3). This difference in absorption enhancement is explained qualitatively by the more intricate and densely distributed lattice in the former type of scales, as observed on SEM and TEM images (Figure 7.3). Also contributing to the blackness is the antireflective device that is provided by micron-sized tapering of the ridges as well as subwavelength tapering of the ridge sides (see inset in Figure 7.3). Although not investigated further by the authors, the luster was attributed to long, hair-like, androconical scales that overlaid the ground scales.

Solar absorption and radiative heat dissipation are expected to be intimately related biological functions of black butterfly wings. The black wings of the male birdwings butterflies *Troides magellanus* and *Troides aeacus* exhibit not only strong (98%) broadband absorption in the visible range [18,19], but those of *T. magellanus* were also found to display two distinct absorption peaks at 3 μm and 6 μm in the spectral region where the black body emits at 313 K [18]. Using a 3D periodic model of the nanostructured scale and comparing it with a planar slab model of equivalent volume, a structural effect on the absorption was demonstrated through numerical simulations, revealing approximately 10% solar absorption enhancement and about 17% emissivity enhancement at 40°C [18].

The West African Gaboon viper (*Bitis rhinoceros*) has a geometrically patterned skin with velvet black spots alternating with light brown and pale colored areas, providing perfect camouflage in its native forest floor habitat, right down to the canopy modulated variations of light and shade observed in that environment (Figure 7.4) [20]. Although these areas differ in pigmentation as confirmed by cross section optical microscopy, the blackness—hence, the high contrast it provides to the skin pattern—originates predominantly from the hierarchical structure found in the velvet black scales. This structure was described from SEM observations as densely packed, leaf-like microstructures striated with nanoridges; the areas between the black scales being also covered with nanoridge striations (Figure 7.4(d–j)). Reflectance spectra of black and pale scales are nearly flat across

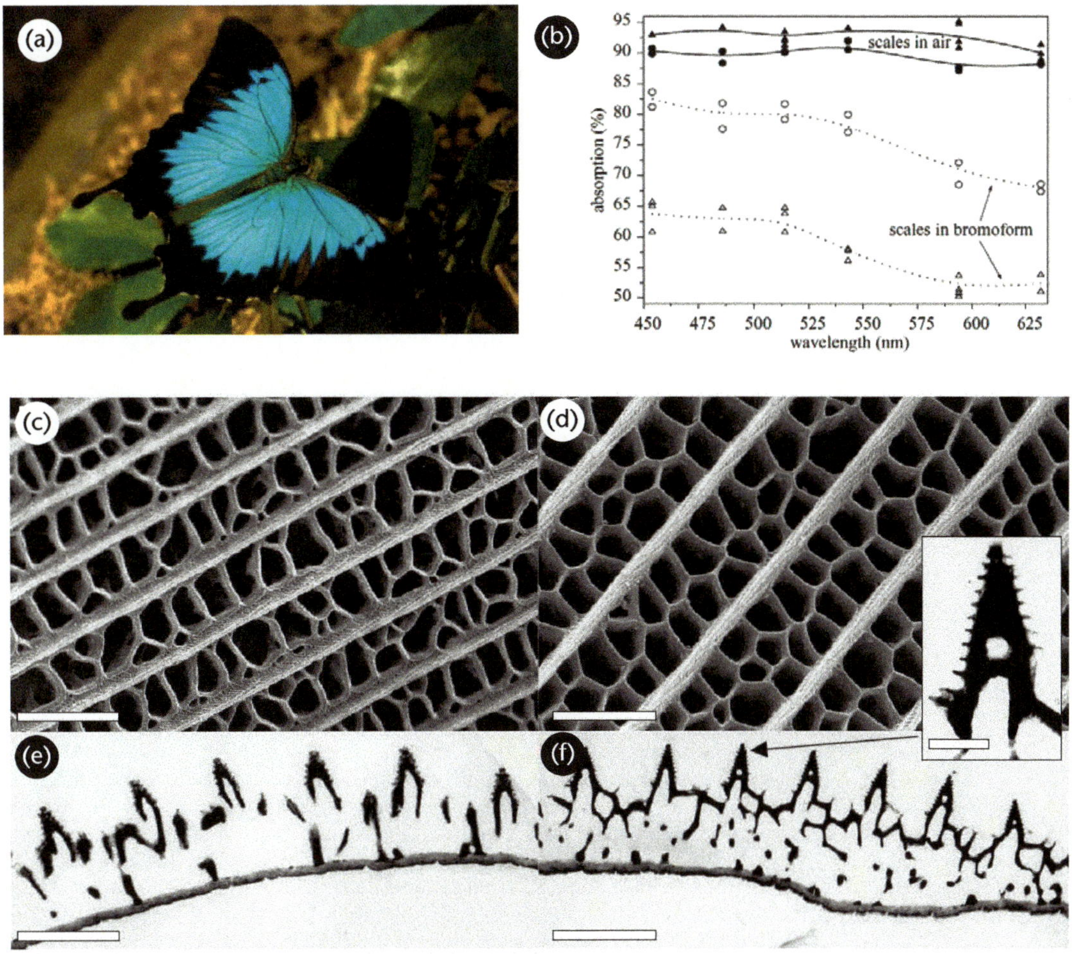

Figure 7.3 The male *P. ulysses* butterfly (a) enhances light absorption in its wings by structuration of their constituent material. Absorption spectra (b) measured at normal incidence on single lustrous black scales and matte black scales; filled and open symbols are for data collected for scales in air and in bromoform (index-matching fluid), respectively. SEM images (c,d) and TEM images (e,f) of lustrous (c,e) and matte (d,f) black scales. Scale bars: 3 μm (c), 2 μm (d); 2 μm (e,f); inset, 300 nm. (Reproduced from W. van Aken (a), https://commons.wikimedia.org/wiki/File:CSIRO_ScienceImage_3831_Ulysses_Butterfly.jpg, License CC-BY-3.0, and from [17] (b–f), with permission from the Royal Society.)

the visible range and increase beyond, with a peak at 880 nm. At 600 nm wavelength, the reflectance is lower than approximately 10% and approximately 30% in the black and pale scales, respectively (Figure 7.4(b)). Most interestingly, coating the back scales with a metallic layer (Au-Pd) not only maintains the black coloration, but also further reduces the reflectance level (Figure 7.4(b)). This observation brings experimental evidence of the structural origin of blackness, which largely prevails over pigmentary absorption. Our interpretation is that the original (uncoated) surface *already* functions as an efficient light trapping device through multiple reflections of incident rays on the leaf-like microstructure. The coating further improves light trapping through reflections on metallic surfaces. The authors' modeling of the diffuse reflection from the surface with Lambertian symmetric

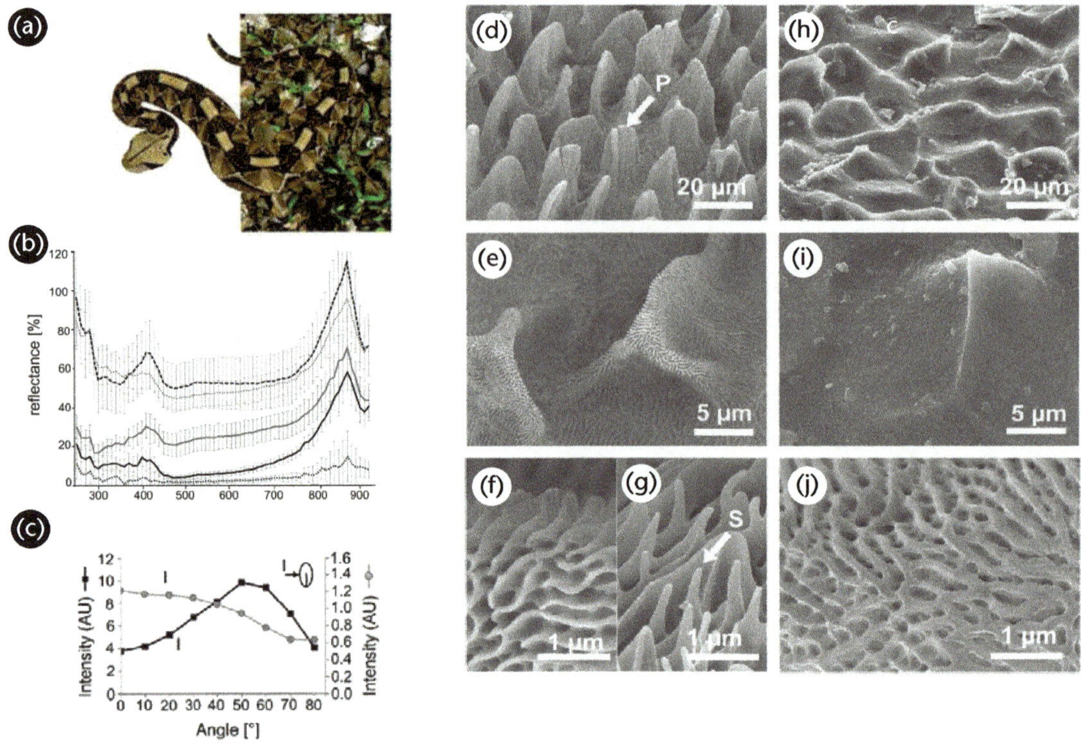

Figure 7.4 The viper *Bitis rhinoceros* (a) has velvet black spots alternating with light brown and pale color on its skin, which provide it with perfect camouflage. Reflectance spectra (b) of black dorsal scales (black line), Au-Pd coated black dorsal scales (dotted black line), pale dorsal scales (gray line), Au-Pd coated pale dorsal scales (dotted gray line), and ventral scales (dashed black line). Scattered light intensities from a pale scale (black line) and a black scale (gray line) (c) at 700 nm as a function of detection angle (illumination at 45°). SEM images (d–j) of the scale microstructures: the black dorsal scales feature densely packed, leaf-like, ornamentation (d), with branched ridges at their surfaces (e-g); pale dorsal scales are comparatively less structured (h), with pits on the surface (i,j). (Reproduced from [20], with permission from Springer Nature.)

V-shaped cavities confirms the light trapping mechanism. It additionally explains quite well the angular dependence of the reflectance (except for the specular peak) for the pale scales, which have similar, but shallower, structures than those of the black scales. Contrary to pale scales, black scales do not exhibit a specular reflection peak, but rather have a slowly decreasing reflectance with increasing detection angle (Figure 7.4(c)). This angular characteristic provides the velvet black with a nonglossy aspect. Suppression of the specular reflection may be regarded as a consequence of the more or less isotropic arrangement of the *B. rhinoceros* viper's scale structure at both the micrometre and nanometre scales. The situation differs from the layered and more regular scale structure of the butterfly *P. ulysses*.

Searching for structurally enhanced blackness may be disappointing in some cases. For instance, the stick insect *Peruphasma schultei* is almost completely matte black (Figure 7.5), with a reflectance as low as approximately 2% across the whole visible range, a level that is comparable to that reported by Vukusic in the black scales of *P. ulysses* butterfly. The insect cuticle has tall denticles (18 μm height) and small papillae (4 μm height), which form rather nice microstructures (Figure 7.5).

Figure 7.5 All-black stick insects: the male of *P. schultei* (a) with its matte black body and the female (b) of *A. paromalus*, with its glossy black body. SEM images of *P. schultei* cuticle morphology: long setae (c) and detail of the microstructure (d) composed of tall denticles and small papillae. Scale bars: 5 mm (a,b). (Reproduced from [21], with permission from Elsevier.)

In spite of their promising appearances, the contribution of these microstructures to the blackness was found to be negligible experimentally after index-matching fluid tests were preformed [21]. A plausible explanation, which is not evoked in [21] but which we will propose here, is that the relatively large distance between the denticles (10 μm) and their vertical positions do not foster light trapping by multiple reflections of rays toward the cuticle. The reason why the much shorter papillae do not provide efficient light trapping deserves further study. By contrast, bleaching experiments using oxidizing or saponificating agents to destroy pigments were conclusive: the black cuticle turned to white once it was bleached! The blackness most probably originates from very effective light-absorbing pigments that have not yet been identified. The matte appearance of the black color, on the other hand, does turn out to arise from the microstructures, as a control stick insect (*Anisomorpha paromalus*), which has a smooth cuticle, exhibits a glossy black appearance (Figure 7.5). Indeed, reflectance measurements as function of the viewing angle showed that the matte cuticle scatters light evenly for all angles, while the glossy cuticle has noticeable reflection directionality between viewing angles of 20° and 60°.

Super-black plumage in five species of birds of paradise (Paradisaeidae) has been recently attributed to structural absorption by McCoy and coworkers, who reported absorption up to 99.95%. This is an extremely high value, approaching those of artificial ultra-absorbent materials [22]. Actually, super-black feathers appear profoundly darker than normal black feathers because of considerably reduced specular reflection, as attested by very low (0.05–0.31%) measured directional reflectance (Figure 7.6). Moreover, super-black feathers appear darkest when viewed from the distal direction, as observed from measurements and confirmed by ray-tracing simulations. Interestingly, this corresponds to the viewing direction of a female observing

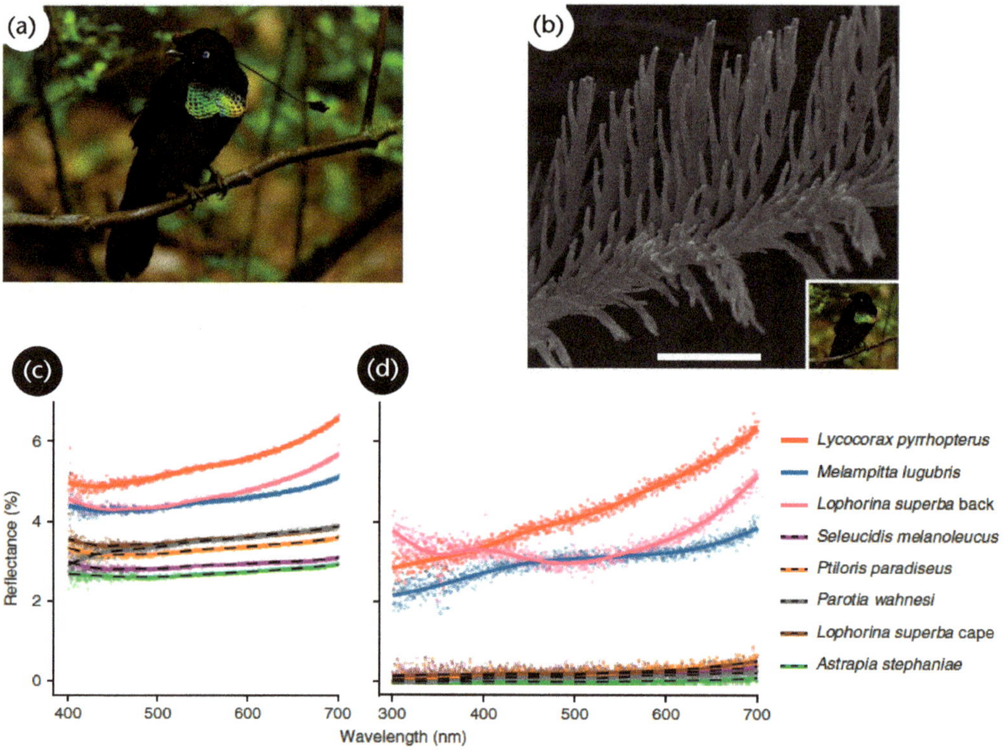

Figure 7.6 The bird of paradise *Parotia wahnesi* (a) exhibits a super-black plumage. SEM image (b) of modified barbules arrays. Reflectance spectra (c,d) of normal black (solid lines) and super-black (dotted lines) plumage: total (diffuse and specular) reflectance (c) and normal, directional reflectance (d). Scale bar: 50 μm (b). (Reproduced from [22], with permission from Springer Nature.)

a displaying male. The origin of the structurally enhanced blackness lies in arrays of curved barbules (Figure 7.6) that are tilted vertically (30°) from the normal toward the distal tip of the feather and increase multiple scattering of light. The overall microstructure can be regarded as regularly spaced cavities with a width of about 5–30 μm and a depth of about 200–400 μm. Looking at the openings of these deep cavities from the distal direction introduces a directional reflectance bias in which the distal viewing quadrant is the darkest under all light illumination directions [22].

7.3 UV Protection Strategies

As explained earlier in this chapter, efficient harvesting of solar energy is crucial for organisms such as plants, whose lives rely on photosynthesis. At the same time as useful radiation is harvested (i.e., visible range of the solar spectrum), protection must be sought from harmful radiation (e.g., UV range), too. It aims, for instance, to protect the organism from DNA damage induced by cells' exposure to UV radiation.

7.3.1 Waveguide Coupling and Energy Dissipation in Edelweiss Flower

This protection is especially critical for plants living at high altitude where exposure to UV is more intense due to rarefied atmosphere. The case of Edelweiss

Figure 7.7 UV protection is an important feature for Edelweiss (*L. nivale*) flowers (a) with their silver-white bracts facing the sky; the bracts are covered by a cottony layer made of criss-crossed transparent filaments (b) of about 10 μm in diameter (optical microscope image), which exibit submicron-sized surface corrugations (c) (SEM image, inset), made of thin parallel fibers of about 180 nm in diameter. The total reflectance (d) (measured at normal incidence on a bract) is high and almost constant over the whole visible range, but drops to zero in the near UV (the solid line corresponds to total reflectance and the dashed line to specular component). Cross-sectional model (e) of a hollow filament and the surface of its wall when its curvature is neglected, so that it becomes a corrugated slab. Absorption spectrum (f) of the filament slab model for TE-polarized light incident at $\theta = 45°$ and $\phi = 30°$. (Reproduced from [23], with permission from the American Physical Society.)

(*Leontopodium nivale*), a perennial herbaceous plant that can be found in the Alps (Europe) up to 3400m-altitude (Figure 7.7(a)), is interesting because it has evolved a waveguiding filamentary structure in the wooly cover layer of the plant's bracts; the wooly layer selectively enhances absorption of UV light before it can reach cellular tissue [23].

The wooly layer covering the edelweiss bracts is made of white, cotton-like pads which, when observed under an optical microscope, appear to be a disordered criss-cross structure of slightly iridescent transparent filaments (Figure 7.7(b)). SEM observations show that the filaments have slowly varying diameters and cross-section shapes along their lengths (Figure 7.7(c)) and are hollow (not shown here), with their walls being much thinner than their diameters (see also the filament cross section model, Figure 7.7(e)). Moreover, their surface is corrugated by an array of thin, parallel fibers, the spatial period (spacing) of which is approximately $p = 180$ nm (see inset in Figure 7.7(c)). The reflectance spectrum of a whole bract is high and flat in the visible range, but drops suddenly to almost zero below 400 nm (Figure 7.7(d)). Since bracts are thick and opaque in transmission, this indicates a strong absorption occurring in the UVA (320–400 nm). The origin of this absorption lies in the submicron features of the filaments' structures. Since the spatial period is much larger than each filament's curvature radius, the filament's surface can be approximated well by a two-dimensional corrugated dielectric slab (Figure 7.7(e)). Because of its short period (corresponding to a product $p \times n \simeq 252$ nm, where

$n \simeq 1.4$ is the filament refractive index), the slab was described as a UV-selective waveguide coupling device, which enables energy transfer of the incident ultraviolet waves into filament waveguide modes (Fano resonances) [23]. As a result, UV light can propagate along much longer distances than would be the case if it were propagating directly across the 10-μm-thick wooly layer. Assuming the filament wall material contains UV-absorbing pigments, the UV photon energy is efficiently dissipated along the hollow guides, hence protecting the cellular tissue underneath. Meanwhile, visible light is not affected by such high losses, as attested by the white color of the wooly layer. A simulation of the absorption spectrum of the filament slab model (Figure 7.7(f)) confirms this mechanism (a much higher imaginary part was added to the refractive index in the UV than in the visible in order to account for pigmentary absorption). Although the displayed simulation was obtained at fixed polar and azimuth angles with respect to the corrugated surface, the criss-cross feature of the filamentary structure guarantees a broadband angular response of the device. The interesting point here is the fact that UV protection is not obtained by reflection, but by absorption through dissipation of quasi-guided modes in the filamentary structure, thanks to a grating-type coupler operating in the UV range.

7.3.2 Backscattering in Avian Eggshells

The optical properties of eggshells are most important for birds because of their role in, for example, recognition by parents, heat control, camouflage, brood parasitim, and UV protection [24–33]. UV-absorbing pigments and fluorophores are naturally embedded in the cuticle of many eggshells. Despite this absorption in the UV, many eggshells were found to give rise to backscattering in the UV range [24,25,34–37]. Due to avian visual sensation in the UV range (see Chapter 1), such optical behavior in the UV range is most probably perceived by birds [38,39]. Spherical pores located in the calcified shell—a volume of the shell deeper than the cuticle and ranging from about 20 μm and about 240 μm—backscatter UV incoherently (Figure 7.8) [38,39]. The pores have a distribution of radii centred around 175 nm and give rise to two reflectance peaks in the UV range through Mie backscattering. Part of the light backscattered by these pores is returned to the incident medium and part of this (backscattered) light is absorbed by the pigments and fluorophores on its way out.

7.4 Thermoregulation and Solar Concentration in Butterflies

Black wings of many butterflies are a good example of an efficient natural solar collector, where the scale's structure contributes in producing very high broadband absorption (98%–99%) of the solar spectrum, such as in *T. magellanus* [18] or *T. aeacus* [19], as described ealier in this chapter. However, natural photonic structures can contribute to thermoregulation of organisms. For instance, butterflies use solar energy in order to heat their muscles when at rest. When they fly, steady-state flight requires a body temperature of approximately 36°C–40°C. Yet, above 40°C, they need to eliminate heat efficiently, as all flight ceases above 42°C [40]. Efficient solar-thermal energy conversion requires the butterfly wings (at least the wing parts closest to the body) to behave as a selective solar collector: reflectance should be

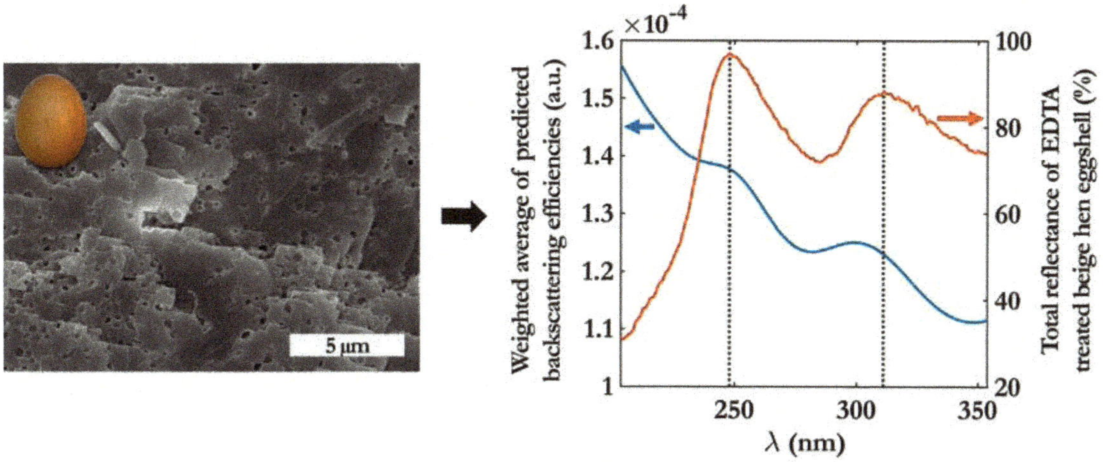

Figure 7.8 Two reflectance peaks (located in the UV range at approximately 250 nm and 310 nm) can be measured from the eggshells of some birds. They arise from pores occurring in the calcified shells of avian eggs. These pores have radii distributed around 175 nm. Predictions of backscattering efficiencies relying on Mie theory and taking into account the pore size distribution have shown that these peaks are due to backscattering. (Reproduced from [38], with permission from the Royal Society of Chemistry.)

low in the visible range (high solar absorptivity), but high in the near-infrared (low thermal emissivity), in order to achieve thermal balance. Moreover, the wings' emissivity should vary according to the temperature, in order to avoid overheating. Berthier showed that such a temperature autoregulation mechanism exists in the tropical butterfly *Prepona meander* [40]. The black parts of the wings harvest solar light with high efficiency (95% absorption in the visible range) not only due to pigments (melanin), but also due to the scales' dendritic structures (Figure 7.9). In the near infrared, absorption drops to less than 2% (which corresponds to a low thermal emission), except around $\lambda \simeq 3\ \mu m$ and $\lambda \simeq 6\ \mu m$, where chitin exhibits absorption peaks arising from $C = O$ vibrations (Figure 7.9). The former peak is not involved in the thermal balance at 40°C. The latter peak, however, is responsible for the temperature autoregulation. For temperatures lower than 40°C, the 6-μm emissivity peak lies at a shorter wavelength than the black body spectrum and the wing behaves as an efficient heat collector (keeping thermal emission low). As temperature rises above 40°C, the black body spectrum starts to overlap with this peak, which increases the wing's thermal emission and helps in stabilizing the butterfly's body temperature.

In butterfly wings, solar absorption aiming at heating may compete with structural coloration in terms of vital biological functions. A striking example is the discoloration phenomenon (due to partial loss of the photonic structure responsible for color) in a Lycaenid butterfly sister species pair [41]. The male *Polyommatus daphnis*, who lives in lower altitude, exhibits blue-violet structural color due to a pepper-pot structure between the scale's ridges. High reflection of UV and blue light may become a disadvantage as far as reduced heating by solar energy absorption is concerned. In this view, losing this structural color function may become an evolutionary advantage. This is exactly what is thought to have happened with the male *Polyommatus marcidus* who lives at higher altitude (>2500m). Its brown color is

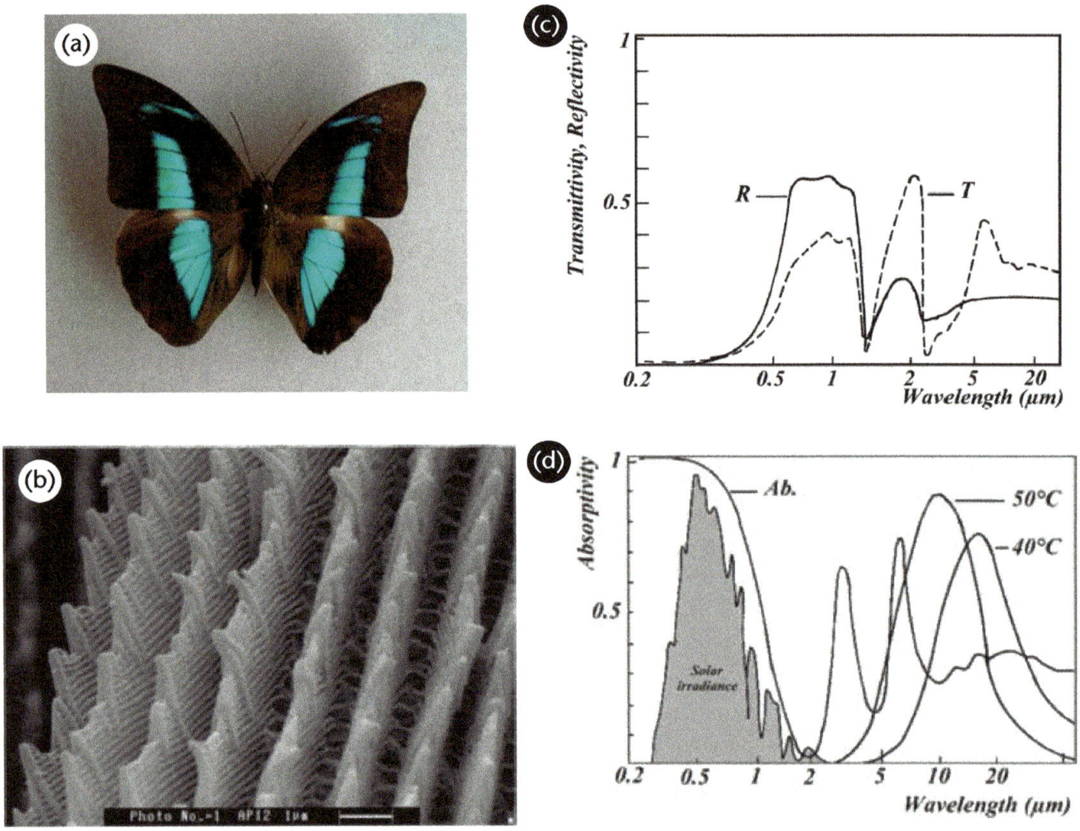

Figure 7.9 Thermoregulation occurs in some butterfly wings, such as the butterfly *Archaeprepona amphimacus* (a), a species closely related to *P. meander*. SEM image (b) of a black scale of *P. meander*. Reflection and transmission factors (c) showing spectral selectivity between visible and near-infrared ranges. Absorptivity (base of wings), solar irradiance, and black body spectra at 40°C and 50°C (d) showing the temperature autoregulation provided by the 6-μm absorption peak. (Reproduced from Notafly (a), https://commons.wikimedia.org/wiki/File:Archaeoprepona_amphimachus.JPG, public domain, and from [40] (b–d), with permission from Springer Nature.)

due to melanin and arises from an almost complete loss of the pepper-pot structure. This adaptative modification enables better heat accumulation by harvesting the UV-blue part of the solar spectrum, which tends to dominate at high altitudes. Unlike mammals, butterflies do not possess any known internal thermal regulation mechanisms, so heating quickly in the early morning is a crucial vital aspect for species living in high altitudes, and thus, harvesting UV-blue photons matters!

While black wings are excellent solar absorbers, white wings are excellent solar reflectors, making them suitable for solar energy concentration. White butterflies of the *Pieris* genus are known to take flight before other butterflies on cloudy days. This capability is believed to be due to thermal basking prior to flight: the V-shaped posture of their wings acts as a V-trough solar concentrator, which helps to heat flight muscles quickly [42]. It is also known that a disordered arrangement of pterin beads, the size of which is optimized for visible light scattering, is responsible for the high wing reflectance in *Pieris* species. Using an infrared camera, Shanks and coworkers measured temperature underneath two detached wings, which were positioned at a varying angle and exposed to light from a solar simulator source [42]. They found a temperature increase of 7.3°C (at the "thorax" position) with respect to flat wings, for an optimal wing angle of 27°. They also showed that a simple monolayer of scale cells was highly reflective and could inspire the design of low weight coatings for solar concentrators.

References

[1] Lee, D. W., and J. B. Lowry, "Physical Basis and Ecological Significance of Iridescence in Blue Plants," *Nature*, Vol. 254, No. 5495, 1975, pp. 50–51.

[2] Jacobs, M., et al., "Photonic Multilayer Structure of *Begonia* Chloroplasts Enhances Photosynthetic Efficiency," *Nature Plants*, Vol. 2, No. 11, 2016, 16162.

[3] Masters, N. J, et al., "Characterization of Chloroplast Iridescence in *Selaginella erythropus*," *Journal of the Royal Society Interface*, Vol. 15, No. 148, 2018, 20180559.

[4] Castillo, M. A., W. P. Wardley, and M. Lopez-Garcia, "Light Dependent Morphological Changes Can Tune Light Absorption in Iridescent Plant Chloroplasts," *bioRxiv*, 2020.09.01.277616.

[5] Gould, K. S., and D. W. Lee, "Physical and Ultrastructural Basis of Blue Leaf Iridescence in Four Malaysian Understory Plants," *American Journal of Botany*, Vol. 83, No. 1, 1996, pp. 45–50.

[6] Chaffey, N., "Iridoplasts, Out of the Shadows at Last," *Botany One*, https://www.botany.one/2017/03/iridoplasts-shadows-last/, 2017.

[7] Sheue, C.-R., et al., "Bizonoplast, a Unique Chloroplast in the Epidermal Cells of Microphylls in the Shade Plant *Selaginella erythropus* (Selaginellaceae)," *American Journal of Botany*, Vol. 94, No. 12, 2007, pp. 1922–1929.

[8] Sheue, C.-R., et al., "A Variation on Chloroplast Development: The Bizonoplast and Photosynthetic Efficiency in the Deep-Shade Plant *Selaginella erythropus*," *American Journal of Botany*, Vol. 102, No. 4, 2015, pp. 500–511.

[9] Gorton, H. L., and T. C. Vogelmann, "Effects of Epidermal Cell Shape and Pigmentation on Optical Properties of Antirrhinum Petals at Visible and Ultraviolet Wavelengths," *Plant Physiology*, Vol. 112, No. 3, 1996, pp. 879–888.

[10] Gkikas, D., A. Argiropoulos, and S. Rhizopoulou, "Epidermal Focusing of Light and Modelling of Reflectance in Floral-Petals with Conically Shaped Epidermal Cells," *Flora*, Vol. 212, 2015, pp. 38–45.

[11] Hünig, R., et al., "Flower Power: Exploiting Plants' Epidermal Structures for Enhanced Light Harvesting in Thin-Film Solar Cells," *Advanced Optical Materials*, Vol. 4, No. 10, pp. 1487–1493.

[12] Schmager, R., et al., "Texture of the Viola Flower for Light Harvesting in Photovoltaics," *ACS Photonics*, Vol. 4, No. 11, 2017, pp. 2687–2692.

[13] Schulte, A. J., et al., "Ultraviolet Patterns of Flowers Revealed in Polymer Replica – Caused by Surface Architecture," *Beilstein Journal of Nanotechnology*, Vol. 10, 2019, pp. 459–466.

[14] Barthlott, W., and C. Neinhuis, "Purity of the Sacred Lotus, or Escape from Contamination in Biological Surfaces," *Planta*, Vol. 202, No. 1, 1997, pp. 1–8.

[15] Neinhuis, C., and W. Barthlott, "Characterization and Distribution of Water-repellent, Self-cleaning Plant Surfaces," *Annals of Botany*, Vol. 79, No. 6, 1997, pp. 667–677.

[16] Chen, X., et al., "Numerical and Experimental Investigation of Light Trapping Effect of Nanostructured Diatom Frustules," *Scientific Reports*, Vol. 5, 2015, 11977.

[17] Vukusic, P., J. R. Sambles, and C. R. Lawrence, "Structurally Assisted Blackness in Butterfly Scales," *Proceedings of the Royal Society B*, Vol. 271, 2004, pp. S237–S239.

[18] Herman, A., et al., "Nanoarchitecture in the Black Wings of *Troides magellanus*: A Natural Case of Absorption Enhancement in Photonic Materials," *SPIE Proceedings*, Vol. 8094, 2011, 8094OH.

[19] Zhao, Q., et al., "Art of Blackness in Butterfly Wings as Natural Solar Collector," *Soft Matter*, Vol. 11, No. 24, 2011, pp. 11433–11439.

[20] Spinner, M., et al., "Snake Velvet Black: Hierarchical Micro- and Nanostructure Enhances Dark Colouration in *Bitis rhinoceros*," *Scientific Reports*, Vol. 3, 2012, 1846.

[21] Maurer, D. L., T. Khol, and M. J. Gebhardt, "Cuticular Microstructures Turn Specular Black into Matt Black in a Stick Insect," *Arthropod Structure and Development*, Vol. 46, No. 2, 2017, pp. 147–155.

[22] McCoy, D. E., et al., "Structural Absorption by Barbule Microstructures of Super Black Bird of Paradise Feathers," *Nature Communications*, Vol. 9, No. 1, 2018, pp. 1–8.

[23] Vigneron, J.-P., et al., "Optical Structure and Function of the White Filamentary Hair Covering the Edelweiss Bracts," *Physical Review E*, Vol. 71, No. 1, 2005, 011906.

[24] Cherry, M. I., and T. D. Bennett, "Egg Colour Matching in an African Cuckoo, as Revealed by Ultraviolet-Visible Reflectance Spectrophotometry," *Proceedings of the Royal Society B*, Vol. 206, No. 1467, 2001, pp. 565–571.

[25] Moreno, J., and J. L. Osorno, "Avian Egg Colour and Sexual Selection: Does Eggshell Pigmentation Reflect Female Condition and Genetic Quality?," *Ecology Letters*, Vol. 6, No. 9, 2003, pp. 803–806.

[26] Avilés, J. M., J. J. Soler, and T. Pérez-Contreras, "Dark Nests and Egg Colour in Birds: A Possible Functional Role of Ultraviolet Reflectance in Egg Detectability," *Proceedings of the Royal Society B*, Vol. 273, No. 1603, pp. 2821–2829.

[27] Kilner, R. M., "The Evolution of Egg Colour and Patterning in Birds," *Biological Reviews*, Vol. 81, No. 3, 2006, pp. 383–406.

[28] Cassey, P., et al., "The Modelling of Avian Visual Perception Predicts Behavioural Rejection Responses to Foreign Egg Colours," *Biology Letters*, Vol. 4, No. 5, 2008, pp. 515–517.

[29] Stoddard, M. C., and M. Stevens, "Pattern Mimicry of Host Eggs by the Common Cuckoo, as Seen Through a Bird's Eye," *Proceedings of the Royal Society B*, Vol. 277, No. 1686, pp. 1387–1393.

[30] Stoddard, M. C., and M. Stevens, "Avian Vision and the Evolution of Egg Color Mimicry in the Common Cuckoo," *Evolution*, Vol. 65, No. 7, 2011, pp. 2004–2013.

[31] Cassey, P., et al., "Why Are Birds' Eggs Colourful? Eggshell Pigments Co-vary with Life-History and Nesting Ecology among British Breeding Non-Passerine Birds," *Biological Journal of the Linnean Society*, Vol. 106, No. 3, 2012, pp. 657–672.

[32] Maurer, G., et al., "First Light for Avian Embryos: Eggshell Thickness and Pigmentation Mediate Variation in Development and UV Exposure in Wild Bird Eggs," *Functional Ecology*, Vol. 29, No. 2, 2015, pp. 209–218.

[33] Hauber, M. E., et al., "The Chemical Basis of a Signal of Individual Identity: Shell Pigment Concentrations Track the Unique Appearance of Common Murre Eggs," *Journal of the Royal Society Interface*, Vol. 16, No. 153, 2019, 20190115.

[34] Fecheyr-Lippens, D. C., et al., "The Cuticle Modulates Ultraviolet Reflectance of Avian Eggshells," *Biology Open*, Vol. 4, No. 7, 2015, pp. 753–759.

[35] Igic, B., et al., "A Nanostructural Basis for Gloss of Avian Eggshells," *Journal of the Royal Society Interface*, Vol. 12, No. 103, 2015, 20141210.

[36] D'Alba, L., et al., "What Does the Eggshell Cuticle Do? A Functional Comparison of Avian Eggshell Cuticles," *Physiological and Biochemical Zoology*, Vol. 90, No. 5, 2017, pp. 588–599.

[37] Ostertag, E., et al., "Pigmentation of White, Brown, and Green Chicken Eggshells Analyzed by Reflectance, Transmittance, and Fluorescence Spectroscopy," *ChemistryOpen*, Vol. 8, No. 8, 2019, pp. 1084–1093.

[38] Ladouce, M., et al., "Scattering of Ultraviolet Light by Avian Eggshells," *Faraday Discussions*, Vol. 223, 2020, pp. 63–80.

[39] Ladouce, M., et al., "UV Scattering by Pores in Avian Eggshells," *Proceedings of SPIE*, Vol. 11481, 2020, 114810G.

[40] Berthier, S., "Thermoregulation and Spectral Selectivity of the Tropical Butterfly *Prepona meander*: A Remarkable Example of Temperature Auto-Regulation," *Applied Physics A*, Vol. 80, 2005, pp. 1397–1400.

[41] Biro, L., et al., "Role of Photonic-Crystal-Type Structures in the Thermal Regulation of a Lycaenid Butterfly Sister Species Pair," *Physical Review E*, Vol. 67, No. 2, 2003, 021907.

[42] Shanks, K., et al. "White Butterflies as Solar Photovoltaic Concentrators," *Scientific Reports*, Vol. 5, 2015, 12267.

CHAPTER 8

Light Emission Management Devices

Emission of visible light occurs in nature in various ways. One origin of light emission is often fluorescence, which is found scattered widely throughout the tree of life: from mammals and insects to marine organisms. It is due to fluorophores and gives rise to colors ranging from blue to red. In addition, nature has evolved light emission management devices. For instance, by embedding fluorophores within a photonic structure, their fluorescence properties can be modified from both intensity and spatial distribution points of view. The case of the beetle *Hoplia coerulea*, described in Chapters 5 and 6 and for which the fluorescence properties have been extensively investigated, is reviewed in this chapter. Although the biological role of this light emission is still far from being understood, control of the emission from the fluorophores by the photonic structure arises from the system's DOS. As a consequence of the structure's porosity, the emission spectrum can be modified upon contact with water. The emission peak is redshifted and the displayed color changes from turquoise to navy blue as water infiltrates the structure. Moreover, the decay time measured at an emission wavelength lying in the structure's photonic bandgap is significantly reduced. In addition to fluorescence, quite a few marine species inhabiting oceans emit light, for different biological purposes, through a different path (i.e., they are bioluminescent). This phenomenon occurs in *Galiteuthis*, a midwater squid, which will be presented in this chapter because it has evolved an interesting camouflage strategy relying on counterillumination of its shadow cast by downwelling sunlight. The light emission management device consists of bioluminescent cells, which have the form of fibers surrounding the eye and bending downward. These fibers guide the emitted light in a way similar to cylindrical Bragg waveguides, yet they are imperfect ones. As often in nature, such imperfections, which cause light to leak laterally from the fibers, are precisely what is interesting in this device, as will be explained hereafter. Another example of light emission management is the structurally induced enhancement of light extraction from bioluminescent organs. Probably the best-known example of bioluminescent terrestrial species are fireflies. Let us make a comparison with our everyday LED lightening sources. We may think that LEDs are so efficient (in terms of internal quantum efficiency) that they do not face photon energy extraction issues. However, this idea is almost entirely untrue and the issue LED designers face is the same as that which bioluminescent living organisms have dealt with over millions years of evolution: poor *external* efficiency. Indeed, extraction of photons from an inner medium into an outer medium of lower refractive index is limited by total internal reflection. Since any medium has a lower index than air, this effect unavoidably impedes light extraction from *all* sources. Actually, for a flat interface between a biological luminescent material ($n_i \simeq 1.56$) and air ($n_e = 1$), the critical angle for total internal reflection is $\theta_c = \arcsin(n_e/n_i) \simeq 40°$, which means that all rays with internal incident angles above 40° will not escape from the medium. Translated in terms of energy extraction, this means that the fraction of light energy that is transmitted to air is only 20%. In this chapter, we will see how a firefly species from Panama has overcome that limit thanks to photonic structures in the firefly lantern (i.e., its light-emitting organ).

8.1 Fluorescence Emission Control in a Beetle's Scales

The integuments of a large variety of natural organisms, both terrestrial and aquatic, have been reported to fluoresce [1,2] (see Chapter 1). Examples have been found in the tissues of some birds (including adult parrots [3], penguins [4], and eggs [5–7]), marine animals (such as some corals [8–10], sea anemones [11], sponges [12], eels [13], and sharks [14]), plants [15–17], amphibians [18,19], reptiles [20], mammals [21] (e.g., tooth enamel, nails, and white hairs), and arthropods [22,23] (such as butterflies [24–32], beetles [33–36], and scorpions [37]). This light emission arises from fluorophores, such as biopterin, papiliochrome II, resilin, psittacofulvin, protoporphyrin IX, biliverdin IXα, and the green fluorescent protein (GFP), which are naturally embedded in the organism's biological tissues. Under illumination by UV light, these organisms display hues ranging from blue, green, and yellow, to orange and red, depending on the embedded fluorophores. This phenomenon was described as being involved in inter- and intraspecific visual communication (e.g., species recognition, mate choice, agonistic behavior, prey detection and camouflage) [1,2]. In spite of its significant roles, fluorescence emission in biological integuments still remains underexplored from physical, biological, and chemical points of view. Living organisms are invariably subject to multiple evolutionary forces simultaneously and some of these evolutionary pressures have resulted in modifications to aspects of their appearances over the millennia. The overall appearance of any living creature can, therefore, be seen as a compromise between multiple evolutionary forces, with each acting on different facets of the appearance and for different reasons. In many instances, it can be hard to discern the biological functionality of an organism's appearance—to what extent the evolution of a given feature was driven by its visual aspect alone and to what extent it resulted from evolution of another functionality. This is specifically the case for the visual aspect, since both the incident UV intensity and the quantum efficiency of the embedded fluorophores can be rather low. For instance, there is no known purpose for the fluorescence of mammalian nails, white hairs, and tooth enamel. Similarly, the contribution of fluorescent emission to the appearance of several butterflies of the *nireus* group (i.e., *Papilio bromius*, *Papilio epiphorbas*, *Papilio nireus*, and *Papilio oribazus*) was shown to be minor [38]. In contrast, UV absorption by fluorophores may provide organisms with protection against damage by harmful UV radiation [39], helping avoiding sunburn, premature aging of integuments, or cancer. So far, only a few species have been shown to have a clear association between fluorescence emission and biological functionalities [1,2].

When fluorophores are embedded within photonic structures, their fluorescence emission may be controlled by the structure through its density of optical states

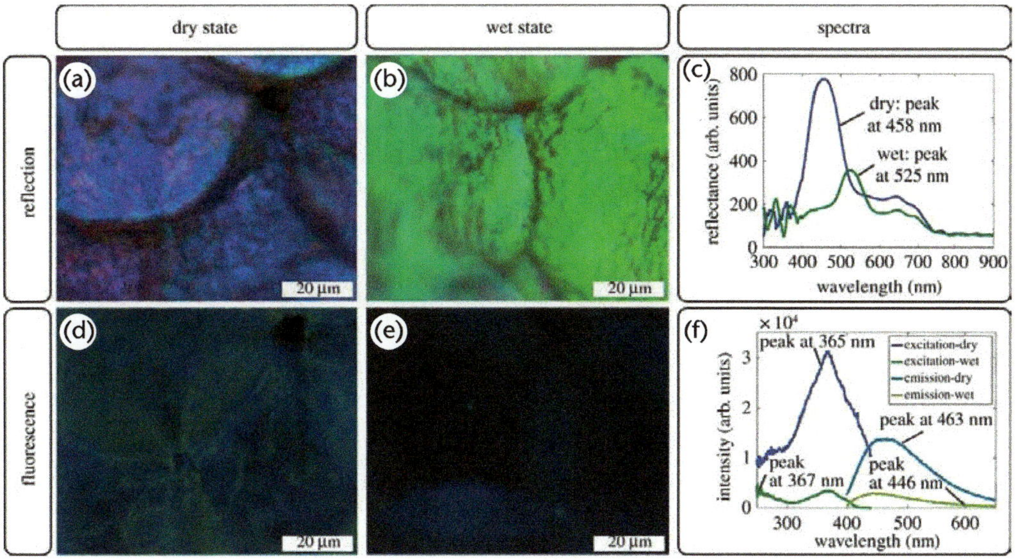

Figure 8.1 Modification of color and fluorescence emission from the scales of the male *H. coerulea* beetle upon contact with water. With incident white light, the scales appear violet-blue in the dry state (a) and green in the wet state (b). The peak reflectance wavelength redshifts from 458 nm to 525 nm (c). With incident UV light, the scales display a turquoise coloration in the dry state (d) and a navy blue color in the wet state (e). Although there is almost no water-induced change in the excitation spectrum's peak wavelength and full width at half maximum, the emission spectrum's peak wavelength blueshifts from 463 nm to 446 nm upon contact with water. The related FWHM decreases from 121 nm to 105 nm (f). (Reproduced from [34], with permission from the Royal Society.)

(DOS) [40–42]. This gives rise to modifications of the associated emission in terms of intensity (enhancement or inhibition) or spatial distribution (i.e., direction of emission) with respect to the emission properties in free space [40–42]. For instance, when the emission wavelength lies within the photonic bandgap of the structure, the emission intensity is reduced. This is the result of the increase in decay time of the excited states [40–42]. Such photonic confinements have been observed in several natural organisms [29,32,34,36,38,43–51].

One example of fluorescence control by a photonic structure is the male *H. coerulea* beetle described in the Chapters 5 and 6. In addition to its structurally colored blue-violet appearance [52] and its fluid-induced color changes [53–56], fluorophores were found to be naturally embedded in its photonic structure [34,48, 57]. Their emission arising from the one-photon absorption process was observed to be modified upon contact with water. This occurred following modifications of the optical system's local density of optical states (LDOS) induced by the penetration of the structure's pores by water (Figure 8.1) [34,57]. The emitted color was shifted reversibly from turquoise to navy blue and the fluorescence decay time related to a wavelength located within the photonic bandgap was reduced. Interestingly, very similar fluorescence properties and photonic structures were observed in the related *Hoplia argentea* [58]. In addition to a slightly thicker crystal period (195 nm in the case of *H. argentea* instead of 175 nm for *H. coerulea*), the main difference is a filamentous layer covering the scales. This filamentous layer acts as a light diffuser that suppresses the iridescence. Fluorescence emission was observed in both the filaments and the scale cores, but was more intense in the latter.

Differences were revealed between the excited states of the embedded fluorophores following one-photon and two-photon absorption in terms of emission spectra, using one-photon and two-photon excited fluorescence (OPEF and TPEF, respectively) microscopy and spectroscopy (Figure 8.2(a–d)) [59, 60]. These differences suggest that different electronic states are excited. In 2020, upconversion luminescence of lanthanide-doped nanoparticles ($NaYF^4$: Yb^{3+}, Er^{3+}) was shown to be tuned by the photonic structures of the butterfly *Cymothoe sangaris* [61]. Upon 980-nm continuous wave (CW) laser incidence, the red and green emission bands of $NaYF^4$: Yb^{3+}, Er^{3+} were selectively controlled, giving rise to various luminescent colorations. In addition, the optical role of local form anisotropy (i.e., a local direction-dependent morphology that displays structures smaller than incident light wavelengths) was highlighted in the optical response of *H. coerulea*. This was investigated by co- and cross-circularly polarized (CP) reflectance spectrophotometry as well as third-harmonic generation (THG) spectroscopy (Figure 8.2(e,f)) [59, 60]. This anisotropy, which influences both reflected and emitted light, arises from the rod-like spacers comprised in the mixed air-cuticle layers (Figure 5.3, Chapter 5). The rod-like subwavelength spacers are locally parallel in groups of about 10. They had previously been neglected and approximated by an effective medium, as discussed in Chapters 5 and 6.

8.2 Light Guides for Counterillumination of Shadows Cast by Squid Eyes

In ocean water, predators looking upward can detect prey because of the shadows cast by downwelling sunlight. One camouflage strategy some bioluminescent species have evolved in order to increase their chances of survival is to counterilluminate their shadow, making them virtually invisible to predators [62]. This management of light emission was nicely described and analyzed in detail by Holt and Sweeney in the case of the midwater squid *Galiteuthis* sp. [63]. The body of this squid is almost transparent except its eyes, which cast shadows downward (Figure 8.3(a,b)). In this bioluminescent squid, photophore cells surround the eyes and have the form of fibers bending vertically at their extremity for the purpose of counterillumination (Figure 8.3(c)). These fibers act as cylindrical Bragg-type leaky waveguides, meaning that light is guided by Bragg reflection from only a few concentric, protein-dense layers of higher refractive index surrounding the inner core (cytoplasm), a guiding mechanism that is radically different from total internal reflection. Waveguiding is not uncommon in bioluminescent marine organisms, including ctenophores [64–66]. For the sake of simplicity, the authors considered a straight model of such fibers (Figure 8.3(d)), aligned vertically with the bioluminescent emission (peaking at 506 nm) located on the fibers' top ends so that light travels downward. Because of lateral openings in the cladding layers, whose cross sections may depart considerably from those of perfect concentric rings, light leaks from the fibers to the sides, hence producing lateral, instead of vertical, illumination—as confirmed by numerical FDTD simulations of different fiber type models (Figure 8.3(e))

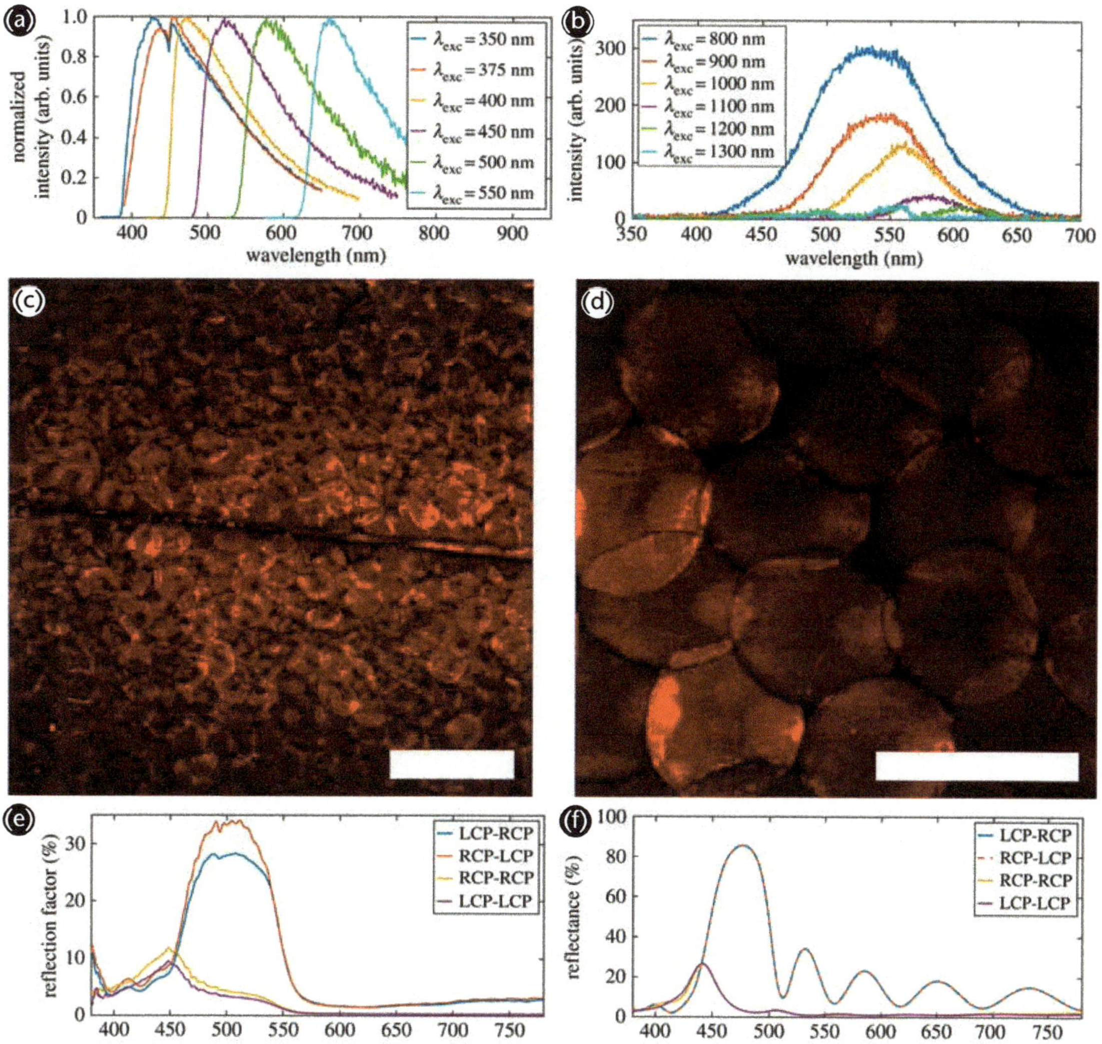

Figure 8.2 OPEF and TPEF spectroscopy (a,b) and microscopy (c,d) investigation of the male *H. coerulea* beetle revealed differences, in terms of emission spectra, between the excited states of the embedded fluorophores following one-photon and two-photon absorption. Both OPEF (a) and TPEF (b) emission are significantly modified as the excitation wavelength is tuned. The differences in shape, peak position, and FWHM between OPEF and related TPEF suggest that different electronic states are excited in OPEF and TPEF [60]. The strong TPEF signal with a 900-nm (c) and a 800-nm (d) excitation light allowed the scales covering the beetle's elytra to be observed. In addition to the strong cross-circularly polarized (i.e., either left-handed CP (LCP) light reflected from right-handed CP (RCP) incident light or vice versa: LCP-RCP and RCP-LCP) reflection at normal incidence, nonnegligible cocircularly polarized (i.e., incident and reflected CP components with the same handedness: LCP-LCP and RCP-RCP) reflection is observed experimentally (e). The spectra predicted by numerical simulations demonstrate that this optical response is due to the presence of local form anisotropy in the photonic structure (f). (Reproduced from [60], with permission from the Royal Society.)

extracted from TEM images [63]. These realistic simulations allowed the authors to calculate the ratio of light exiting the ends of a fiber relative to the sides as a function of the wavelength (Figure 8.3(f)). The striking, yet expected, result is that this ratio depends greatly on the arrangement of the cladding layers. Turning on and off different fiber types would allow the squid to tune its reshaped emission to

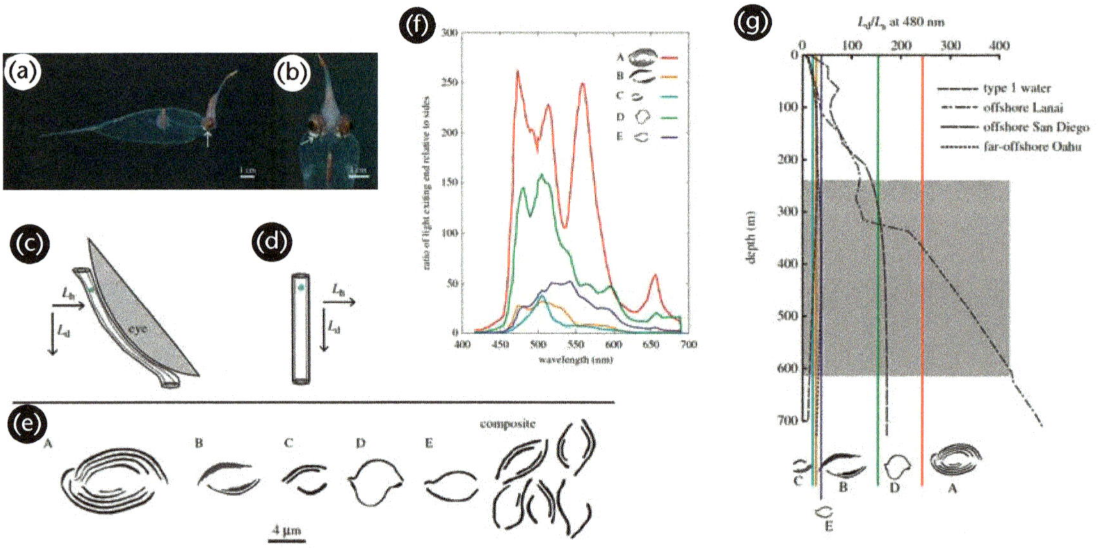

Figure 8.3 Light guides in squid eyes. In situ photographs (a,b) of *Galiteuthis*: the photophore organs are visible on the ventral surfaces of the eyes and indicated by white arrows. Schematic (c) showing the photophore cells surrounding the squid's larger eye (not to scale). Fibers run parallel to the outside of the eye, except for those on the ventral-most surface, which turn 90° to point down into the ocean. Vectors for downwelling (L_d) and sidewelling (L_h) radiances are shown for reference. Blue circles show an approximate representation of the bioluminescent emission. Simplified situation (d) in which light guides on average point vertically down in the ocean. Typical fiber cross sections (e), called A-E, extracted from TEM images. Ratio of light exiting the ends of a fiber, relative to that exiting via the sides, as a function of wavelength (f), given the emission spectrum of the photophores. Predicted downwelling to sidewelling radiance ratio (g) of observed fiber structures compared with ocean radiance at 480 nm for two ocean locations, as well as a generic radiative transfer model of ocean water. Colored vertical lines indicate the L_d/L_h ratio at 480 nm calculated for each fibre taken from (g). Gray box shows the vertical range of *Galiteuthis* in ocean water. (Reproduced from [63], with permission from the Royal Society.)

the range of downwelling to sidewelling radiance ratios (L_d/L_h) found at the ocean depths and locations it inhabits. These authors' speculations tend to be supported by the comparison between predicted L_d/L_h values for the different fiber types and those calculated for relevant oceanic habitats as a function of water column depth (Figure 8.3(f)).

8.3 Light Extraction from Fireflies

A specific firefly species of the genus *Photuris* (Lampyridae) has evolved photonic structures in different parts of its lantern (Figure 8.4), which enable more efficient extraction of light from its natural bioluminescence [67–70]. Through numerical simulations using a detailed photonic model of the lantern morphology (Figure 8.4), Bay and coworkers showed that the global transmission into air reached 39.6%. This amounts to a substantial gain in *external* energy conversion efficiency, with respect to the 20% limit set by total internal reflection for a flat interface [67,68,70]. By evaluating the impact on light scattering of each of the seven lantern parts separately, they found that the most significant contribution arises from the corrugation

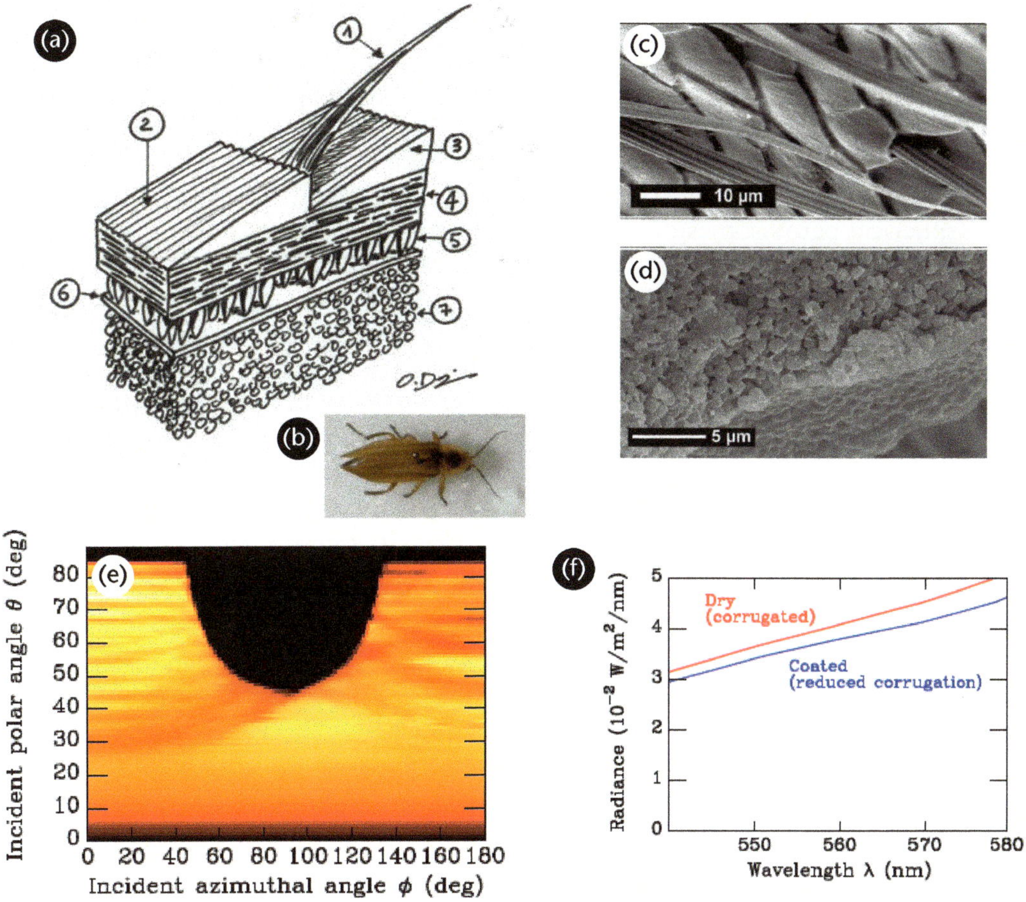

Figure 8.4 Photonic model of the whole lantern structure of *Photuris sp.* firefly derived from SEM observations; the sketch (a) is not to scale and shows (1) a seta (hair), (2) a scale's rib-like corrugations (250-nm line spacing), (3) scales arranged like tiles on a roof (tiles of 10 μm length and 3 μm vertical offset), (4) textured cuticle (2.4 μm thick), (5) muscular layer (tentative interpretation), (6) membrane (60 nm thick), and (7) layer of photocytes (clusters of spherical peroxisomes). Specimen (b) collected in The Darien Forest, Republic of Panama. SEM images of substructures n°3 (c) and n°7 (d), which contribute significantly to enhanced light extraction. Hemispherical transmission (e) on logarithmic scale of the full lantern model as a function of incidence (polar and azimuth) angles. Hemispherical radiance (f) of the firefly's abdomen measured with (red) and without (blue) index-matching liquid on the surface. (Original drawing (a) from O. Deparis, March 2021. Reproduced from [70], with permission from the Optical Society (b–f).)

of the insect's abdomenal surface, which they described as having a "tiles-on-a-roof" structure (labeled n°3 in Figure 8.4). The regular arrangement of slanted scales (i.e., tiles) on the cuticle surface forms a large-step grating (with a period of circa 10 μm) with sharp edges protruding approximately 3 μm above the surface. Scattering at these edges probably explains the increase in light extraction observed, not only in simulations, but also experimentally. Indeed, the calculated hemispherical transmission (at 560 nm—the typical wavelength of yellow-green color flashes emitted by the firefly) shows, for the full lantern model, a large angular redistribution of the scattered light well beyond the critical angle ($\theta \simeq 40°$) relative to a planar interface (Figure 8.4). When only the grating corrugation is taken into account in the lantern model, the integration over the emergence hemispherical solid angle already gives a global transmission of 29.5%, which surpasses the planar interface limit. Additional contribution to light extraction (of the order of 5%) was found to come from the structure of the photocyte layer, where the bioluminescence chemical reaction takes place. The structure (labeled 7 in Figure 8.4) consists of a disordered, three-dimensional array of densely packed (filling fraction of approximately 0.74), spherical peroxisomes of refractive index ($n \simeq 1.4$) lower than the bulk chitinous material. This layer helps to reduce the lantern's effective refractive index and, therefore, to improve the external efficiency, by lowering the critical angle limit. In some firefly species, the bioluminescent tissues were described as an irregular stratum, lying over a layer comprising packed hollow granules that act as a reflector layer; in such cases, reflectance reached 82% at circa 550 nm [69,71]. Spectral hemispheric radiance measurements by Bay and coworkers confirmed their numerical predictions about the effect of the tiles-on-a-roof corrugation. Specifically, they proved that it gave a 9.5% transmission increase, with an enhancement of about 7% (at 560 nm) when comparing the measured results obtained for the insect abdomen with and without index matching liquid on the cuticle surface [70].

References

[1] Lagorio, M. G., G. B. Cordon, and A. Iriel, "Reviewing the Relevance of Fluorescence in Biological Systems," *Photochemical & Photobiological Sciences*, Vol. 14, No. 9, 2015, pp. 1538–1559.

[2] Marshall, J., and S. Johnsen, "Fluorescence as a Means of Colour Signal Enhancement," *Philosophical Transactions of the Royal Society B*, Vol. 372, No. 1724, 2017, 20160335.

[3] Arnold, K. E., I. P. F. Owens, and N. J. Marshall, "Fluorescent Signaling in Parrots," *Science*, Vol. 295, No. 5552, 2002, p. 92.

[4] McGraw, K. J., et al., "A Description of Unique Fluorescent Yellow Pigments in Penguin Feathers," *Pigment Cell Research*, Vol. 20, No. 4, 2007, pp. 301–304.

[5] Keneddy, G. Y., and H. G. Vevers, "Eggshell Pigments of the Araucano Fowl," *Comparative Biochemistry and Physiology Part B*, Vol. 44, No. 1, 1973, pp. 11–25.

[6] Keneddy, G. Y., and H. G. Vevers, "A Survey of Avian Eggshell Pigments," *Comparative Biochemistry and Physiology Part B*, Vol. 55, No. 1, 1976, pp. 117–123.

[7] Ladouce, M., et al., "Scattering of Ultraviolet Light by Avian Eggshells," *Faraday Discussions*, Vol. 223, 2020, pp. 63–80.

[8] Catala-Stucki, R., "Fluorescence Effects from Corals Irradiated with Ultra-Violet Rays," *Nature*, Vol. 183, No. 4666, 1959, p. 949.

[9] Cox, G., and A. Salih, "Confocal Microscopy of GFP-Like Pigments in Corals." In *Multi-Modality Microscopy*, pp. 238–245, H. Yu, P.-C. Cheng, P.-C. Lin, and F.-J. Kao (eds.), Singapore: World Scientific Publishing Co, 2006.

[10] Salih A., "Screening Reef Corals for Novel GFP-Type Fluorescent Proteins by Confocal Imaging." *In In Vivo Cellular Imaging Using Fluorescent Proteins. Methods in Molecular Biology (Methods and Protocols)*, pp. 217–233, R. Hoffman (ed.), Totowa, NJ: Humana Press, 2012.

[11] Phillips, C. E. S., "Fluorescence of Sea Anemones," *Nature*, Vol. 119, No. 3003, 1927, p. 747.

[12] Read, K. R. H., J. M. Davidson, B. M. Twarog, "Fluorescence of Sponges and Coelenterates in Blue Light," *Comparative Biochemistry and Physiology*, Vol. 25, No. 3, 1968, pp. 873–876.

[13] Gruber, D. F., et al., "Adaptive Evolution of Eel Fluorescent Proteins from Fatty Acid Binding Proteins Produces Bright Fluorescence in the Marine Environment," *PLoS ONE*, Vol. 10, No. 11, 2015, e0140972.

[14] Gruber, D. F., et al., "Biofluorescence in Catsharks (Scyliorhinidae): Fundamental Description and Relevance for Elasmobranch Visual Ecology," *Scientific Reports*, Vol. 6, 2016, 24751.

[15] Goodwin, R. H., "Fluorescent Substances in Plants," *Annual Review of Plant Physiology*, Vol. 4, No. 1, 1953, pp. 283–304.

[16] Iriel, A. and M. G. Lagorio, "Is the Flower Fluorescence Relevant in Biocommunication?," *Naturwissenschaften*, Vol. 97, No. 10, 2010, pp. 915–924.

[17] Iriel, A. and M. G. Lagorio, "Implications of Reflectance and Fluorescence of Rhododendron Indicum Flowers in Biosignaling," *Photochemical & Photobiological Sciences*, Vol. 9, No. 3, 2010, pp. 342–348.

[18] Taboada, C., et al., "Naturally Occurring Fluorescence in Frogs," *Proceedings of the National Academy of Sciences of the United States of America*, Vol. 114, No. 14, 2017, pp. 3672–3677.

[19] Deschepper, P., B. Jonckheere, and J. Matthys, "A Light in the Dark: The Discovery of Another Fluorescent Frog in the Costa Rican Rainforests," *Wilderness & Environmental Medicine*, Vol. 29, No. 3, 2018, pp. 421–422.

[20] Mohd Top, M., et al., "First Record of Ultraviolet Fluorescence in the Bent-Toed Gecko *Cyrtodactylus quadrivirgatus* Taylor, 1962 (Gekkonidae: Sauria)," *Herpetology Notes*, Vol. 13, 2020, pp. 211–212.

[21] Tani, K., et al., "Fluorescent Properties of Porcelain-Restored Teeth and Their Discrimination," *Materials Transactions*, Vol. 45, No. 4, 2004, pp. 1010–1014.

[22] Lawrence, R. F., "Fluorescence in Arthropoda," *Journal of the Entomological Society of Southern Africa*, Vol. 17, No. 2, 1954, pp. 167–170.

[23] Welch, V. L., et al., "Fluorescence in Insects," *Proceedings of SPIE*, Vol. 8480, 2012, 848004.

[24] Cockayne, E. A., "The Distribution of Fluorescent Pigments in Lepidoptera," *Transactions of the Royal Entomological Society of London*, Vol. 72, No. 1-2, 1924, pp. 1–19.

[25] Umebachi, Y., and K. Yoshida, "Some Chemical and Physical Properties of Papiliochrome II in the Wings of *Papilio xuthus*," *Journal of Insect Physiology*, Vol. 16, No. 6, 1970, pp. 1203–1228.

[26] Umebachi, Y., "Papiliochrome, a New Pigment Group of Butterfly," *Zoological Science*, Vol. 2, 1985, pp. 163–174.

[27] Gilbert, L. E., et al., "Correlations of Ultrastructure and Pigmentation Suggest How Genes Control Development of Wing Scales of *Heliconius* Butterflies," *Journal of Research on the Lepidoptera*, Vol. 26, No. 1-4, 1988, pp. 141–160.

[28] Umebachi, Y., and M. Osanai, "Perturbation of the Wing Color Pattern of a Swallowtail Butterfly, *Papilio xuthus*, Induced by Acid Carboxypeptidase," *Zoological Science*, Vol. 20, 2003, pp. 325–331.

[29] Vukusic, P., and I. Hooper, "Directionally Controlled Fluorescence Emission in Butterflies," *Science*, Vol. 310, No. 5751, 2005, p. 1151.

[30] Reed, R. D., et al., "Gene Expression Underlying Adaptive Variation in *Heliconius* Wing Patterns: Non-Modular Regulation of Overlapping Cinnabar and Vermilion Prepatterns," *Proceedings of the Royal Society B*, Vol. 275, No. 1630, 2008, pp. 37–45.

[31] Briscoe, A. D., et al., "Positive Selection of a Duplicated UV-Sensitive Visual Pigment Coincides with Wing Pigment Evolution in *Heliconius* Butterflies," *Proceedings of the National Academy of Sciences of the United States of America*, Vol. 107, No. 8, 2010, pp. 3628–3633.

[32] Van Hooijdonk, E., S. Berthier, and J.-P. Vigneron, "Contribution of Both the Upperside and the Underside of the Wing on the Iridescence in the Male Butterfly *Troïdes magellanus* (Papilionidae)," *Journal of Applied Physics*, Vol. 112, No. 7, 2012, 074702.

[33] Israelowitz, M., S. H. W. Rizvi, and H. P. von Schroeder, "Fluorescence of the "Fire-Chaser" Beetle, *Melanophila acuminate*," *Journal of Luminescence*, Vol. 126, No. 1, 2007, pp. 149–154.

[34] Mouchet, S. R., et al., "Controlled Fluorescence in a Beetle's Photonic Structure and Its Sensitivity to Environmentally Induced Changes," *Proceedings of the Royal Society B*, Vol. 283, No. 1845, 2016, 20162334.

[35] Mouchet, S. R., et al., "Colour and Fluorescence Emission of *Euchroea auripigmenta* Beetle," *Proceedings of SPIE*, Vol. 10965, 2019, pp. 72–82.

[36] Mouchet, S. R., et al., "Unveiling the Non-Linear Optical Response of *Trictenotoma childreni* Longhorn Beetle," *Journal of Biophotonics*, Vol. 12, No. 9, 2019, e201800470.

[37] Pavan, M., and M. Vachon, "Sur l'existence d'une substance fluorescente dans les téguments des Scorpions (Arachnides)," *Comptes Rendus de l'Académie des Sciences*, Vol. 239, 1954, pp. 1700–1702.

[38] Wilts, B. D., et al., "Papiliochrome II Pigment Reduces the Angle Dependency of Structural Wing Colouration in *nireus* Group Papilionids," *Journal of Experimental Biology*, Vol. 215, No. 5, 2012, pp. 796–805.

[39] Berthier, S., *La couleur des papillons ou l'impérative beauté – Propriétés optiques des ailes de papillons*, Paris: Springer, 2000.

[40] Purcell, E. M., "Spontaneous Emission Probabilities at Radio Frequencies," *Physical Review*, Vol. 69, No. 11-12, 1946, p. 681.

[41] Yablonovitch, E., "Inhibited Spontaneous Emission in Solid-State Physics and Electronics," *Physical Review Letters*, Vol. 58, No. 20, 1987, pp. 2059–2062.

[42] John, S., "Strong Localization of Photons in Certain Disordered Dielectric Superlattices," *Physical Review Letters*, Vol. 58, No. 23, 1987, pp. 2486–2489.

[43] Kumazawa, K., et al., "Fluorescence from wing of *Morpho sulkowskyi* Butterfly," *Japanese Journal of Applied Physics*, Vol. 33, No. 4A, 1994, pp. 2119–2122.

[44] Lawrence, C., et al., "Grazing-Incidence Iridescence from a Butterfly Wing," *Applied Optics*, Vol. 41, No. 3, 2002, pp. 437–441.

[45] Vigneron, J.-P., et al., "Correlated Diffraction and Fluorescence in the Backscattering Iridescence of the Male Butterfly *Troïdes magellanus* (Papilionidae)," *Physical Review E*, Vol. 78, No. 2, 2008, 021903.

[46] Van Hooijdonk, E., et al., "Detailed Experimental Analysis of the Structural Fluorescence in the Butterfly *Morpho sulkowskyi* (Nymphalidae)," *Journal of Nanophotonics*, Vol. 5, No. 1, 2011, 053525.

[47] Trzeciak, T. M., et al., "Variable Multilayer Reflection Together with Long-Pass Filtering Pigment Determines the Wing Coloration of Papilionid Butterflies of the *Nireus* Group," *Optics Express*, Vol. 20, No. 8, 2012, pp. 8877–8890.

[48] Van Hooijdonk, E., S. Berthier, and J.-P. Vigneron, "Bio-Inspired Approach of the Fluorescence Emission Properties in the Scarabaeid Beetle *Hoplia coerulea* (Coleoptera): Modeling by Transfer-Matrix Optical Simulations," *Journal of Applied Physics*, Vol. 112, No. 11, 2012, 114701.

[49] Van Hooijdonk, E., et al., "Bi-Functional Photonic Structure in the *Papilio nireus* (Papilionidae): Modeling by Scattering-Matrix Optical Simulations," *Optics Express*, Vol. 20, No. 20, 2012, pp. 22001–22011.

[50] Van Hooijdonk, E., et al., "Angular Dependence of Structural Fluorescent Emission from the Scales of the Male Butterfly *Troïdes magellanus* (Papilionidae)," *Journal of the Optical Society of America B*, Vol. 29, No. 5, 2012, pp. 1104–1111.

[51] Van Hooijdonk, E., et al., "Yellow Structurally Modified Fluorescence in the Longhorn Beetles *Celosterna pollinosa sulfurea* and *Phosphorus virescens* (Cerambycidae)," *Journal of Luminescence*, Vol. 136, 2013, pp. 313–321.

[52] Vigneron J.-P., et al., "Natural Layer-by-Layer Photonic Structure in the Squamae of *Hoplia coerulea* (Coleoptera)," *Physical Review E*, Vol. 72, No. 6, 2005, 061904.

[53] Rassart M., et al., "Scale Coloration Change Following Water Absorption in the Beetle *Hoplia coerulea* (Coleoptera)," *Physical Review E*, Vol. 80, No.3, 2009, 31910.

[54] Mouchet S. R., et al., "Vapor Sensing with a Natural Photonic Cell," *Optics Express*, Vol. 24, No. 11, 2016, pp. 12267–12280.

[55] Mouchet S. R., et al., "Liquid-Induced Colour Change in a Beetle: The Concept of a Photonic Cell," *Scientific Reports*, Vol. 6, 2016, 19322.

[56] Mouchet S. R., et al., "Assessment of Environmental Spectral Ellipsometry for Characterising Fluid-Induced Colour Changes in Natural Photonic Structures," *Material Today: Proceedings*, Vol. 4, No. 4, 2017, pp. 4987–4997.

[57] Mouchet S. R., et al., "Photonic Scales of *Hoplia coerulea* Beetle: Any Colour you Like," *Material Today: Proceedings*, Vol. 4, No. 4, 2017, pp. 4979–4986.

[58] Kilchoer, C., et al., "Diffusive Structural Colour in *Hoplia argentea*," *Journal of Experimental Biology*, Vol. 222, No. 24, 2019, jeb213306.

[59] Verstraete, C., et al., "Linear and Nonlinear Optical Effects in Biophotonic Structures Using Classical and Nonclassical Light," *Journal of Biophotonics*, Vol. 12, No. 1, 2018, e201800262.

[60] Mouchet, S. R., et al., "Nonlinear Optical Spectroscopy and Two-Photon Excited Fluorescence Spectroscopy Reveal the Excited States of Fluorophores Embedded in a Beetle'S Elytra," *Interface Focus*, Vol. 9, No. 1, 2019, 20180052.

[61] Gao, T., et al., "Selectively Manipulating Upconversion Emission Channels with Tunable Biological Photonic Crystals," *Journal of Physical Chemistry* C, Vol. 125, No. 1, 2021, pp. 732–739.

[62] Mäthger, L. M., et al., "Mechanisms and Behavioural Functions of Structural Coloration in Cephalopods," *Journal of the Royal Society Interface*, Vol. 6, No. S2, 2009, pp. S149–S163.

[63] Holt, A. L., and A. M. Sweeney, "Open Water Camouflage Via Leaky Light Guides in the Midwater Squid *Galiteuthis*," *Journal of the Royal Society Interface*, Vol. 13, No. 119, 2016, 20160230.

[64] Welch, V. L, *Structural Colouration in Jellyfish, Fish and Ctenophores*, PhD Thesis, University of Oxford, UK, 2003.

[65] Welch, V. L., J.-P. Vigneron, and A. R. Parker, "The Cause of Colouration in the Ctenophore *Beroe cucumis*," *Current Biology*, Vol. 15, No. 24, 2005, pp. R985–R986.

[66] Welch, V. L., et al., "Optical Properties of the Iridescent Organ of the Comb-Jellyfish, *Beroe cucumis* (Ctenophora)," *Physical Review E*, Vol. 73, No. 4, 2006, 041916.

[67] Bay, A., and J.-P. Vigneron, "Light Extraction from the Bioluminescent Organs of Fireflies," *Proceedings of SPIE*, Vol. 7401, 2009, 740108.

[68] Bay, A., M. Sarrazin, and J.-P. Vigneron, "Light Extraction: What We Can Learn from Fireflies," *Proceedings of SPIE*, Vol. 8480, 2012, 84800G.

[69] Kim, J.-J., et al., "Biologically Inspired LED Lens from Cuticular Nanostructures of Firefly Lantern," *Proceedings of the National Academy of Sciences of the United States of America*, Vol. 109, No. 46, 2012, pp. 18674–18678.

[70] Bay, A., et al., "Improved Light Extraction in the Bioluminescent Lantern of a *Photuris* Firefly (Lampyridae)," *Optics Express*, Vol. 21, No. 1, 2013, pp. 764–780.

[71] Chen, L., et al., "Bioinspired Photonic Structures by the Reflector Layer of Firefly Lantern for Highly Efficient Chemiluminescence," *Scientific Reports*, Vol. 5, 2015, 12965.

CHAPTER 9

Design of Bioinspired Photonic Devices

Since human imagination can be said to have no limit and creativity often goes along unpredictable tracks, it is hard to elaborate general design rules for bioinspired photonic devices. To claim the design relies on Mother Nature is nice, but does not help engineers much. To conceive innovative photonic devices from nature's templates is a case-by-case task, in which we let ourselves be guided, sometimes tightly, sometimes loosely, by what we have learned from the analysis of a potentially interesting property exhibited by the template. Actually, such inspiration typically stems more from the way this property is produced rather than the property itself, which is often well-known or described. At the same time, it is necessary to keep in mind technological advantages, economical profits, and ecological benefits scientists can derive from a tight or loose implementation of the said property into an artificial device. This chapter presents a global methodology that is based on learning from nature, but also goes beyond biomimetism, or straightforward transfer of ideas, from the natural world to technology. As electromagnetic simulations are of ubiquitous importance in the design of bioinspired devices, the numerical methods that are most commonly used for computing the devices' optical responses will be introduced. Bioinspiration is, ultimately, a matter of concepts. The concept of switchable Bragg mirrors will be discussed in more detail at the end of this chapter, where we will briefly take a look at bioinspiration at work through a few examples.

9.1 Methodology: From Field Observations to Modeling

Learning from nature has long been and still is the rationale of natural science. Natural photonics, a young branch of natural science that emerged at the dawn of the twenty-first century, is no exception. Everything starts with observations [1]. Visual observations of colors displayed by natural organisms begin with the naked eye—ideally in the field, or, alternatively, within specimen collections, always keeping in mind that human visual perception could differ from other animals' visual perceptions, not to mention those colors *Homo sapiens* do not see, but species with UV- or IR-photoreceptors do detect. In these preliminary observations, particular attention will be paid to sexual dimorphism (male colored, female not, or vice versa), interspecific visual communication (e.g., aposemantism), and the species' habitat, since the latter is often relevant for biological functions related to colors (e.g., camouflage). Color changes deserve specific attention, whether they appear to be controlled by the organism or induced by external stimuli (see Chapters 5 and 6). Observations are conducted in the laboratory under the microscope's objective lens using white

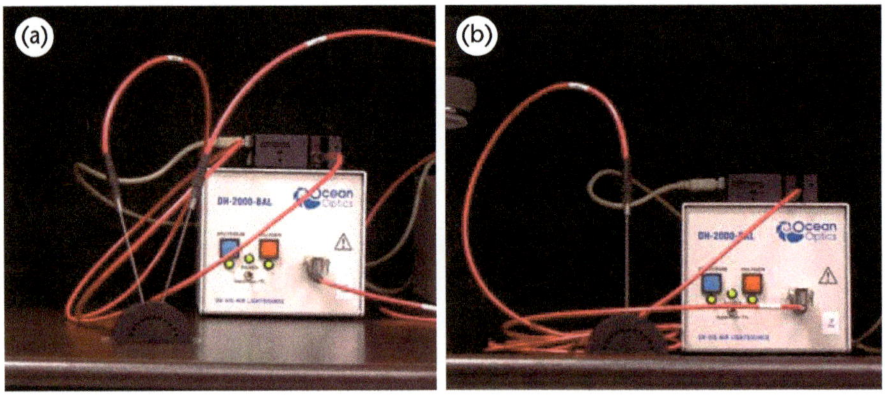

Figure 9.1 A fiber-optic spectrophotometer is a convenient device for measuring reflectance or transmittance from biological samples in the laboratory. An optical fiber that is connected to a white light source illuminates the sample at a chosen angle, using a fiber holder for positioning. The reflected light is collected at another angle by another fiber and directed toward the detector that separates spectral components (a). A bifurcated fiber is used to illuminate the sample and, simultaneously, to collect the reflected light in the backward direction (b). (Pictures were taken by S. R. Mouchet, 2021.)

light illumination. Placing the samples in contact with a refractive-index matching fluid allows researchers to check whether the observed color is structural or not (of course, such a simple test works only if the photonic structure from which the color originates has pores open to the surrounding environment). Using a Wood' lamp for UV illumination can also quickly help with detecting fluorescent body parts and draw further attention to any influence the fluorescence may have on the perceived colors. In the laboratory, a next step can be the quantitative spectral characterization by fiber-optic spectrophotometer (Figure 9.1). This device is particularly well adapted to spectral reflectance measurements with illumination and detection at specific angles, normal incidence backscattering configuration (bifurcated fiber), or diffuse reflection using an integrating sphere. The portability of such an instrument makes it suitable even for measurements on living specimens, such as insects at rest on leaves, depending on the experimentalist's patience.

More sophisticated scattering measurements and observations can be done, if necessary. Using a goniometric setup, an ellipsoidal mirror, or Fourier optics, for instance, the bidirectional reflectance distribution function (BRDF) or a scatterogram of the sample's surface can be measured. In such measurements, the light intensity scattered across the whole hemisphere is resolved, spatially, in terms of azimuth and polar angles. The measurements can be performed at a single wavelength, which is selected typically using filters. Microspectrophotometry and hyperspectral imaging techniques are also highly relevant in the context of natural photonics. By combining an optical microscope with a charge-coupled device (CCD) camera connected to a miniature spectrometer, hyperspectral techniques enable the measurement of optical spectra at each pixel position in the plane of focus. These techniques are very convenient for studying, for example, disordered or polycrystalline photonic structures. On a more practical level, they solve the issue of measuring the optical response of tiny and nonuniform samples, whose dimensions are smaller than the spot size of benchtop spectrophotometers.

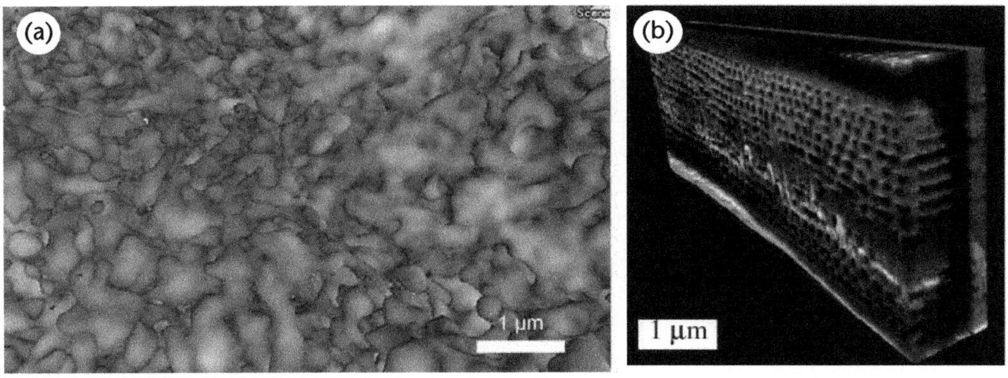

Figure 9.2 Three-dimensional imaging of natural photonic structures. X-ray nanotomography landscape image (a) of the internal structure of a firefly's photocyte, showing disordered and slightly multidispersed peroxisomes. 3D reconstruction (b) from FIB-SEM images of the porous multilayer structure of a scale of the beetle *Hoplia coerulea.* (Reproduced from [2], with permission from the Optical Society (a) and from [3], with permission from the Royal Society (b).)

Once potentially interesting optical properties have been identified in specimens, which may include color changes induced by some stimuli, the next step is to characterize the morphology of body parts (cuticle, scales, etc.) exhibiting those properties. Since the relevant dimensions for optical interference phenomena in the visible spectrum lie in the 10–1000 nm range, both scanning electron microscopy (SEM) and transmission electron microscopy (TEM) are widely used for this purpose. In addition to the specific sample preparation required for these techniques, an unavoidable limitation of SEM or TEM is the fact that they provide two-dimensional projected views of the real structure. SEM observations at various angles of view may compensate for this lack of 3D rendering to some extent. Despite that, one has always to take care (e.g., when estimating characteristic lengths of the observed structures). SEM combined with focused ion beam milling (FIB-SEM) allows one to reveal cross sections progressively and to give rise to reconstructed 3D images of the observed structure. Similarly, X-ray tomography gives access to direct 3D observation of the real structure (Figure 9.2). Such 3D characterizations, however, involve costly and time-consuming experiments. Sometimes it is necessary to investigate the extreme surface morphology of samples by other microscopy techniques, such as atomic force microscopy (AFM).

A geometrical model of the structure is devised on the basis of morphological characterization. Often, if not always, natural photonic structures exhibit disorder and this can take various forms: layer thickness variations in seemingly periodic stacks, surface and interface roughness, pore radius distribution, and so on. In arthropods, the material most often encountered is chitin—more precisely, various chitinous materials as the natural polymer chitin often contains proteins or other molecules (such as melanin). Keratinous materials are also found, for example, in bird feathers, collagen in mammalian skins and bird caruncles, cellulose in plants, as well as guanine crystals such as in fish scales. All these biological materials are well known and it is often sufficient to consider a consensus value for the refractive index. However, the wavelength dependence of the refractive index may matter in some cases, for instance, especially the imaginary part of the refractive index, which is responsible for light absorption and influences much of the optical response. In addition, disorder can have a strong influence on the reflection or transmission spectrum as well (e.g., by suppressing coherence effects such as Fabry-Pérot oscillations on the sides of a Bragg peak). In the end, it is the responsibility of the investigators to decide whether or not, or to which degree, disorder should be taken into account in the modeling. Experience has taught us that the decision to include disorder must often be taken on a case-by-case basis. Another aspect to consider when dealing with natural photonic architectures is the occurrence of different levels of structuring—either in the form of a hierarchical structure or independent organizations at multiple length-scales. This multiscale property, when present, may complicate the modeling, but it also produces interesting effects. Again, the investigators should assess the opportunity to include this aspect in the modeling. For instance, when studying colors from bird feathers, the curvature of a barb may be neglected and its cylindrical geometry can be ignored in such a way as to retain only the photonic crystal within the cortex. To a first-order approximation, the barb model will then consist of a planar photonic crystal slab.

9.2 Simulation Methods

Typical length scales of the photonic structure determine the choice of the simulation method. The ray-tracing method is well suited to cope with large-scale and disordered structures, for which resorting to geometrical optics is justified when the dimensions of interest are much greater than the wavelength of light. Curved surfaces, such as pits on a beetle cuticle or the cortex of a feather barb, are some examples. When the structure has typical feature sizes of the order of the wavelength, a wave optics approach becomes necessary. This is the case for a majority of studies dealing with iridescence, since it originates from wave interference. The most popular computational methods used in the natural photonics community are the one-dimensional transfer-matrix (1D TM) method, the finite difference time domain (FDTD) method, and the rigorous coupled wave analysis (RCWA) method. All three methods solve Maxwell's equations exactly in inhomogeneous optical media and have their own advantages and drawbacks in terms of computational resource requirements and numerical accuracy. Other methods are sometimes used, but their application is restricted to specific cases. For instance, a frequency-domain method based on an eigensolver for Maxwell's equations in a plane wave basis or a Kronig-Penney model approach are used for photonic band structure calculations in order to predict the positions and widths of photonic bandgaps in infinite photonic crystals. The interest in these methods is limited to perfectly ordered crystals of large size, which are rare in natural organisms with a few exceptions. Mie scattering calculations can be useful in some cases, too—for instance, to describe the appearances of structures containing spherical pores.

In this section, we will describe the basics of the main simulation methods used in natural photonics in order to understand the underlying optical mechanisms, as well as to develop bioinspired photonic devices.

9.2.1 One-Dimensional Transfer-Matrix Method

As far as planar multilayer stacks are concerned, whether they are periodic or not, the 1D TM method [4–6] is a well-known method that solves wave propagation exactly in arbitrarily stratified optical media. The method relies on expressing electromagnetic field boundary conditions at each layer's planar interface and connecting forward- and backward-propagating electric field complex amplitudes through 2×2 matrices. The transfer matrix of the whole stack is simply the matrix product of the layers' matrices in the order they succeed one another in the stack. Reflectance and transmittance are calculated in a straightforward fashion from matrix elements by taking squared modulus of complex quantities. Since it is a frequency-domain method that operates on complex field amplitudes, dispersive and absorbing materials are easily taken into account by using wavelength-dependent complex refractive index data $n(\lambda) + ik(\lambda)$, from which the reflectance spectrum $R(\lambda)$ and the transmittance spectrum $T(\lambda)$ are calculated by running the algorithm at each wavelength independently; the absorptance spectrum is deduced from energy conservation; that is, $A(\lambda) = 1 - R(\lambda) - T(\lambda)$. The method also gives access to the spatial evolution of the fields along the direction normal to the layers, from which field intensity or local absorption depth profiles can be obtained [7]. Computer codes are available commercially or in open access. Since they deal with 1D electromagnetic problems, codes run fast on a personal computer and can be written easily by undergraduate physics students. The TM method has been extended to 4×4 matrix formalisms in order to account for optically anisotropic planar layers [8–11]. This version can be used, for example, for calculating circularly polarized reflection (Figure 9.3) of the natural Bouligand-type helicoidal photonic structures found in the cuticle of some beetles [12, 13]. Returning to the isotropic case, it is worth noting that the optical wave dispersion relation (i.e., photonic band diagram) has a closed-form analytical expression for 1D photonic crystals. From this expression, the DOS of the infinite crystal can be calculated numerically (Figure 9.3) and compared, in terms of photonic bandgap location, with the reflectance spectrum obtained by 1D TM calculations for a finite slab [3]. Finally, it is worth mentioning that the transfer-matrix method has also been extended to model fluorescence emission from a single light-emitting planar layer, which mimics fluorescent molecules embedded in a multilayer stack [3, 14].

9.2.2 Ray-Tracing and Beyond

As noted above, the ray-tracing method is relevant when the size and characteristic features of the scattering object are large in comparison with the wavelength of light, so that geometrical optics applies. A three-dimensional polygonal mesh of the actual structure must be built first. Source and detector elements are then placed on the hemispheres surrounding the object. When the detector is placed on the same hemisphere as the source, the detected intensity (number of ray hits) defines the reflectance. Conversely, as one would expect, when the detector is placed on the opposite hemisphere from the source, the detected intensity defines the transmittance. Typically, numerous rays are launched from the light source in order to reach numerical convergence in calculations. All rays initially carry equal amounts of radiant power, and their paths as well as their power evolution are followed through

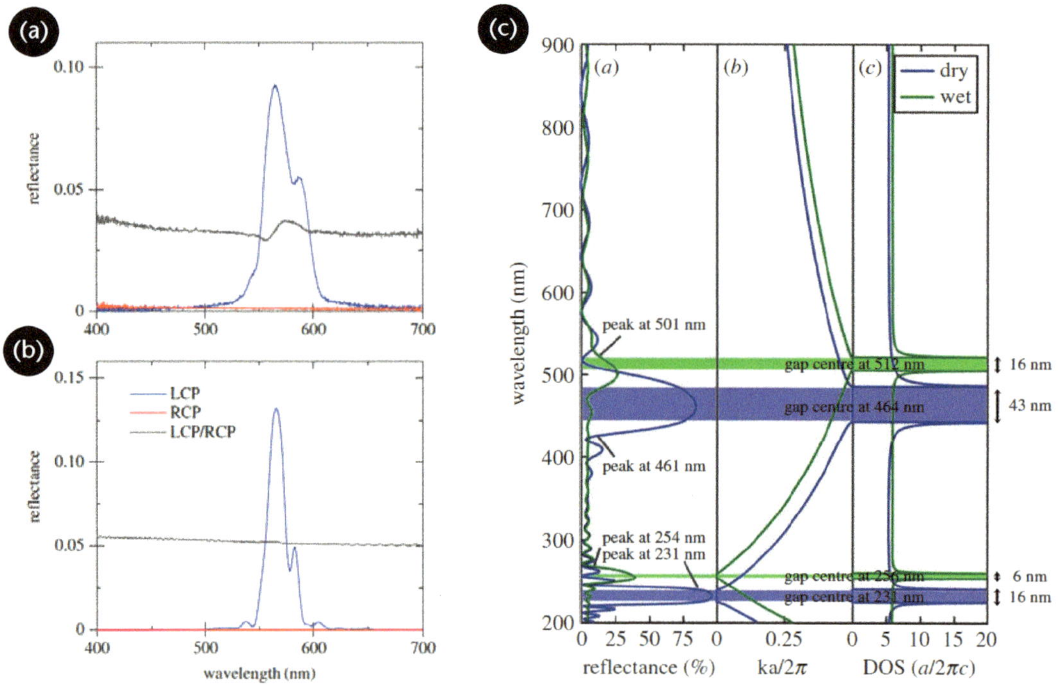

Figure 9.3 1D TM calculations. Circularly polarized reflectance spectra measured (a) and calculated (b) in copolarized (LCP and RCP) and cross-polarized (LCP/RCP) configurations for the beetle *Chalcothea smaragdina*. A 1D TM code for anisotropic layer stacks was used for reflectance calculation. Reflectance spectra, photonic band diagram, and DOS calculated for the beetle *H. coerulea* (c). Blue and green curves correspond to dry and wet states of the porous photonic model, respectively. 1D TM code for isotropic layer stacks was used for reflectance calculations. The photonic band diagram was calculated analytically. DOS was calculated by taking numerical derivatives from the band diagram. (Reproduced from [13] (a,b) and [3] (c), with permission from the Royal Society.)

multiple scattering events on the object's surface elements (reflection, refraction, absorption) until they reach detector elements or intersect the hemisphere without being detected. By ascribing a wavelength to each ray, the wavelength-dependent properties of the object constitutive materials are taken into account and the directional reflectance, transmittance, and absorptance spectra of the object can be calculated.

Ray-tracing simulations were used by McCoy and coworkers to account for multiple scattering by barbules in bird of paradise feather [15]. These simulations allowed the researchers to examine the impact of viewing direction and illumination on the appearance of feathers (Figure 9.4). Either omnidirectional or directional light source was employed.

A ray-tracing technique can be combined with, for example, 1D TM or RCWA method, in a *multiscale* approach when both geometrical optics and wave optics are needed to describe interactions separately at different length scales [16, 17]. This situation may be encountered when surface elements are no longer simple dioptres but, for example, are multilayer structures. In this case, instead of calculating Fresnel reflection and transmission coefficients, the 1D TM method is used to calculate the radiant power redistribution between the reflected and transmitted waves at the surface *structured* elements. Vigneron and coworkers applied this

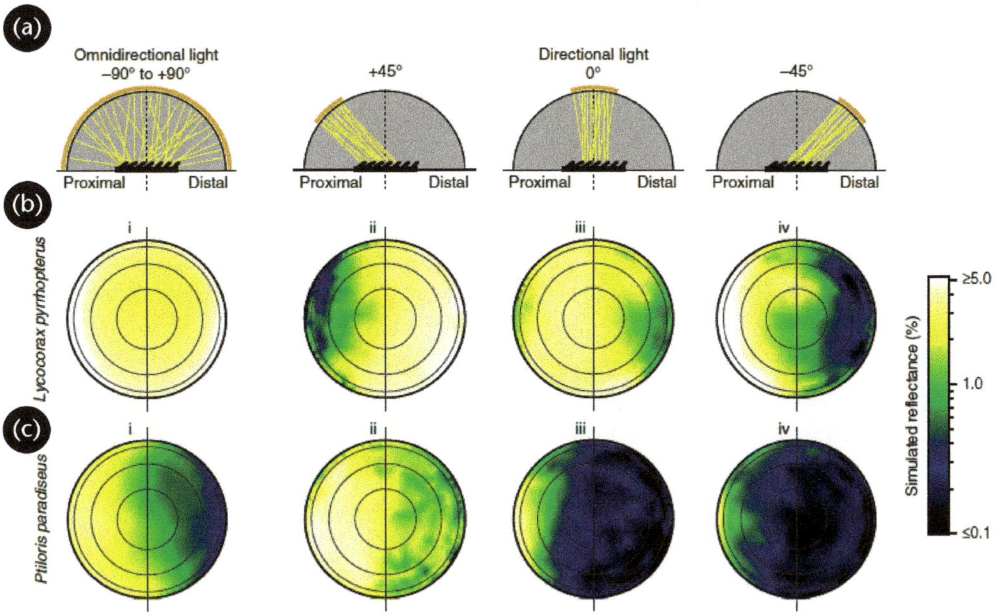

Figure 9.4 Ray-tracing calculations on 3D models of bird feathers. In sketches (a) of source-detector configurations, the reflectance hemisphere is colored gray and the feather model is black, with yellow lines showing the path of incident light from the hemisphere arc. Polar plots of the directional reflectance of *Lycocorax pyrrhopterus* (b) and *Ptiloris paradiseus* (c) feathers under the four lighting conditions (concentric rings correspond to the specific polar angles: 22.5°, 45°, 67.5°, and 90°). (Reproduced from [15], License CC-BY 4.0.)

multiscale approach to the modeling of the reflectance of the African shieldbug *Calidea panaethiopica* [16]. The beetle's cuticle is covered by hemispherical concave dips with a conformal Bragg layer stack underneath the curved surface. In multiscale computations, the 1D TM code accounted for wave interference within the Bragg stack, whereas the ray-tracing code accounted for reflections on the mesoscopic concave surface.

9.2.3 Finite-Difference Time-Domain Method

In 1966, Kane S. Yee derived a solution to Maxwell's equations in the time domain on arbitrary optically isotropic media by resorting to central finite differences in order to approximate both time and spatial partial derivatives and make them numerically workable [18]. This method was later popularized as the FDTD method. The analytical representation of Maxwell's equations is discarded, keeping only a discretized form of the equations on a spatial grid whose cubic voxels are denoted Yee's cells [19]. Discretization of the medium into voxels unavoidably causes numerical errors in the calculated fields (i.e., artifacts) that can be reduced by decreasing the voxel size at the expense of the computation time. In Yee's cell [18, 19], electric field Cartesian components occupy the middle of the edges and magnetic field Cartesian components the center of the faces due to the discretization of the curl operators in Ampere's and Faraday's equations. Gauss's electric and magnetic equations are automatically satisfied. Boundary conditions at the interfaces between regions of different permittivities (or permeabilities) are expressed straightforwardly thanks to this scheme. Starting from their initial values on the grid, fields values are updated recursively, both in discretized space and time, until a steady-state is reached. Fields in the frequency domain are obtained by discrete Fourier transform, which is incremented at each temporal step. Fluxes of the Poynting vectors associated with the incident and scattered waves are calculated across predefined measurement planes in the computational box. The reflectance (or transmittance) spectrum is deduced from normalization of scattered flux with respect to the incident flux. Since the computational box is of finite size and limited by computer memory resources, boundary conditions must be used at the box's edges in order to avoid artifacts such as nonphysical spurious reflections. Bloch periodic conditions allow periodic optical media such as photonic crystals to be emulated, whereas perfectly matched layer (PML) boundary conditions [20] enable emulation of an infinite medium containing a scattering object of finite size and arbitrary structure. The spatial discretization (i.e., grid) offers huge flexibility regarding the material composition of the scattering object, since permittivity and permeability values can be assigned independently in each voxel. Because the FDTD method operates in the time domain, the use of dispersive and absorbing materials (i.e., wavelength-dependent complex refractive indices) in simulations is less straightforward and flexible than in frequency-domain methods, as it requires a theoretical description of the dielectric function $\varepsilon(\omega)$, such as the Drude or Lorentz models. The introduction of $\varepsilon(\omega)$ into Ampere's equation, once back to the time-domain, leads to an auxiliary differential equation that must also be approximated by finite differences. Finally, without going into subtle technical detail, it should be mentioned that the procedure for updating electric and magnetic fields is divided into two interwoven steps in a leapfrog algorithmic scheme in order to keep accuracy in the second order of the numerical calculations [19]. As a consequence, the discrete electric field is not defined at the same time as the discrete magnetic field.

In the context of natural photonics, FDTD simulations are used quite often [3,21–27]. Chandler and coworkers used binarized versions of TEM images to build a model of the cuticular structure of the red alga *Chondrus crispus*, which actually exhibits blue structural coloration at the tips of the thalli [24]. Using a commercial FDTD software (Lumerical Inc.), they calculated the reflectance spectrum and simulated the light scattering pattern through a near-field to far-field transformation (Figure 9.5). Although the main feature of *C. crispus* reflectance spectrum can be captured by one-dimensional transfer-matrix calculations using an idealized planar multilayer model [24], there are many other cases for which the complexity of the natural photonic structure does not permit such a simplification. For instance, the complex anatomy of the breast feather barbules of Lawes's Parotia *Parotia lawesii* was rendered by a realistic three-dimensional model; FDTD modeling enabled direct visualization of light propagation within the photonic structure as well as simulation of the far-field light scattering pattern (Figure 9.5) [23].

9.2.4 Rigorous-Coupled Wave Analysis Method

In 1981, Moharan and Gaylord proposed a rigorous coupled-wave approach to analyze diffraction by arbitrary planar gratings [28], which was later known as the RCWA method. In the original problem, the grating was defined by a sinusoidal

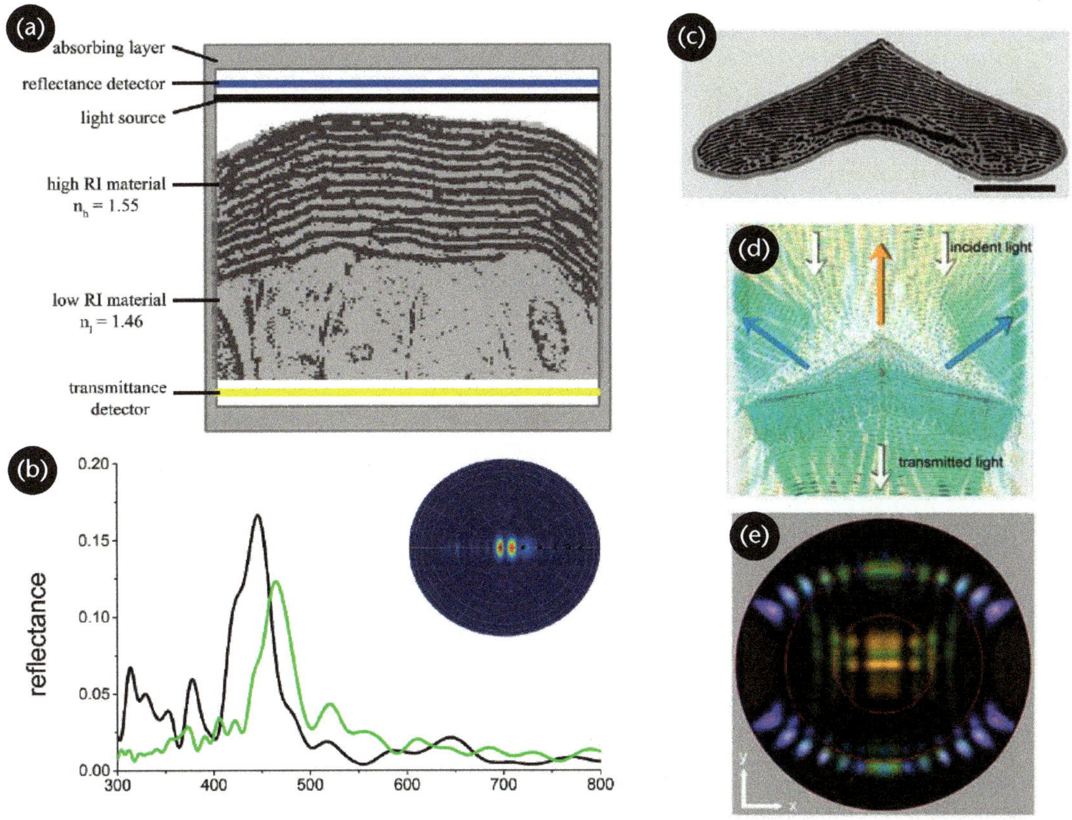

Figure 9.5 FDTD simulations. Binarized TEM image (a) used as model of the cuticle structure of the alga *Chondrus crispus* and reflectance spectra (b) calculated by FDTD from two different TEM images (inset: simulated light-scattering pattern). TEM image (c) of *P. lawesii* breast feather barbule (scale bar: 5 μm) used to build a realistic photonic model; two superimposed video frames (d) of light propagation in the model for normally incident light at 420 nm (blue) and 610 nm (orange); superimposed simulated far-field light-scattering patterns (e) for different incidence angles. (Reproduced from [24], License CC-BY 4.0 (a,b) and from [23], with permission from PNAS (c–e).)

variation of the permittivity ε (single spatial period or grating pitch), yet spatial variations were allowed along two directions: the z-axis (perpendicular to the grating surface) and the x-axis. Translation invariance was assumed along the y-axis. This grating configuration allowed slanted gratings to be studied at oblique or normal incidence. The generalization of Moharam and Gaylord's treatment followed by considering an arbitrarily layered medium (step-wise variations of the dielectric permittivity $\varepsilon(\vec{r})$ along the z-axis), with independent and arbitrary periodic variations of $\varepsilon(\vec{r})$ along the lateral directions (x and y axes) [29]. Although it requires assuming infinite lateral extension of the medium and periodicity in one lateral dimension at least, the generic concept of a laterally periodically stratified medium (Figure 9.6) encompasses a large variety of optical systems of finite thicknesses (i.e., slabs). These are all idealized templates of natural photonic structures: Bragg stacks, two- or three-dimensional photonic crystal slabs and pseudo-disordered photonic slabs [30]. The Bragg stack is a trivial case, where there is no lateral periodicity, only longitudinal periodicity over a finite length. The case of a pseudo-disordered photonic slab is treated by introducing a fictive lateral periodicity to satisfy the

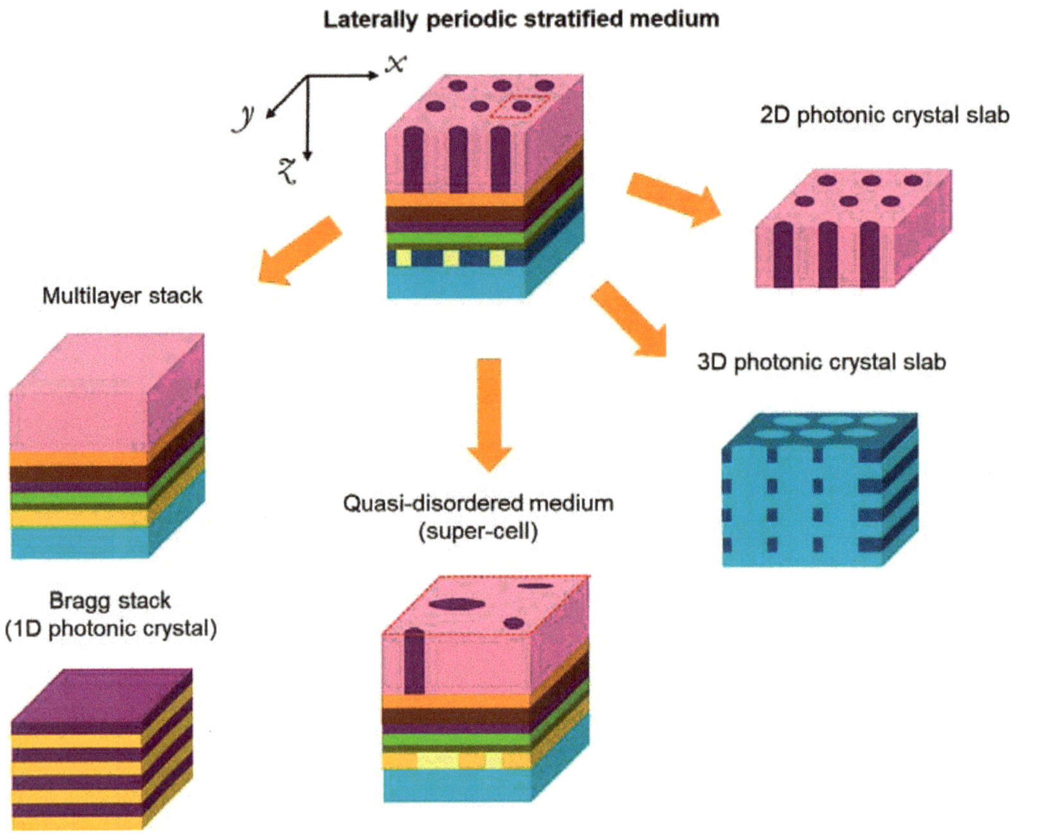

Figure 9.6 The generic concept of a laterally periodically stratified medium encompasses a large variety of photonic structures found in nature. The unit cell is highlighted in the dashed red box for the generic and the pseudo-disordered cases.

RCWA theoretical framework (hence, the prefix "pseudo" was used here), while allowing arbitrary permittivity variations (i.e., disorder) within the unit cell, which is called a super-cell, for this reason [31].

In contrast with FDTD, RCWA is a rigorous method in the sense that it solves Maxwell's equations exactly (i.e., keeping analytical expressions of temporal and spatial derivatives). Therefore, unlike in FDTD, no spatial mesh is needed, except for the division of the slab into layers, in which $\varepsilon(\vec{r})$ is assumed to be constant along the z-axis. More layers can always be taken to obtain a finer medium geometrical description; this comes at the expense of the computational time, though. The principle of RCWA relies on the Bloch-Floquet theorem: the spatial periodicity of the medium in lateral directions translates the electromagnetic field decomposition into lateral Bloch waves [29]. The Fourier components of the laterally varying permittivity is associated with a two-dimensional reciprocal lattice and the corresponding diffraction directions. By expressing electromagnetic boundary conditions at layer interfaces, fields on both layer sides can be connected and a layer transfer matrix can be defined. The latter involves Fourier components of forward- and backward-propagating complex amplitudes, which are associated with either radiative or evanescent Bloch waves. The size of the matrix is determined by the number

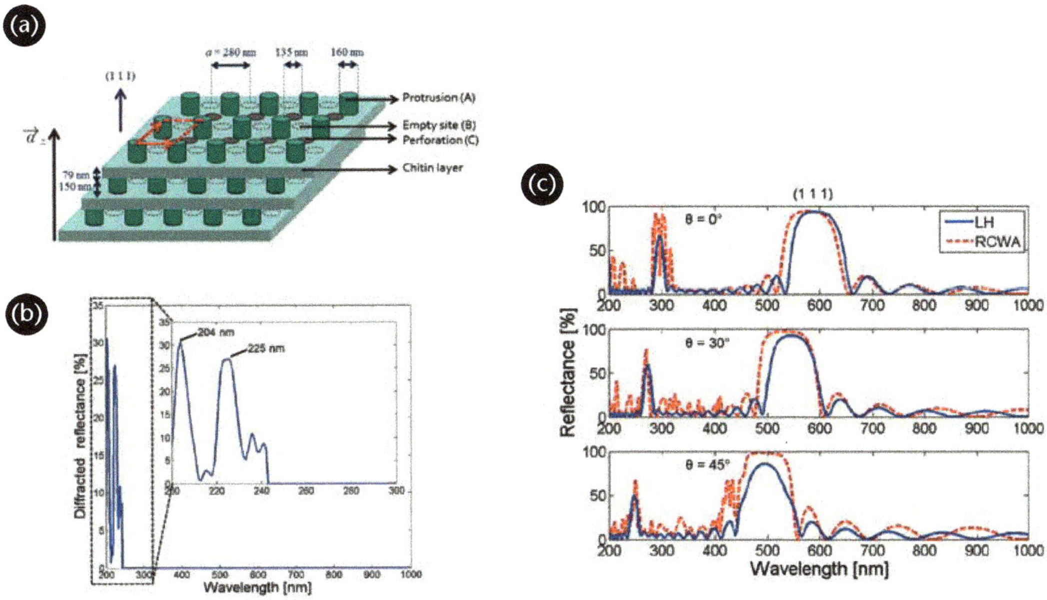

Figure 9.7 RCWA simulations. Model of the natural 3D photonic crytal (a) and diffracted reflectance spectrum calculated by RCWA (b) showing all diffraction orders except the specular component. Reflectance spectra (c) calculated by RCWA and layer homogeneization (LH) methods at different incidence angles for (111) crystal orientation. (Reproduced from [34], with permission from the Optical Society.)

of Fourier components that are retained in the two-dimensional Fourier expansion of the permittivity (or equally the number of plane wave components in the field expansion into Bloch waves). This unavoidable Fourier series truncation leads to numerical errors, which must be evaluated through convergence tests, in order to find a trade-off between the number of plane waves and the required computational resources. For reasons of numerical stability, the layer is divided into identical, subwavelength-thick slices and the transfer-matrix of a slice is converted into a scattering matrix. Scattering matrices are assembled using Pendry's formulas [33] (nonassociative operation) in order to obtain the scattering matrix of the layer. The same operation is performed on all the layers and leads to the scattering matrix of the photonic slab.

At the slice level, scattering matrices allow researchers to reconstruct the electromagnetic field spatial distribution with subwavelength resolution along the z-axis (slice thickness) and arbitrarily high resolution along the x- and y-axes (analytical expression of Bloch waves). At the slab level, the scattering matrix gives direct access to scattered electromagnetic fields on both sides of the slab for given polarization and directions of the incident wave. From Poynting vector fluxes calculated in the incidence and emergence media, hemispherical (total) reflectance and transmittance are obtained and the absorptance of the slab is deduced from energy conservation. By retaining only one Bloch wave component in such calculations, one has access to the diffraction efficiency (either in reflection or transmission) of the corresponding diffraction order. In our own RCWA modeling of a natural three-dimensional photonic crystal (Figure 9.7), we calculated, for a given crystal direction, the diffracted reflectance; that is, the hemispherical reflectance from which the specular (zeroth order diffraction) component was excluded [34]. We also compared, at different incidence angles, RCWA simulations with 1D TM simulations that had been performed using a layer homogenization approximation of the 3D photonic crystal [34].

RCWA is a frequency-domain method. It implies that computation can be performed independently at each wavelength. Therefore, in contrast to FDTD, experimentally obtained refractive index spectral data can be used directly without the need to use dielectric function models. Decreasing computational resources is possible thanks to the frequency-domain operation of the RCWA method: a smaller number of plane waves can be afforded at some wavelengths depending on the slab structure and the dispersive properties of the material constituents. This adaptation and its advantage has, unfortunately, not been highlighted sufficiently in the literature on RCWA.

Finally, it is worth mentioning that, in early publications by Vigneron and coworkers, the electromagnetic computational method described above was referred to as a transfer-matrix method, although it does implement the scattering matrix (S-matrix) formulation (i.e., it is truly RCWA). This might be unfortunate terminology with respect to the one-dimensional transfer-matrix method that is used extensively in natural photonics. Moreover, it is also distinct from Pendry's transfer-matrix method [32], which relies on an original k-space discretization and S-matrix formalism to calculate the scattering of photons by three-dimensionally periodic optical media. Actually, the original idea of using an S-matrix (instead of a T-matrix) to solve numerical stability issues belongs to Pendry [32, 33]. As far as the authors remember, Vigneron and Pendry met in the late 1990s, in the context of emerging natural photonic studies—this seems a likely reason why Jean-Pol Vigneron developed his own code using the S-matrix "trick." It is our personal interpretation that Jean-Pol, as a solid-state physicist, chose the "transfer-matrix" terminology spontaneously, rather than RCWA, which has become more popular in the optics community.

9.2.5 Other Numerical Prediction Methods

Although it uses scattering matrices, the transfer-matrix method developed by Pendry [32, 33] differs from RCWA. To the best of our knowledge, it has not been used much in the context of natural photonics. Zi's group cites Pendry's method with regard to their calculations of photonic band diagrams and reflectance spectra for the 2D photonic crystal structure found in the brown barbules of the male peacock's tail feathers. However, unfortunately, they do not give any computational details [35].

In the late 1990s, Prum and coworkers developed a Fourier method for predicting coherent scattering by biological structures and their related reflectance spectra [36–41]. This method is based on Benedek's relationship between the Fourier transform of spatial variation in refractive index and the intensity of coherent scattering [42]. It requires computation of the two-dimensional fast Fourier transform of electron micrographs or two-dimensional slices of three-dimensional tomographic reconstructions of natural structures that are thought to give rise to coherent scattering. The resulting 2D Fourier components are then squared to obtain the 2D Fourier power spectrum that may be expressed as a function of spatial frequencies (e.g., in nm^{-1}). Each value of this 2D Fourier power spectrum corresponds to the magnitude of the related periodicity in the original micrograph for a specific direction. The radial average of a computed 2D Fourier spectrum can be performed over all directions, but only within a specific quadrant or over restricted angles. A normalization with respect to the total Fourier power should ideally be carried out, specifically when calculating radial averages of 2D power spectra from several micrographs. When the normalized radial average is restricted to the visible range and expressed as a function of the inverse of the spatial frequencies multiplied by twice the average refractive index of the biological structures, it can be compared directly with reflectance spectra measured by spectrophotometry or simulated by other computational methods. The average refractive index is usually assessed on the basis of the micrograph(s). This technique has proven very effective for quasi-ordered structures (i.e., structure with partial absence of organization in their physical morphology), such as those occurring in bird caruncles [37,39,40], bird feathers [36,38,40,41], mammalian skins [43], dragonfly integuments [44], butterfly scales [45], or salamander egg sacs [46].

9.3 Examples of Bioinspired Design Approaches

In this section, we present two examples of concepts in photonics that have emerged from bioinspiration.

9.3.1 The Concept of Spectral Richness

When considering the iridescence properties of natural 1D photonic crystal slabs from a bioinspired perspective, one may wonder how to maximize the color change as the incidence angle is varied. In other words, how to maximize the Bragg peak shift for a given couple of layer materials (fixed refractive indices) that are compatible with layer-by-layer fabrication requirements. This idea was rationalized through the concept of *spectral richness* introduced by Deparis [47].

Let us consider a periodic multilayer coating. The Bragg peak spectral position is the first design parameter that determines the coating's dominant reflected wavelength (i.e., its hue). The normalized photonic bandgap width is the second design parameter and that determines the coating's color saturation. The spectral richness, noted Θ, is a dimensionless quantity that expresses the relative Bragg peak shift as function of the incidence angle, noted θ (the reference being the Bragg peak wavelength at normal incidence, $\lambda_{\theta=0°}$). The interesting point is that Θ is independent of the absolute physical dimensions of the multilayer structure. Because multilayers exhibit form birefringence, we distinguish those cases where the electric field of the incident wave is parallel or perpendicular to the layer interfaces. The corresponding refractive indices are given by $n_{\|} = n_1 \left(\frac{1+rc^2}{1+r}\right)^{\frac{1}{2}}$ and $n_{\perp} = n_1 \left(\frac{1+r}{1+r/c^2}\right)^{\frac{1}{2}}$, where $c = n_2/n_1$ is the refractive index contrast, with n_1 (n_2) being the refractive index of the least (most) refringent layer, and $r = d_2/d_1$ is the ratio between layer thicknesses. The spectral richness is defined by $\Theta = \frac{\lambda_\theta}{\lambda_{\theta=0°}}$, where $\lambda_\theta = 2a\sqrt{n^2 - \sin^2\theta}$

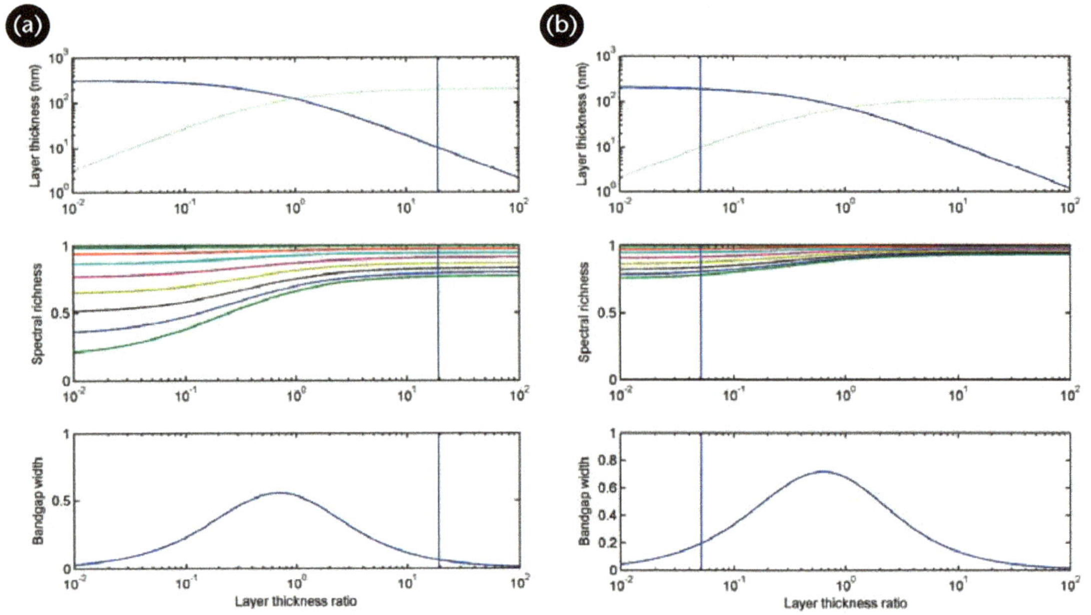

Figure 9.8 Design of iridescent periodic multilayer coatings for a Bragg peak at normal incidence set to $\lambda_{\theta=0^\circ} = 640$ nm. Layer thicknesses, spectral richness, and bandgap width are plotted as functions of the layer thickness ratio. The spectral richness was calculated at incidence angles varying from 0° to 80° in steps of 10°. Results are shown for natural air/chitin multilayers (a) ($n_1 = 1.0, n_2 = 1.56$) and artificial SiO_2/TiO_2 multilayers (b) ($n_1 = 1.5, n_2 = 2.7$). The vertical lines on the charts (a) correspond to the beetle *C. vittata*'s air/chitin design ($d_1 = 10$ nm, $d_2 = 194$ nm) and those on the other charts (b) to the dual SiO_2/TiO_2 design ($d_1 = 194$ nm, $d_2 = 10$ nm). All calculations were done for TE polarization. (Reproduced from [47], with permission from the Optical Society.)

is the dominant reflected wavelength (i.e., Bragg peak associated with the spatial period $a = d_1 + d_2$). The spectral richness Θ can be expressed, for both linear polarizations, as a function of the dimensionless parameters r, c, the incidence angle θ, and the refractive index n_1 [47]. The normalized bandgap width at normal incidence, which determines the spectral selectivity, can be expressed analytically as a function of r and c as well [47].

In order to achieve a high spectral richness, the design has to rely on a strong imbalance between layer thicknesses; that is, r must be very different from unity, as shown in Figure 9.8 in the case of transverse-electric (TE) polarization (electric field normal to the plane of incidence) and for chitin/air and TiO_2/SiO_2 refractive index pairs. This design rule is a lesson taken from nature, as illustrated by the iridescent beetle *Chrysochroa vittata* for which $r \simeq 20$ [47]. For design purposes, it is convenient to plot the spectral richness as a function of the layer thickness ratio for a given baseline (lowest) refractive index and refractive index contrast. The ability of the Bragg peak to exhibit a large angular shift (i.e., strong iridescence) is expressed by the great separation of $\Theta(r)$ curves calculated for set of θ values. Interestingly, more unbalanced designs also come with a higher spectral selectivity (i.e., more saturated color), as shown by the normalized bandgap width's dependence on the layer thickness ratio. This makes the iridescence a more striking visual effect. A last point to mention is that, for a fixed refractive index contrast, the smaller the

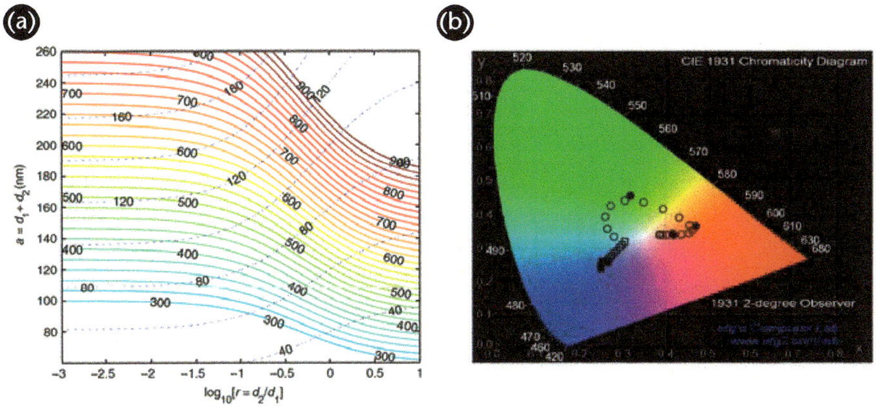

Figure 9.9 Design map (a) displaying $\lambda_{\theta=0^\circ}$ (solid lines) and $\delta\lambda_\theta$ (dotted lines) in (r, a) parameter space for SiO_2/TiO_2 multilayer coatings (values are displayed in nanometers along the lines). The spectral richness (calculated at $\theta = 80^\circ$) is noted as $\delta\lambda_\theta$ here. Calculations were done for transverse-electric (TE) polarization, using $n_1 = 1.5$ and $n_2 = 2.5$. Chromaticity coordinates (b) of two types of SiO_2/TiO_2 multilayer coatings fabricated by magnetron sputtering techniques (circles and squares). Full and open symbols correspond to measurements ($\theta = 20^\circ, 40^\circ, 60^\circ$) and predictions (from $\theta = 0^\circ$ to 80° by steps of 5°), respectively. (Reproduced from [48], with permission from IOP Publishing.)

baseline index, the larger the spectral richness [47]. The fact that air gaps occur in the lower refractive index layers in *C. vittata*'s cuticle tends to suggest that the beetle has evolved an optimal design for spectral richness in terms of layer thickness ratio and baseline index. In a bioinspired design approach, we can generate high spectral richness similar to the beetle by substituting air and chitin with SiO_2 and TiO_2, respectively (Figure 9.8). However, the higher index contrast ($c = 1.8$), as well as the higher baseline index ($n_1 = 1.5$) of the artificial design, leave us no choice but to select thin high-index layers and thick low-index layers, as illustrated by the dual design with respect to *C. vittata*.

Replacing biological materials with oxides such as SiO_2 and TiO_2 allows us to fabricate bioinspired iridescent coatings for commercial applications using, for instance, magnetron sputtering techniques, as we demonstrated it in 2008 [48]. One has to be careful in designing the reflector, however, as illustrated above, by the swapping between the high and low index layers in the unbalanced design. The spectral richness value at $\theta = 80^\circ$, noted $\delta\lambda_\theta$ and calculated as a difference instead of a ratio in our aforementioned study [48], gives a good indication of the maximum Bragg peak shift when the angle varies from normal to grazing incidences. Plotted in the (r, a) parameter space (Figure 9.9), it enables the selection of the layers thicknesses, $d_1 = a/(1+r)$ and $d_2 = ar/(1+r)$, for tailored reflector responses [48]. Using silicon oxide and titanium oxide as layer materials, for instance, and choosing a period $a = d_1 + d_2$ around $200 - 230$ nm, with $r = d_2/d_1$ of the order of 0.01 should produce a red hue at normal incidence, which turns to green at grazing incidence. On the other hand, using the same materials, decreasing the period down to approximately 80–100 nm and increasing $r = d_2/d_1$ up to 10 should produce a blue hue at normal incidence, which turns to violet at grazing incidence. As highlighted above, the layer thickness ratio is a key design parameter: the former choice ($r \ll 1$) leads to a large $\delta\lambda_\theta$ value, whereas the latter design ($r \gg 1$) leads

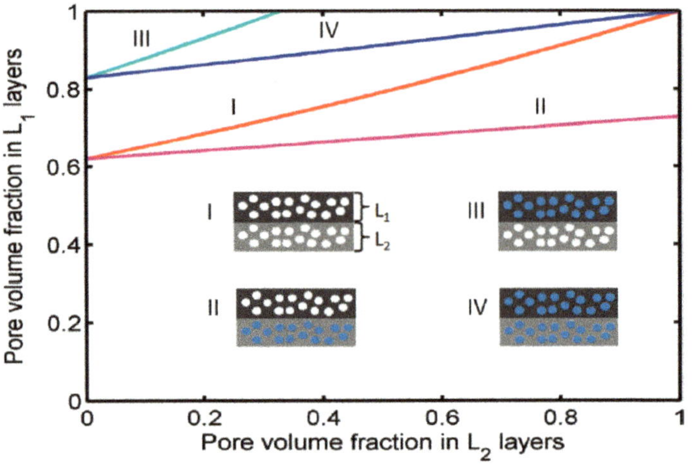

Figure 9.10 Transparency relationships corresponding to porous bilayers made from TiO_2 (L_1 layer) and SiO_2 (L_2 layer), whose pores are either empty (white circles), or filled with water (blue circles). (Reproduced from [50], with permission from AIP Publishing.)

to a small $\delta\lambda_\theta$ value. These particular designs produce an appearance (Figure 9.9) mimicking *C. vittata* and *H. coerulea*, respectively, as the incidence angle varies [48].

9.3.2 The Concept of a Color-Switchable Mirror

In Chapter 5, the changes of structural color induced by contact with liquids in some insect cuticles were explained by the displacement of a fluid inside a porous multilayer structure, which could be periodic or not. The case of *Charidotella egregia* is particularly interesting since, as the liquid is drained out of the structure, the multiple coherent scattering effect responsible for the structural gold appearance vanishes [49]. If the effective refractive indices of all layers match each other, the emptied structure would become transparent. Actually, it would transmit light in a diffuse way because the empty pores scatter light randomly. From a bioinspiration perspective, it is instructive to examine the theoretical conditions that need to be imposed on a Bragg mirror and on the fluid displacement so that the multilayer can switch between a colored state and a transparent state [50]. Although in the examples found in nature so far, the fluid was assumed to be either infiltrated in or removed from the pores of the whole multilayer structure, we will not restrict ourselves to these two cases. Instead, we will also consider the two other cases where the fluid is selectively displaced between the adjacent layers that form the periodic Bragg stack (Figure 9.10, inserts).

Let us consider a periodic multilayer stack (i.e., a one-dimensional photonic crystal) made of two arbitrary porous layers, the host materials of which are assumed to be transparent in the spectral range of interest. The size of the pores are assumed to be small enough that an effective medium approximation can be used to describe the refractive indices of both constitutive layers. Consequently, the porous structure can be regarded as an *effective* multilayer and its optical response can be calculated using the Abeles formalism of thin-film optics. Let us further assume that the porous network is interconnected so that the fluid can be displaced from layer to layer (hereafter, the pore volume filling fractions refer to the open porosity, i.e., with the pores accessible to the fluid). Let us examine the specific case in which the effective refractive index contrast between adjacent layers cancels out and, consequently, the photonic crystal behaves as an effective homogeneous transparent material. For a porous layer made of dielectric material with a given pore volume filling fraction, the effective permittivity can be calculated using the Bruggeman effective medium theory [51]. In spite of the fact that both constituents of the medium (pores and dielectric material) are treated on an equal basis in Bruggeman's theory, we will refer to the dielectric material as the host and the pores as the inclusions, as they would be named in the Maxwell Garnett effective medium theory. The advantage of using Bruggeman's theory, however, is that the full range of pore filling fraction values can be considered without affecting the accuracy of the results. In this formalism, the effective permittivity of a porous layer is the solution of the second-degree equation:

$$f_{\mathrm{p}} \frac{\varepsilon_{\mathrm{p}} - \varepsilon_{\mathrm{eff}}}{\varepsilon_{\mathrm{eff}} + (\varepsilon_{\mathrm{p}} - \varepsilon_{\mathrm{eff}})L} + f_{\mathrm{h}} \frac{\varepsilon_{\mathrm{h}} - \varepsilon_{\mathrm{eff}}}{\varepsilon_{\mathrm{eff}} + (\varepsilon_{\mathrm{h}} - \varepsilon_{\mathrm{eff}})L} = 0, \tag{9.1}$$

where f_{p} ($f_{\mathrm{h}} = 1 - f_{\mathrm{p}}$) is the volume fraction of the pores (host material); ε_{p} (ε_{h}), the dielectric permittivity of the pores (host material), and L, the depolarization factor, which depends on the pores' geometrical shape. The refractive indices are obtained by taking the square root of the permittivities (i.e., $n = \sqrt{\varepsilon}$). In the following, the refractive indices are assumed to be constant and real over the whole spectral range of interest. For spherical pores ($L = \frac{1}{3}$), the solution of (9.1) is

$$\varepsilon_{\mathrm{eff}} = \frac{1}{4}\left(\beta + \sqrt{\beta^2 + 8\varepsilon_{\mathrm{p}}\varepsilon_{\mathrm{h}}}\right), \tag{9.2}$$

with $\beta = (3f_{\mathrm{h}} - 1)\varepsilon_{\mathrm{h}} + (3f_{\mathrm{p}} - 1)\varepsilon_{\mathrm{p}}$. Now, if we make a bilayer from two different materials with two different pore filling fractions, it can behave as a single homogenous slab, provided that $\varepsilon_{\mathrm{eff},1} = \varepsilon_{\mathrm{eff},2}$, with both permittivity expressions given by (9.2). If we fix, for instance, the porosity in the first layer ($f_{\mathrm{p},1}$), we can determine the porosity in the second layer ($f_{\mathrm{p},2}$) so that the above equality is fulfilled ($\varepsilon_{\mathrm{eff},1} = \varepsilon_{\mathrm{eff},2}$). This condition, called hereafter the *transparency condition*, is expressed by [50]:

$$f_{\mathrm{p},2} = f_{\mathrm{p},1} \frac{g(u_{\mathrm{p},1}) - g(u_{\mathrm{h},1})}{g(u_{\mathrm{p},2}) - g(u_{\mathrm{h},2})} + \frac{g(u_{\mathrm{h},1}) - g(u_{\mathrm{h},2})}{g(u_{\mathrm{p},2}) - g(u_{\mathrm{h},2})}, \tag{9.3}$$

where $u = \frac{\varepsilon_{\mathrm{eff}}}{\varepsilon}$ is a dimensionless variable and $g(u) = \frac{1-u}{1+(\frac{1}{L}-1)u}$.

The relationships between $f_{\mathrm{p},2}$ and $f_{\mathrm{p},1}$ at transparency are shown in Figure 9.10 in the case of a bilayer made of porous titanium dioxide and silicon dioxide for the four configurations of water distribution inside the pores.

Although it may seem easier to achieve transparency with empty pores in the high-index layer (here, TiO_2) and liquid-filled pores in the low-index layer (here, SiO_2), the three other configurations allow transparency to be reached as well [50].

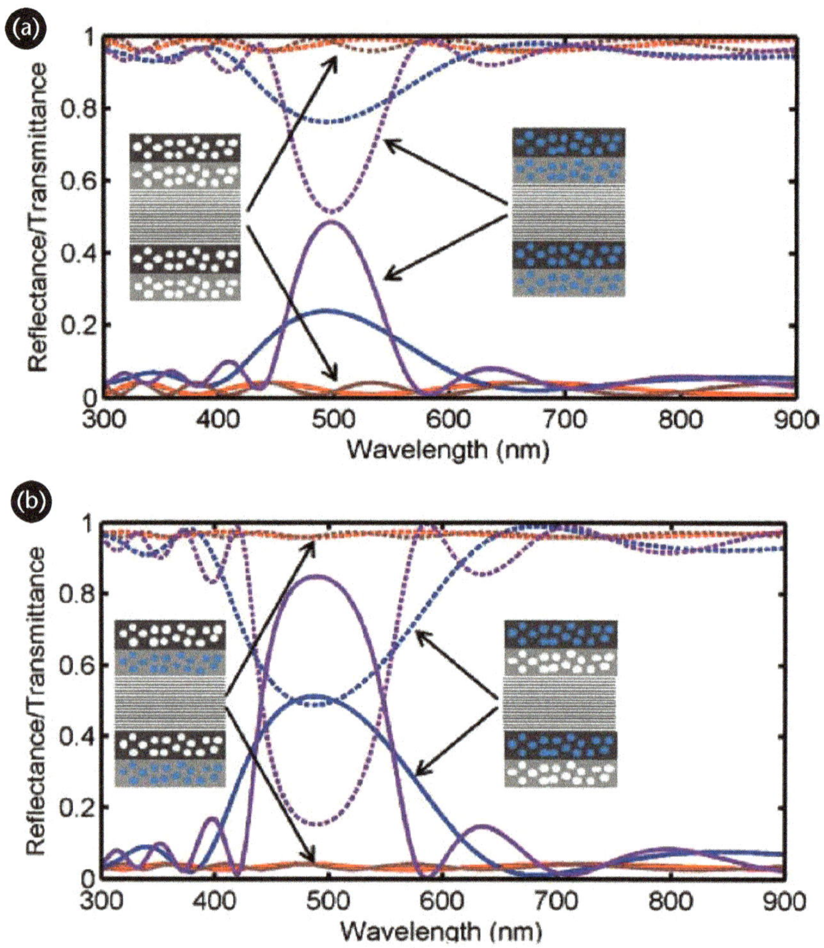

Figure 9.11 Reflectance spectra (solid curves) and transmittance spectra (dashed curves) of porous TiO_2/SiO_2 Bragg mirrors on a glass substrate ($n_{sub} = 1.51$). The thicknesses of the layers were chosen to be equal to 65 nm (TiO_2) and 100 nm (SiO_2) and the number of bilayers equal to $N = 3$ (blue and red curves) and $N = 6$ (violet and brown curves). Blue and violet (red and brown) curves relate to (a) pores filled with water (air) in both TiO_2 and SiO_2 layers, or (b) pores filled with water (air) in TiO_2 layers and air (water) in SiO_2 layers. The pores' volume fractions were equal to $f_{p,TiO_2} = 0.72$ (a), $f_{p,TiO_2} = 0.65$ (b), and $f_{p,SiO_2} = 0.30$ (a,b). (Reproduced from [50], with permission from AIP Publishing.)

It is worth noting that the opposite, counterintuitive configuration (curve III in Figure 9.10) also works, but is limited by an upper bound on the pore filling fraction in the low-index layer.

Let us now examine the switching behavior of a periodic multilayer made of N bilayers, whose pore filling fractions were chosen in order to reach transparency in a given configuration of liquid distribution (Figure 9.11). Starting with empty pores in both layers, the filling of the pores with liquid leads to the appearance of a Bragg peak in the reflectance spectrum. The same switching behavior, between transparency and Bragg mirror states, occurs when starting with empty pores in one layer and filled pores in the other and reversing the liquid distribution within the bilayer. The latter scenario is preferable to the former, as far as maximization of the reflectance (or transmittance) contrast between transparency and Bragg mirror state is concerned.

References

[1] Vukusic, P., and D. G. Stavenga, "Physical Methods for Investigating Structural Colours in Biological Systems," *Journal of the Royal Society Interface*, Vol. 6, No. S2, pp. S133–S148.

[2] Bay, A., et al., "Improved Light Extraction in the Bioluminescent Lantern of a *Photuris* Firefly (Lampyridae)," *Optics Express*, Vol. 21, No. 1, 2013, pp. 764–780.

[3] Mouchet, S. R., et al., "Controlled Fluorescence in a Beetle's Photonic Structure and Its Sensitivity to Environmentally Induced Changes," *Proceedings of the Royal Society B*, Vol. 283, No. 1845, 2016, 20162334.

[4] Born, M., and E. Wolf, *Principles of Optics*, Oxford, UK: Pergamon, 1964.

[5] Lekner, J., and M. C. Dorf, "Matrix Methods for the Calculation of Reflection Amplitudes," *Journal of the Optical Society of America A*, Vol. 4, No. 11, 1987, pp. 2092–2095.

[6] Yeh, P., *Optical Waves in Layered Media*, Hoboken, NJ: Wiley, 2005.

[7] Deparis, O., "Poynting Vector in Transfer-Matrix Formalism for the Calculation of Light Absorption Profile in Stratified Isotropic Optical Media," *Optics Letters*, Vol. 36, No. 20, 2011, pp. 3960–3962.

[8] Teitler, S., and B. W. Henvis, "Refraction in Stratified, Anisotropic Media," *Journal of the Optical Society of America*, Vol. 60, No. 6, 1970, pp. 830–834.

[9] Berreman, D. W., "Optics in Stratified and Anisotropic Media: 4 × 4-Matrix Formulation," *Journal of the Optical Society of America A*, Vol. 62, No. 4, 1972, pp. 502–510.

[10] Lin-Chung, P. J., and S. Teitler, "4 × 4 Matrix Formalisms for Optics in Stratified Anisotropic Media," *Journal of the Optical Society of America A*, Vol. 1, No. 7, 1984, pp. 703–705.

[11] Ko, D. Y. K., and J. R. Sambles, "Scattering Matrix Method for Propagation of Radiation in Stratified Media: Attenuated Total Reflection Studies of Liquid Crystals," *Journal of the Optical Society of America A*, Vol. 5, No. 11, 1988, pp. 1863–1866.

[12] Lowrey, S., et al., "Observation and Modeling of Polarized Light from Scarab Beetles," *Journal of the Optical Society of America A*, Vol. 24, No. 8, 2007, pp. 2418–2425.

[13] McDonald, L. T., et al., "Circularly Polarized Reflection from the Scarab Beetle *Chalcothea smaragdina*: Light Scattering by a Dual Photonic Structure," *Interface Focus*, Vol.7, No. 4, 2017, 20160129.

[14] Van Hooijdonk, E., S. Berthier, and J.-P. Vigneron, "Bio-Inspired Approach of the Fluorescence Emission Properties in the Scarabaeid Beetle *Hoplia coerulea* (Coleoptera): Modeling by Transfer-Matrix Optical Simulations," *Journal of Applied Physics*, Vol. 112, No. 11, 2012, 114702.

[15] McCoy, D. E., et al., "Structural Absorption by Barbule Microstructures of Super Black Bird of Paradise Feathers," *Nature Communications*, Vol. 9, No. 1, 2018, pp. 1–8.

[16] Vigneron, J.-P., et al, "Spectral Sideband Produced by a Hemispherical Concave Multilayer on the African Shield-Bug *Calidea panaethiopica* (Scutelleridae)," *Physical Review E*, Vol. 79, No. 2, 2009, 021907.

[17] Vigneron, J.-P., et al., "Optical Reflectance and Transmittance of Photonic Polycrystalline Structures from Living Organisms," *Proceedings of SPIE*, Vol. 8480, 2012, 84800H.

[18] Yee, K. S., "Numerical Solution of Initial Boundary Value Problems Involving Maxwell's Equations in Isotropic Media," *IEEE Transactions on Antenna and Propagation*, Vol. 14, No. 3, 1966, pp. 302–307.

[19] Taflove, A., and S. C. Hagness (eds.), *Computational Electrodynamics: the Finite-Difference Time-Domain Method*, Norwood, MA: Artech House, 2005.

[20] Berenger, J.-P., "A Perfectly Matched Layer for the Absorption of Electromagnetic Waves," *Journal of Computational Physics*, Vol. 114, No. 2, 1994, pp. 185–200.

[21] Plattner, L., "Optical Properties of the Scales of *Morpho rhetenor* Butterflies: Theoretical and Experimental Investigation of the Back-Scattering of Light in the Visible Spectrum," *Journal of the Royal Society Interface*, Vol. 1, 2004, pp. 149–159.

[22] Wilts, B. D., et al., "Brilliant Camouflage: Photonic Crystals in the Diamond Weevil, *Entimus imperialis*," *Proceedings of the Royal Society B*, Vol. 279, No. 1738, 2012, pp. 2524–2530.

[23] Wilts, B. D., et al., "Sparkling Feather Reflections of a Bird-of-Paradise Explained by Finite-Difference Time-Domain Modeling," *Proceedings of the National Academy of Sciences of the United States of America*, Vol. 111, No. 12, 2014, pp. 4363–4368.

[24] Chandler, C. J., et al., "Structural Colour in *Chondrus crispus*," *Scientific Reports*, Vol. 5, 2015, 11645.

[25] Mouchet, S. R., et al., "Photonic Scales of *Hoplia coerulea* Beetle: Any Colour You Like," *Materials Today: Proceedings*, Vol. 4, No. 4, 2017, pp. 4979–4986.

[26] Feyer, P., B. D. Wilts, and D. G. Stavenga, "Reflections on Iridescent Neck and Breast Feathers of the Peacock, *Pavo cristatus*," *Interface Focus*, Vol. 9, No. 1, 2018, 20180043.

[27] Wu, X., et al., "Polarization Conversion Effect in Biological and Synthetic Photonic Diamond Structures," *Advanced Optical Materials*, Vol. 6, No. 24, 2018, pp. 1–11.

[28] Moharam, M. G., and T. K. Gaylord, "Rigorous Coupled-Wave Analysis of Planar-Grating Diffraction," *Journal of the Optical Society of America*, Vol. 71, No. 7, 1981, pp. 811–818.

[29] Vigneron, J.-P., and V. Lousse, "Variation of a Photonic Crystal Color with the Miller Indices of the Exposed Surface," *SPIE Proceedings*, Vol. 6128, 2006, 61281G.

[30] Deparis, O., and J.-P. Vigneron, "Modeling the Photonic Response of Biological Nanostructures Using the Concept of Stratified Medium: The Case of a Natural Three-Dimensional Photonic Crystal," *Materials Science and Engineering B*, Vol. 169, No. 1–3, 2010, pp. 12–15.

[31] Muller, J., et al., "A Fair Comparison Between Ultrathin Crystalline-Silicon Solar Cells with Either Periodic or Correlated Disorder Inverted Pyramid Textures," *Optics Express*, Vol. 23, No. 11, 2015, pp. A657–A670.

[32] Pendry, J. B., and A. MacKinnon, "Calculation of Photon Dispersion Relations," *Physical Review Letters*, Vol. 69, No. 19, 1992, pp. 2772–2775.

[33] Pendry, J. B., "Photonic Band Structures," *Journal of Modern Optics*, Vol. 41, No. 2, 1994, pp. 209–229.

[34] Mouchet, S. R., et al., "Method for Modeling Additive Color Effect in Photonic Polycrystals with Form Anisotropic Elements: The Case of *Entimus imperialis* Weevil," *Optics Express*, Vol. 21, No. 11, 2013, pp. 13228–13240.

[35] Li, Y., et al., "Structural Origin of the Brown Color of Barbules in Male Peacock Tail Feathers," *Physical Review E*, Vol. 72, No. 1, 2005, 010902(R).

[36] Prum, R. O., et al., "Coherent Light Scattering by Blue Feather Barbs," *Nature*, Vol. 396, No. 706, 1998, pp. 28–29.

[37] Prum, R. O., et al., "Coherent Light Scattering by Nanostructured Collagen Arrays in the Caruncles of the Malagasy Asities (Eurylaimidae: Aves)," *Journal of Experimental Biology*, Vol. 202, No. 24, 1999, pp. 3507–3522.

[38] Prum, R. O., et al., "Two-Dimensional Fourier Analysis of the Spongy Medullary Keratin of Structurally Coloured Feather Barbs," *Proceedings of the Royal Society B*, Vol. 266, No. 1414, 1999, pp. 13–22.

[39] Prum, R. O., and R. H. Torres, "Structural Colouration of Avian Skin: Convergent Evolution of Coherently Scattering Dermal Collagen Arrays," *Journal of Experimental Biology*, Vol. 206, No. 14, 2003, pp. 2409–2429.

[40] Prum, R. O., and R. H. Torres, "A Fourier Tool for the Analysis of Coherent Light Scattering by Bio-Optical Nanostructures," *Integrative and Comparative Biology*, Vol. 43, No. 4, 2003, pp. 591–602.

[41] Shawkey, M. D., et al., "Electron Tomography, Three-Dimensional Fourier Analysis and Colour Prediction of a Three-Dimensional Amorphous Biophotonic Nanostructure," *Journal of the Royal Society Interface*, Vol. 6, No. S2, 2009, pp. S213–S220.

[42] Benedek, G. B., "Theory of Transparency of the Eye," *Applied Optics*, Vol. 10, No. 3, 1971, pp. 459–473.

[43] Prum, R. O., and R. H. Torres, "Structural Colouration of Mammalian Skin: Convergent Evolution of Coherently Scattering Dermal Collagen Arrays," *Journal of Experimental Biology*, Vol. 207, No. 12, 2004, pp. 2157–2172.

[44] Prum, R. O., J. A. Cole, and R. H. Torres, "Blue Integumentary Structural Colours in Dragonflies (Odonata) Are Not Produced by Incoherent Tyndall Scattering," *Journal of Experimental Biology*, Vol. 207, No. 22, 2004, pp. 3999–4009.

[45] Prum, R. O., T. Quinn, and R. H. Torres, "Anatomically Diverse Butterfly Scales All Produce Structural Colours by Coherent Scattering," *Journal of Experimental Biology*, Vol. 209, No. 4, 2006, pp. 748–765.

[46] Zabuga, A. V., et al., "Translucent in Air and Iridescent in Water: Structural Analysis of a Salamander Egg Sac," *Soft Matter*, Vol. 16, No. 7, 2020, pp. 1714–1721.

[47] Deparis, O., et al., "Color-Selecting Reflectors Inspired from Biological Periodic Multilayer Structures," *Optics Express*, Vol. 14, No. 8, 2006, pp. 3547–3555.

[48] Deparis, O., et al., "Structurally Tuned Iridescent Surfaces Inspired by Nature," *New Journal of Physics*, Vol. 10, 2008, 013032.

[49] Vigneron, J.-P., et al., "Switchable Reflector in the Panamanian Tortoise Beetle *Charidotella egregia* (Chrysomelidae: Cassidinae)," *Physical Review E*, Vol. 76, No. 3, 2007, 031907.

[50] Deparis, O., et al., "Theoretical Condition for Transparency in Mesoporous Layered Optical Media: Application to Switching of Hygrochromic Coatings," *Applied Physics Letters*, Vol. 104, No. 2, 2014, 023704.

[51] Choy, T. C., *Effective Medium Theory: Principles and Applications*, Oxford, UK: Oxford University Press, 1999.

CHAPTER 10

Fabrication of Bioinspired Photonic Devices

The fabrication of bioinspired photonic devices faces challenges common to any microfabrication at the nanoscale. Natural photonic structures have typical features, such as pores or layer thicknesses, of the order of tens to hundreds of nanometers. Moreover, body parts where such structures are present (butterfly scales, bird feather barbules, beetle elytra, etc.) may spread over lengths ranging from a few to hundreds of microns. Photonic devices designed for technological applications, however, usually require good uniformity over areas from hundreds of mm^2 to several cm^2, depending on the application (optical sensors, smart coatings, etc.) [1]. Top-down and bottom-up approaches have been employed, both with success, either to produce a faithful copy of the biological template (this practice is called biomimetism) or to fabricate a device that shares selected structural features with the template (i.e., bioinspiration). First described in this chapter are nanoimprint lithography, self-assembly of spheres, and sol-gel methods, which all resort to bottom-up approach techniques. Nanoimprint lithography uses natural samples directly, as templates, whereas self-assembly of spheres allows one to synthesize artificial templates that closely resemble some natural ones. Sol-gel methods can be used to transfer natural templates into synthetic ones, thanks to the infiltration of liquid solutions into the structure. A hybrid sol-gel process, combined with spin coating, can be used to deposit porous oxide layers on a substrate in order to synthesize mesoporous oxide multilayered coatings with hygrochromic properties similar to those found in some insect cuticles. On the other side of the portfolio of methods, silicon processing technologies, atomic layer deposition, and magnetron sputtering constitute top-down approach techniques. Photolithography, wet etching, and dry etching are all familiar techniques for silicon processing that can be employed to produce sophisticated bioinspired devices on silicon. Atomic layer deposition allows the production of conformal organic-inorganic structures from intricate natural samples. It also permits the tuning of their optical properties precisely through atomic-level accuracy on layer thickness. Magnetron sputtering is a widely used coating technique in the glass and steel industries that can be applied directly to the fabrication of bioinspired smart coatings. Different methods, including layer-by-layer assembly and evaporation-induced self-assembly, have gained in popularity for synthesizing cellulose nanocrystal (CNC) films. This chapter does not pretend to be exhaustive. It is humbly hoped to be a source of inspiration for engineers and researchers who are considering transferring and adapting ideas from nature to their fabrication processes, with a possibly innovative combination of those.

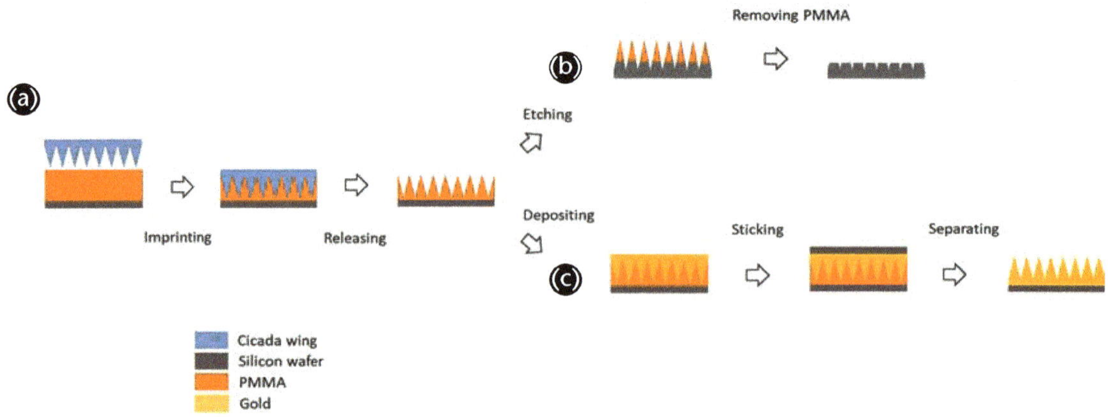

Figure 10.1 Nanoimprint lithography process applied to the replication of cicada wing nanostructures in PMMA (a), followed either by direct transfer of the negative pattern onto a silicon substrate (b) or indirect transfer onto a deposited gold layer, in order to obtain a positive replicate (c).

10.1 Nanoimprint Lithography

Nanoimprint lithography (NIL) has become a widespread method for fabricating nanostructures on large-scale samples [2–6]. For instance, artificial moth-eye antireflective structures have been fabricated on silicon wafers using a soft roller nanoimprint technique [4]. In this case, a UV-sensitive resist was first spin-coated on the silicon wafer and then imprinted by rolling a flexible master on it. The flexible master was prepared from a nickel master which was, itself, molded in photoresist by two-beam laser interference photolithography (giving rise to a moth-eye pattern). The imprinted resist was cross-linked by UV-curing and its moth-eye pattern was transferred into silicon by dry etching. A hot embossing technique has also previously been used to fabricate a polyvinyl chloride (PVC) transparent template with a moth-eye pattern, which was then used for a UV nanoimprint [5]. The UV imprinted resin pattern on glass acted as an antireflective coating.

Moving to biomimetic approaches, NIL can be used to replicate natural nanostructures, such as the arrays of nipples found on moths' eyes, or similar arrays of protuberances found on the wings of cicadas. In these cases, the biological samples serve directly as stamps for NIL. They therefore need to possess good mechanical properties in order to withstand the imprint process and, ideally, to be reusable. Chitinous structures—such as those found on butterfly wings, for instance—possess stillness and chemical stability, thanks to their crystalline chitin fibers that are embedded in a matrix of proteins and other constituents (Young's modulus of cicada wings has been reported to be as high as 7–9 GPa [2]). In some cases, the low surface tension of the template, due to, for example, a wax layer on cicada wings, permits the use of antiadhesive layer on the stamp to be avoided [2]. The need to clean the template surface to remove contaminants is, in comparison, a small drawback. Zhang's group has used cicada wings as a stamp for NIL [2,3]. The principle of NIL is illustrated in Figure 10.1. In the application described, it involves two steps [2]: the wing is pressed (40 bar, 180 s) on a spin-coated PMMA film heated at 190°C (70°C above its glass transition temperature T_g =120°C); the embossed PMMA film is then released from the stamp (Figure 10.1(a)). The inverted (negative) replica of the wing structure consists of a hexagonal array of nanowells [2]. This negative replica can be used for various purposes. For instance, it can be used as a mask for etching the silicon substrate onto which the PMMA film has already been spin-coated (Figure 10.1(b)). Although the PMMA mask was not ideal for reactive ion etching in the aforementioned study, the thin nanostructured layer etched in silicon exhibited some antireflective capability [2]. Another possibility is to use the negative replica as a mold: after gold is deposited on it, the gold layer sticks to the silicon substrate and is separated from the mold (Figure 10.1(c)). The resulting gold pillar array is a faithful positive replicate of the original cicada wing, which can serve as a substrate for surface enhanced Raman scattering (SERS) [2].

10.2 Self-Assembly of Spheres

Self-assembly of colloidal silica (e.g., 360-nm diameter SiO_2 spherical particles) has been used to create a biomimetic template that reproduced moth eye patterns (Figure 10.2(a)) [7]. This template was used to cast a poly(dimethylsiloxane) (PDMS) mold, which, after solidification and peeling off, was then pressed on a monomer film that had been deposited on glass. Finally, UV-curing was used to polymerize the moth-eye patterned film. The latter was made of poly(ethoxylated trimethylolpropane triacrylate) (PETPTA), which acted as an antireflective coating. An intermediate step (i.e., reactive ion etching) was used to adjust the height of the silica spheres that protrude from the spin-coated polymer nanocomposite [7]. Ultimately, this step controls the protrusion depth of the moth-eye PETPTA replica. Another way to exploit the biomimetic template is to use it as a mask for etching the underlying Si wafer, which offers a nice, bottom-up approach for patterning sub-wavelength nipple arrays on silicon directly [8]. In this method, the self-assembled sphere template is first subject to etching in O_2 plasma in order to adjust its geometry before being used as mask during SF_6 reactive ion etching of silicon (Figure 10.2(b)). A final HF etching step is used to selectively remove the silica spheres.

10.3 Sol-Gel Methods

The sol-gel process is a versatile, wet-chemical method for the synthesis of glassy or ceramic materials at moderate temperatures. It relies on hydrolysis of a colloidal solution (sol) containing precursors (e.g., metal alkoxides) and poly-condensation of hydrolyzed products, leading to the formation of a gel; that is, a solid, inorganic network containing a liquid phase. Since it starts from a liquid solution, it is often combined with spin-coating and subsequent thermal annealing for the deposition of solid films on substrates. Among numerous materials that can be deposited, metal oxides, such as SiO_2, TiO_2, or Al_2O_3, are of particular interest for producing multilayered coatings with optical properties. A sol-gel method was used by Han and coworkers in combination with selective etching in order to produce SiO_2 inverse replica of butterfly wing [9]. The replication process first required a series of pretreatments to be performed on the wing's sample (rinsing, degreasing, gradual dehydration, drying) at room temperature in order to make it suitable for use as

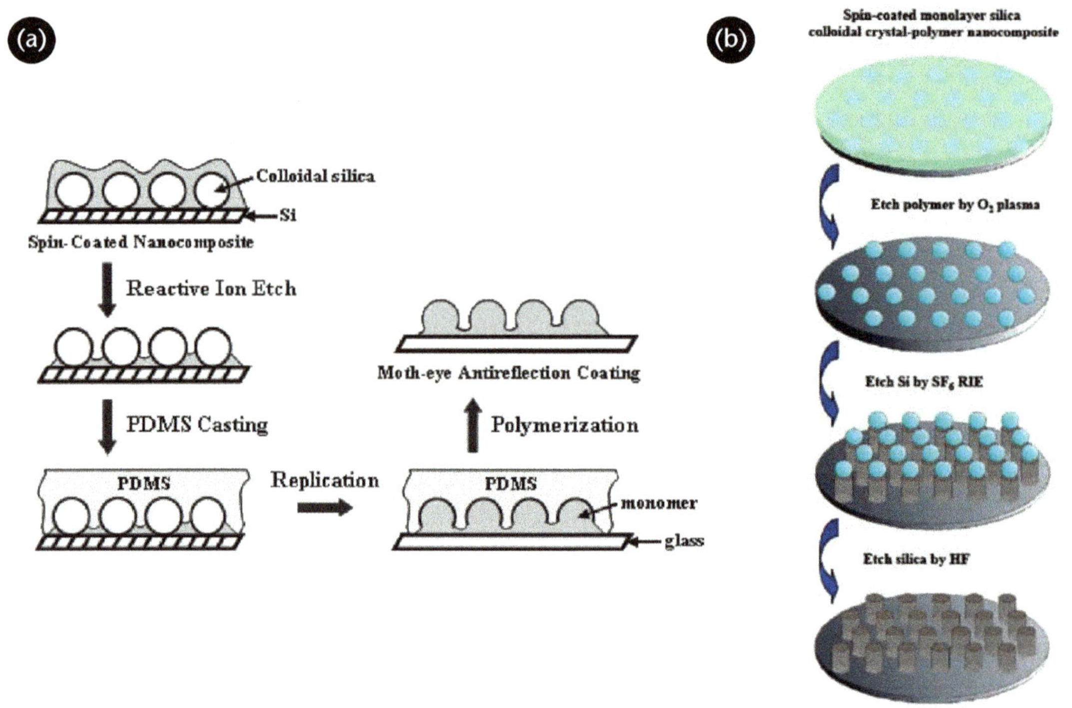

Figure 10.2 Templating procedure (a) for fabrication of biomimetic moth-eye resin patterns on glass, starting from self-assembled silica sphere nanocomposite. Similar templating procedure (b) adapted for fabrication of moth eye patterns on silicon. (Reproduced from [7] (a) and [8] (b), with permission from AIP Publishing.)

a template. The pretreated sample was then sandwiched between two glass slides and the silica precursor solution was added from the edges of this assembly, which was heated at 130°C for 25 min to evaporate any residual solvent (Figure 10.3). The final step consisted of etching in a mixture (1:1 in volume) of nitric acid and perchloric acid (at 120°C for 30 min) in order to remove the biological template. The SiO_2 sample thus obtained was a faithful inverse replica of the butterfly wing at all levels of the original hierarchical structure (Figure 10.3).

Porous materials can also be synthesized using a hybrid organic-inorganic sol-gel process, in which organic micelles are used as porogenic agents. Ghazzal and coworkers fabricated mesoporous dielectric multilayer films using such an hybrid sol-gel preparation and spin-coating deposition [10]. Mesoporous titania and silica layers of high optical quality and well-defined pore size were synthesized. Commercially available organic molecules were added to the polymeric sols in order to serve as template for creating mesopores inside inorganic host materials. The control of the open porosity, which was achieved by using an adequate choice of the structuring agent, allowed them to tailor the optical transmission of periodic Bragg stacks (Figure 10.4) in response to the infiltration of liquids [11]. Briefly, the fabrication proceeded as follows: when dissolved in ethanol, individual organic molecules formed small spherical aggregates (i.e., micelles) and, during the sol-gel process, evaporation of the solvent led to the formation of an hybrid xerogel that contained these spherical micelles. Calcination of the spin-coated sol-gel film (in air,

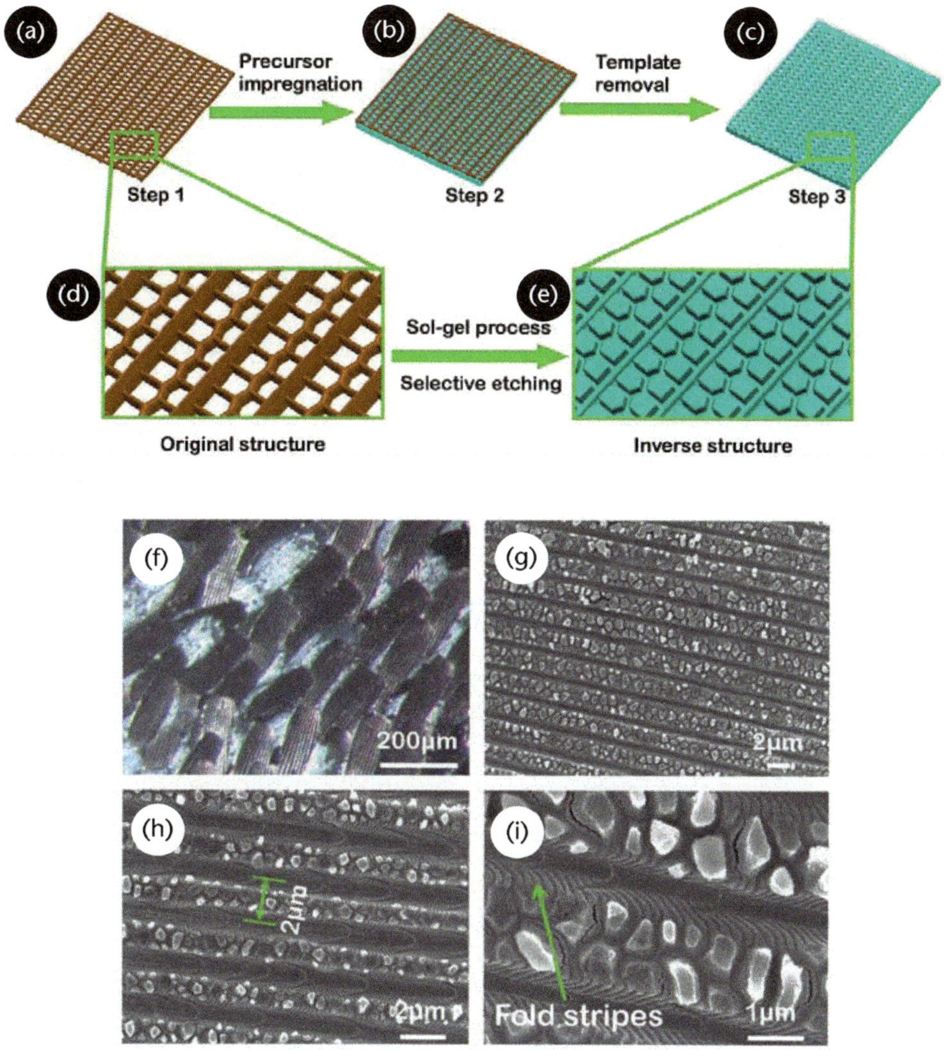

Figure 10.3 Sol-gel based replication process of butterfly wing (a–e). Original template structure (a), solidification of infiltrated sol-gel precursor (b), removal of template by etching (c). Details of original template (d) and inverse replica (e) dual structures. SEM images (f–i) of SiO_2 inverse replica at different magnifications, showing the faithful reproduction of the butterfly wing structure at different length scales. (Reproduced from [9], with permission from the Royal Society of Chemistry.)

at 400°C) ensured complete removal of the organic template, leaving a consolidated inorganic mesoporous layer. Stacking of mesoporous layers made, alternately, of titanium dioxide and silicon dioxide, required consolidation of each layer by gentle successive thermal annealing steps at moderate temperatures (< 200°C). Pore size ranged, typically, from 1 nm to 10 nm, depending on the template used. Open porosities as high as 45% and 41% were obtained in silica films and titania films, respectively. This approach for synthesizing single oxide mesoporous layers was extended later to mixed oxide mesoporous layers, offering an additional parameter to tune the layers' effective refractive index [12]. By using a single porogenic agent while varying molar fractions of TiO_2 and Al_2O_3, or TiO_2 and SiO_2, Ghazzal and

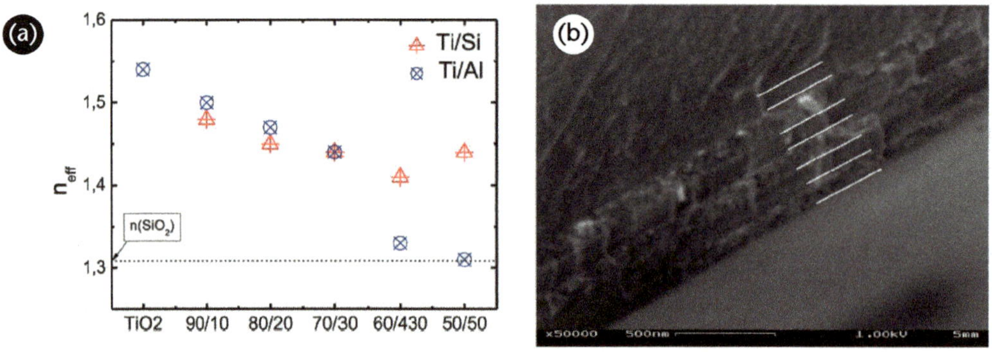

Figure 10.4 Effective refractive index (a) of mesoporous mixed oxide (titania/alumina or titania/silica) layers as function of oxide ratio. Porous layers were prepared by the sol-gel method, using Pluronic P123 as the porogenic agent. Refractive indices of host materials are $n_{TiO_2} = 2.50$, $n_{Al_2O_3} = 1.60$, and $n_{SiO_2} = 1.51$. SEM image (b) of the cross section of a periodic silica-titania mesoporous multilayer that was prepared by the sol-gel method using porogenic agents. (Reproduced from [12] (a) and [10] (b), with permission from the Royal Society of Chemistry.)

coworkers were able to tune the effective refractive index over a range spanning from 1.3 to 1.55 (Figure 10.4). The lowest value matched the effective index of porous silica used as the low-index material and allowed them to achieve optical transparency of the Bragg stack with air-filled (empty) pores [12]. As described in Chapter 9, this transparency condition can be exploited to synthesize a bioinspired switchable Bragg reflector by infiltration of a liquid into the mesoporous multilayer [11].

10.4 Silicon Processing Technologies

Techniques used in microelectronics for silicon processing can be employed to fabricate bioinspired photonic structures. Hereafter, among many other possibilities, we will describe the fabrication of a bioinspired porous silicon photonic crystal. Using standard microfabrication techniques combined with electrochemical etching, Rasson and coworkers fabricated a periodic array of concave multilayers in porous silicon, which mimicked the nanostructure of *Papilio blumei*'s scales [13]. Briefly, a silicon oxide layer was formed on a p-type boron-doped silicon wafer by wet oxidation in order to serve as a hard mask for etching. A periodic stripe pattern was defined on a resist by photolithography and opened using buffered HF. Wet etching of silicon through the mask openings created V-shaped cavities. After removal of the oxide mask, electrochemical etching created porosities in the silicon using two different current densities to produce two layers of different effective refractive index. This cycle was repeated several times to build a periodic porous multilayer. High conformity of the electrochemical etching process enabled the production of alternate layers of high and low porosities, which preserved the original concave profile of the cavity (Figure 10.5). This concave porous silicon multilayer has demonstrated improved performance for ethanol vapor optical sensing in comparison with a flat porous silicon multilayer [13].

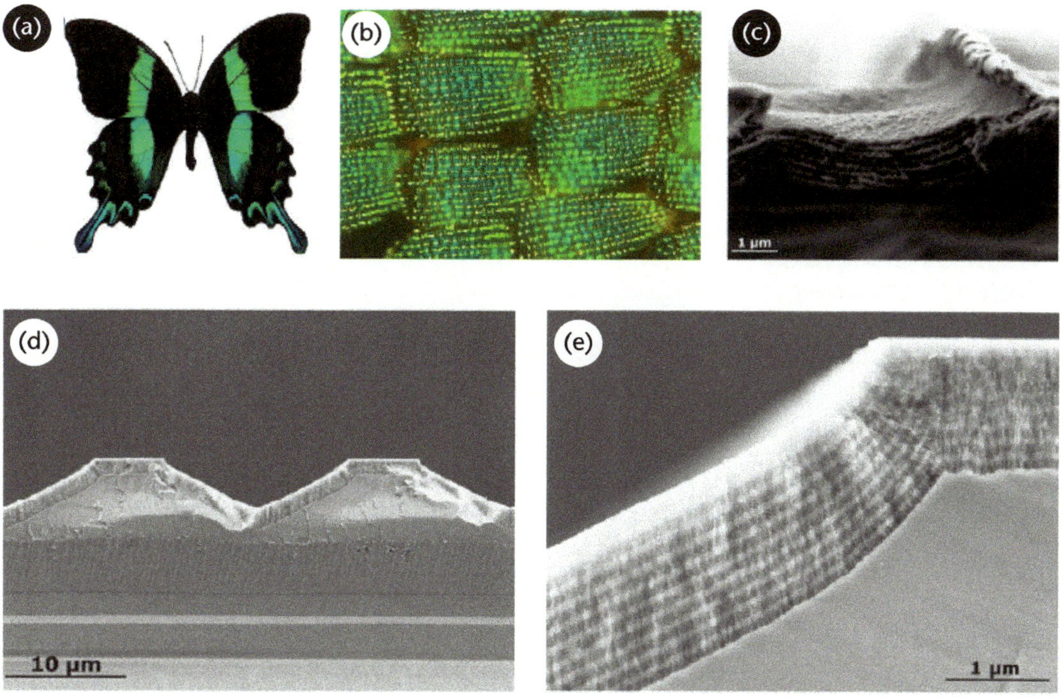

Figure 10.5 Photograph of *P. blumei* (a), light-microscope image of green scales (b), SEM cross-section image of the scale's concave porous multilayer structure (c), and SEM cross-sectional images of the bioinspired concave mutilayer structure in porous silicon (d,e). (Reproduced from [13], with permission from Elsevier.)

10.5 Atomic Layer Deposition

Atomic layer deposition (ALD) is a technique during which the surface to be coated is repeatedly exposed to gas precursors, whose adsorption forms monolayers. Unlike other chemical vapor deposition techniques, the different precursors are never put together, but react with the surface during independent, sequential cycles. The unique advantage of ALD is its ability to coat uneven surfaces with high aspect ratio in a conformal manner.

Gaillot and coworkers have used ALD to fabricate composite organic-inorganic structures from samples (5×5 mm^2) taken from the wing of the butterfly *P. blumei* [14]. Amorphous layers of TiO_2 were deposited conformally onto and into originally green iridescent scales, which were composed of an ordered array of 5–10 μm concavities separated by ridges (Figure 10.6). The scales had a porous structure inside them, made of nanometer-sized, rectangularly shaped voids. Because original scales were sealed, this structure was not accessible to ALD gas precursors and layers could be only deposited on top and all around the scales' surfaces. However, for a few scales on the edge of the sampling area, microcracks or fractures allowed infiltration of gas precursors and deposition of conformal TiO_2 layers on the inner walls of the voids. Sequential ALD cycles of 10-nm thin layers were performed at 100°C, leading to, respectively, coated scales and infiltrated scales. The

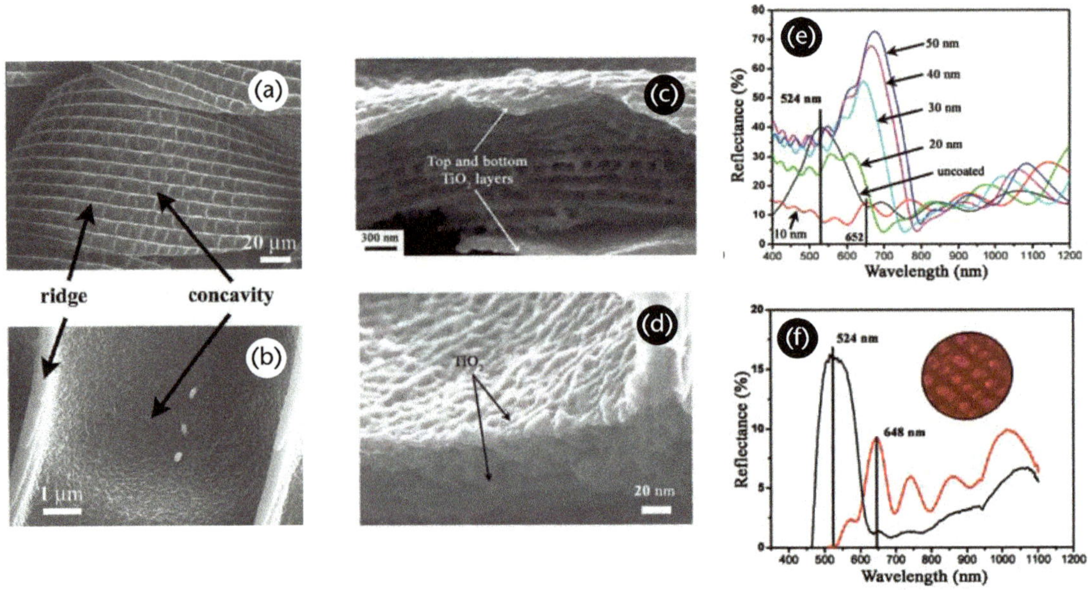

Figure 10.6 SEM images (a,b) of *P. blumei* scales: view of single scale (a) and magnified view of ridges and concavities (b). Cross-sectional SEM images (c,d) taken after ALD of coated scale (c) and infiltrated scale (d). Reflectance spectra of coated scales for increasing thickness of conformal TiO_2 layers (e) and of an infiltrated scale with 10-nm thickness of TiO_2 (f). (Reproduced from [14], with permission from the American Physical Society.)

optical reflectance of these scales varied, according to the thickness of the conformally deposited TiO_2 (Figure 10.6). It was noted that coated sealed scales could be regarded as hybrid organic-inorganic Fabry-Pérot etalons, with partially reflecting TiO_2 mirrors, which explained the observed modifications of the reflectance spectrum (Figure 10.6). By tuning the titanium oxide thickness, the team was able to control the color of the coated scales precisely. Actually, TiO_2 layers as thin as 10 nm were sufficient to modify the scale's color from green to yellow green. On the other hand, with 10 nm of TiO_2, infiltrated scales exhibited a large reflectance peak shift, changing color from green to red (Figure 10.6). This was explained by the reduction in the air filling fraction, from 59% to 37%, due to the conformal coating of the voids' inner walls, which increased the effective refractive index of the infiltrated porous stack. These interpretations were supported by numerical reflectance simulations of coated or infiltrated scale models.

The capability of ALD to achieve accurate thickness control and to use a wide range of oxides has been exploited by Poncelet and coworkers, in order to synthesize *Morpho*-inspired artificial photonic structures for gas sensing applications [15]. In that case, ALD was used to coat silicon wafers with TiO_2/Al_2O_3 and HfO_2/Al_2O_3 Bragg mirrors. Dry- and wet-etching steps were then used to sculpt these stacks into arrays of Christmas-tree-like photonic structures, which were reminiscent of the blue iridescent scales of *Morpho* butterflies. Starting from the ALD-deposited Bragg stack, the process consisted of the following steps: a chromium hard mask was deposited and patterned (array of slits) using photolithography, followed by a Cl_2/O_2 plasma step. The multilayer was etched through the opened slits using CCl_4 plasma, to create V-shaped trenches. The exposed Al_2O_3 layers were slightly

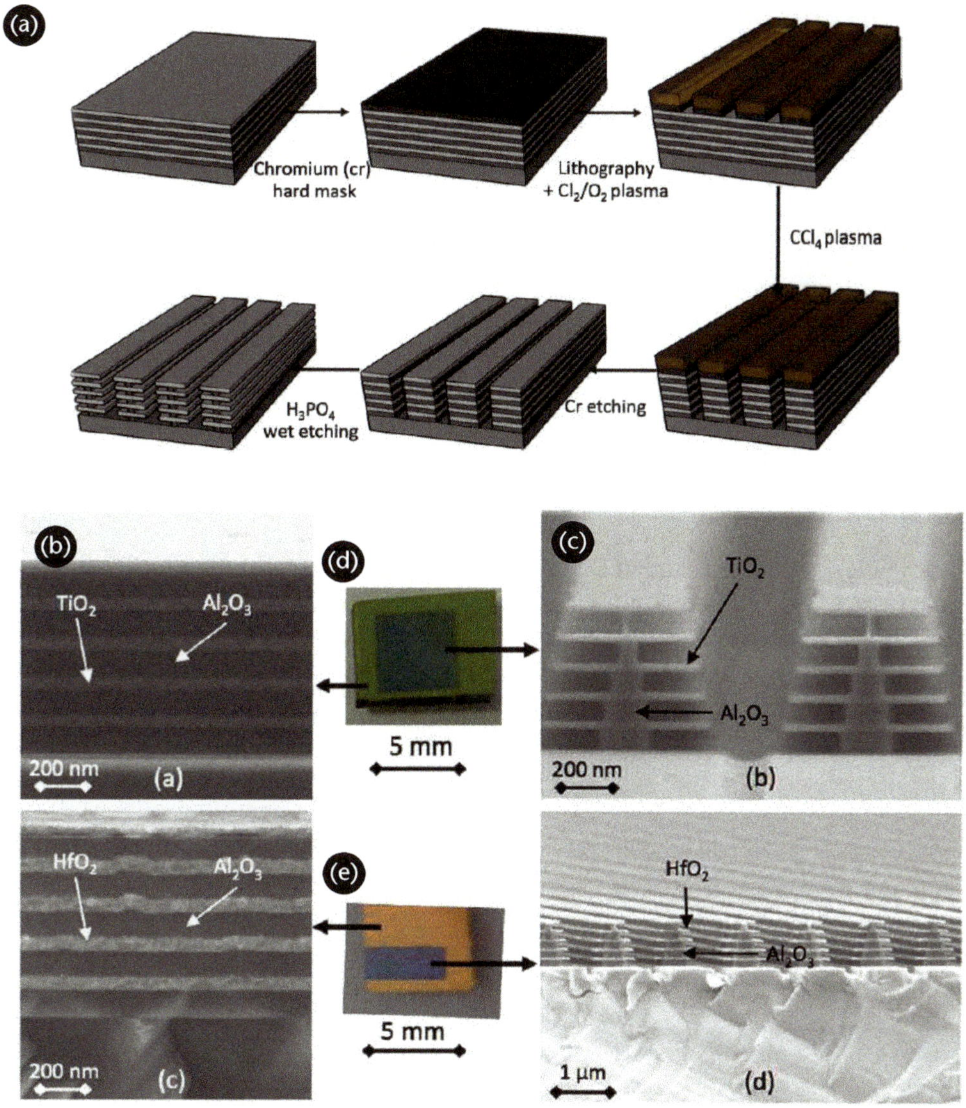

Figure 10.7 Process steps for microfabrication of *Morpho*-like structures, starting from ALD-deposited Bragg mirrors (a). SEM images (b,c) of Bragg mirrors made of TiO_2/Al_2O_3 and HfO_2/Al_2O_3 before (b) and after (c) processing. Photographs of the corresponding samples (d,e). (Reproduced from [15], with permission from IOP Publishing.)

underetched with a hot H_3PO_4 : IPA wet etching solution, whose temperature was carefully chosen to get maximum selectivity between Al_2O_3 and TiO_2. These underetched alumina oxide layers formed the trunk of the Christmas tree structure, while the titania layers formed the branches (Figure 10.7). The resultant structural color effect is striking (Figure 10.7): the planar ALD-deposited Bragg mirrors reflect orange (HfO_2/Al_2O_3) or green (TiO_2/Al_2O_3) colors; whereas, after processing, colors changed to *Morpho* blue in both cases. This changed appearance arose because the effective refractive index of Al_2O_3 layers (underetching introduced air into otherwise solid layers) was precisely tuned by monitoring the underetching.

10.6 Magnetron Sputtering

Magnetron sputtering is a physical vapor deposition (PVD) method for synthesizing thin films on substrates. Thin films of metals, dielectrics, alloys, and various compounds can be deposited with excellent uniformity and reproducibility in terms of both layer thicknesses and compositions. Because the process is scalable to large areas, it is much used in industry for coating glass products, for example. Basically, sputtering is a plasma-assisted process that consists of ejecting ions from a target (cathode) with energetic ions and directing the sputtered ions onto a substrate (anode). Energetic ions bombarding the target are produced by ionization of an inert gas, such as argon, in a vacuum chamber. Increasing the background gas pressure favors collisions of ions with gas atoms, thereby allowing a transition from ballistic to thermalized ion motions. The magnetron serves to confine charged plasma particles to a band close to the target surface, which eventually increases the deposition rate. Reactive direct current (DC) magnetron sputtering can be used to deposit an oxide film from a metal target, thanks to a chemical reaction with a reactive gas (e.g., oxygen or nitrogen) introduced into the chamber. The film stoichiometry can be controlled by tuning the partial pressure and flow of the reactive gas. Radio-frequency (RF) sputtering, however, enables direct deposition of an oxide film from an insulating target; the charge build-up on the latter being avoided by alternating the sign of the anode-cathode bias at high frequency.

Iridescent coatings can be fabricated easily on a glass substrate by magnetron sputtering of periodically alternated layers of titanium oxide and silicon oxide [16,17]. According to the theoretical predictions discussed in the previous chapter, by selecting the layer thicknesses (d_1, d_2) carefully , the iridescence of periodic multilayer coatings can be tuned considerably for a given pair of materials (with refractive indices n_1, n_2). For instance, strongly iridescent coatings can be designed by using a large period ($a = d_1 + d_2$), in combination with a small layer thickness imbalance ($r = d_2/d_1 \ll 1$), whereas weakly iridescent coatings can be designed in just the opposite way [17]. As a proof of principle, both strongly and weakly iridescent bioinspired coatings were deposited on glass by magnetron sputtering and their angle-dependent reflectance spectra were analyzed [17]. In the aforementioned study, alternated deposition of TiO_2 and SiO_2 was performed at room temperature, producing amorphous oxide layers (up to 12 bilayers in total). Titania layers were produced from a titanium target in DC magnetron sputtering mode using a reactive O_2 gas. The flow rate of the O_2 was adjusted to operate beyond the metal-oxide transition of the cathode-voltage-versus-oxygen-flow curve. Silica layers were produced from a silicon oxide target in RF magnetron sputtering mode. Deposition times were selected in order to obtain silica/titania layer thicknesses of 218 nm/9 nm and 10 nm/84 nm for strongly and weakly iridescent multilayer designs, respectively. At normal incidence, the reflectance peaked in the red and blue parts of the visible spectrum, respectively. However, with increasing incidence angle, the reflectance shifted considerably (from red to green) in the former sample, whereas it shifted only slightly (from blue to violet) in the latter sample (Figure 10.8). These iridescence behaviors were reminiscent of the beetles *H. coerulea* and *C. vittata*, respectively, by which the multilayer designs were inspired [16].

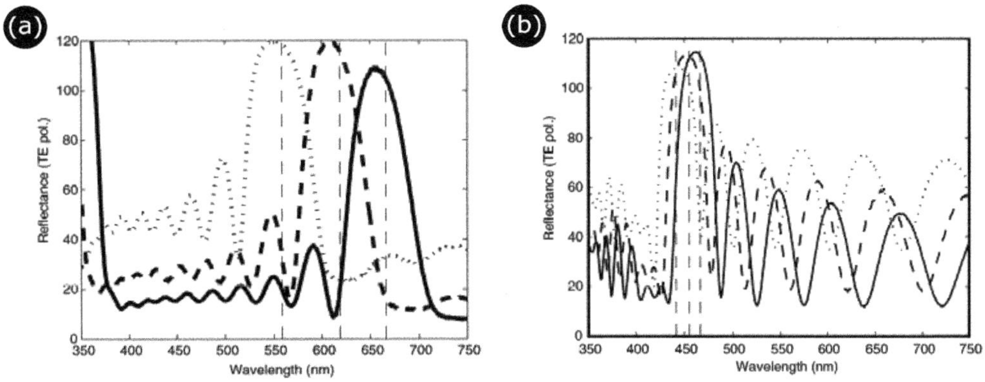

Figure 10.8 Reflectance of strongly (a) and weakly (b) iridescent TiO_2/SiO_2 coatings (12 bilayers) fabricated by magnetron sputtering. Spectra were measured using TE-polarized light in specular configuration at incidence angles equal to 20°(solid curves), 40°(dashed curves), and 60°(dotted curves). (Reproduced from [16], with permission from John Wiley and Sons.)

10.7 Processes Involving Cellulose Nanocrystals

Fabrication methods of bioinspired and biomimetic photonic devices are currently relying more and more on CNCs. CNCs are extracted from plant cell walls (typically wood pulp) by acid hydrolysis. The choice of this material is partly due to its environmentally benign aspects—cellulose is the most abundant biopolymer on earth—and partly due to its chemical properties, which make it simple to employ. One approach is layer-by-layer assembly [18–20]. This technique allowed Tzeng, Hewson, and coworkers to synthesize an analog of the multilayer structure occurring in the beetle *Chrysochroa rajah* (Figure 10.9) [20]. Layer-by-layer deposition like this consists of assembling oppositely charged materials from aqueous solutions. The optical properties of each layer are controlled by combining two materials in defined ratios within each layer. The layer thicknesses are controlled by varying the number of deposition cycles. The higher refractive index layer (labeled "A" in Figure 10.9) consists of 13 bilayers of cationic polyethylenimine (PEI) and anionic vermiculite clay (VMT). The lower refractive index layer (labeled "B" in Figure 10.9) is made of 40 bilayers of cationic colloidal silica (SiO_2) and anionic CNCs.

Evaporation-induced self-assembly (EISA) of CNC thin films is another synthesis method employing CNCs [21–24]. These films were shown to reflect circularly polarized light [23,24]. One of the advantages of this method lies in its simplicity: droplets of an aqueous solution of CNCs are left to dry at ambient temperature and pressure. Droplets with suspended particles are known to form a characteristic, ring-shaped deposit when drying due to water flow carrying the particles to the edge of the droplets [25]. The confinement of CNCs at the edge leads to its self-assembly into a Bouligand structure (giving rise to circularly polarized light reflection) and the displayed colors. Usually, the evaporation process results in an inhomogeneous distribution of the deposited films. By maintaining constant relative humidity (at 50%) and temperature in a custom-built steel chamber, Wilts and coworkers were able to synthesize homogeneous films. With this method, they were also able to minimize defects and formation of several domains by implementing a slow evaporation of the droplets [24].

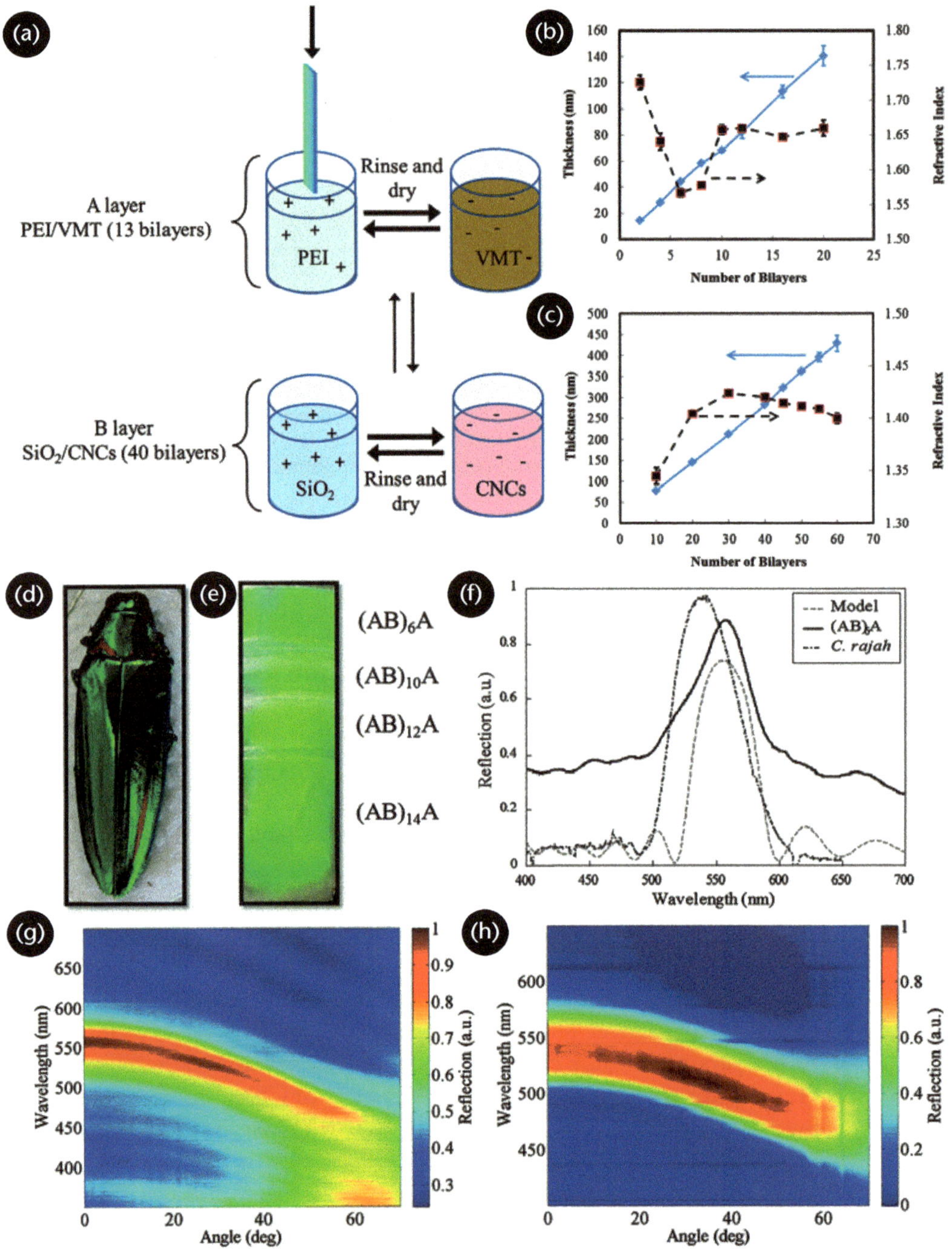

Figure 10.9 Layer-by-layer assembly process applied to the replication of the Bragg mirror occurring in the elytra of the beetle *C. rajah* (a). Both the thicknesses and the refractive indices of layer A (in PEI/VMT) and layer B (in SiO_2/CNC) vary as functions of the number of deposited bilayers. The green area of *C. rajah*'s elytra (d) is well mimicked by the layer-by-layer assembled Bragg mirror (e). Four different multilayers are shown, with 6, 10, 12, and 14 (AB) periods. They all have a final A layer. The reflection spectra measured from the green area of *C. rajah*'s elytra and one synthesized Bragg mirror, as well as a simulated spectrum, are in agreement (f). In addition, the optical behaviors of both the 6-period sample (g) and the beetle's elytra (h) are similar with respect to change in incidence angle. (Reproduced from [20], with permission from the Royal Society of Chemistry.)

References

[1] Mouchet, S. R., et al., "Optical Costs and Benefits of Disorder in Biological Photonic Crystals," *Faraday Discussions*, Vol. 223, 2020, pp. 9–48.

[2] Zhang, G., et al., "Cicada Wings: A Stamp from Nature for Nanoimprint Lithography," *Small*, Vol. 2, No. 12, 2006, pp. 1440–1443.

[3] Xie, G., et al., "The Fabrication of Subwavelength Anti-Reflective Nanostructures Using a Bio-Template," *Nanotechnology*, Vol. 19, No. 9, 2008, 095605.

[4] Chen, Q., et al., "Broadband Moth-Eye Antireflection Coatings Fabricated by Low-Cost Nanoimprinting," *Applied Physics Letters*, Vol. 94, No. 26, 2009, 263118.

[5] Hong, S.-H., et al., "Imprinted Moth-Eye Antireflection Patterns on Glass Substrate," *Electronic Materials Letters*, Vol. 5, No. 1, 2009, pp. 39–42.

[6] Van Opdenbosch, D., et al., "Fabrication of Three-Dimensional Photonic Crystals with Tunable Photonic Properties by Biotemplating," *Photonics and Nanostructures – Fundamentals and Applications*, Vol. 10, No. 4, 2012, pp. 516–522.

[7] Linn, N. C., et al., "Self-Assembled Biomimetic Antireflection Coatings," *Applied Physics Letters*, Vol. 91, No. 10, 2007, 101108.

[8] Sun, C.-H., et al., "Broadband Moth-Eye Antireflection Coatings," *Applied Physics Letters*, Vol. 92, No. 6, 2008, 061112.

[9] Han, Z., et al., "An Ingenious Replica Templated from the Light Trapping Structure in Butterfly Wing Scales," *Nanoscale*, Vol. 5, No. 18, pp. 8500–8506.

[10] Ghazzal, M. N., et al., "Porosity Control and Surface Sensitivity of Titania/Silica Mesoporous Multilayer Coatings: Applications to Optical Bragg Resonance Tuning and Molecular Sensing," *Journal of Materials Chemistry*, Vol. 22, No. 48, 2012, pp. 25302–25310.

[11] Deparis, O., et al., "Theoretical Condition for Transparency in Mesoporous Layered Optical Media: Application to Switching of Hygrochromic Coatings," *Applied Physics Letters*, Vol. 104, No. 2, 2014, 023704.

[12] Ghazzal, M. N., et al., "Tailored Refractive Index of Inorganic Mesoporous Mixed-Oxide Bragg Stacks with Bio-Inspired Hygrochromic Properties," *Journal of Materials Chemistry C*, Vol. 1, 2013, pp. 6202–6209.

[13] Rasson, J., et al., "Vapor Sensing Using a Bio-Inspired Porous Silicon Photonic Crystal," *Materials Today Proceedings*, Vol. 4, No. 4A, 2017, pp. 5006–5012.

[14] Gaillot, D. P., et al., "Composite Organic-Inorganic Butterfly Scales: Production of Photonic Structures with Atomic Layer Deposition," *Physical Review E*, Vol. 78, No. 3, 2008, 031922.

[15] Poncelet, O., et al., "Vapour Sensitivity of an ALD Hierarchical Photonic Structure Inspired by *Morpho*," *Bioinspiration and Biomimetism*, Vol. 11, No. 3, 2016, 036011.

[16] Deparis, O., et al., "Dielectric Multilayer Films Fabricated by Magnetron Sputtering: How Far Can the Iridescence be Tuned?," *Plasma Processes and Polymers*, Vol. 6, 2009, pp. S746–S750.

[17] Deparis, O., et al., "Structurally Tuned Iridescent Surfaces Inspired by Nature," *New Journal of Physics*, Vol. 10, 2009, 013032.

[18] Kurt, P., et al., "Structural Color *via* Layer-by-Layer Deposition: Layered Nanoparticle Arrays with Near-UV and Visible Reflectivity Bands," *Journal of Materials Chemistry*, Vol. 19, No. 47, 2009, pp. 8920–8927.

[19] Nogueira, G. M., et al., "Spray-Layer-by-Layer Assembly Can More Rapidly Produce Optical-Quality Multistack Heterostructures," *Langmuir*, Vol. 27, No. 12, 2011, pp. 7860–7867.

[20] Tzeng, P., et al., "Bio-Inspired Iridescent Layer-by-Layer Assembled Cellulose Nanocrystal Bragg Stacks," *Journal of Materials Chemistry C*, Vol. 3, No. 17, 2015, pp. 4260–4264.

[21] Dumanli, A. G., et al., "Digital Color in Cellulose Nanocrystal Films," *ACS Applied Materials & Interfaces*, Vol. 6, No. 15, 2014, pp. 12302–12306.

[22] Mu, X. Y., and D. G. Gray, "Droplets of Cellulose Nanocrystal Suspensions on Drying Give Iridescent 3-D 'Coffee-Stain' Rings," *Cellulose*, Vol. 22, No. 2, 2015, pp. 1103–1107.

[23] Hewson, D., P. Vukusic, and S. J. Eichhorn, "Reflection of Circularly Polarized Light and the Effect of Particle Distribution on Circular Dichroismin Evaporation Induced Self-Assembled Cellulose Nanocrystal Thin Films," *AIP Advances*, Vol. 7, No. 6, 2017, 065308.

[24] Wilts, B., et al., "Invited Article: Chiral Optics of Helicoidal Cellulose Nanocrystal Films," *APL Photonics*, Vol. 2, No. 4, 2017, 040801.

[25] Deegan, D., et al., "Capillary Flow as the Cause of Ring Stains from Dried Liquid Drops," *Nature*, Vol. 389, No. 6653, 1997, pp. 827–829.

CHAPTER 11

Ideas at Work: Bioinspired Hygrochromic Devices

Most of the liquid-induced color changes observed in insects and discussed in this book (see Chapter 5) involve a porous structure that can be described as an effective multilayer system (i.e., Bragg stack). Depending on the presence or absence of liquid in the pores, the pore filling fraction, and the refractive indices of each layer, the Bragg stack may be either highly reflective or transparent, as illustrated nicely, for example, in Vigneron's study of the switchable reflector found in the tortoise beetle *Charidotella egregia* [1]. In Chapter 9, a bioinspired design concept was presented for porous multilayer systems that are capable of switching between a reflective (colored) state and a transparent state [2]. After describing the design process and the tools used to develop bioinspired optical applications (Chapters 9 and 10), we will describe some ideas at work through selected examples relating to bioinspired hygrochromic devices. First, we will briefly review some color-tunable porous multilayer coatings. We will then describe an original implementation of the transparency theory for the synthesis of porous dielectric coatings that are able to switch, reversibly, from transparency to colored visual appearances, upon contact with a liquid. Other bioinspired devices will be reviewed in the next chapter.

Natural photonic structures that change color upon contact with liquids are a source of inspiration for designing hygrochromic materials. In the examples described in Chapter 5, such a property was generally achieved thanks to two characteristics of the structure: porosity and multilayer assembly. The first of these allows liquids to be accommodated into the structure and, thereby, changes the effective index, whereas the second of these produces structural color. As an option, the porosity of the layers can be such that, for a certain distribution of the liquid inside the structure, the effective refractive index is homogenized. This means that the multilayer becomes transparent [2], as described theoretically in Chapter 9 (or, if it attenuates light, e.g., through intrinsic material absorption, the transmission loss has no wavelength-selective contribution due to multiple scattering at the layer interfaces).

Ozin's group reported on color-tunable spin-coated multilayer stacks of mesoporous TiO_2 and mesoporous SiO_2 that exhibited Bragg wavelength shifts upon infiltration with alcohols and alkanes into their pores [3]. Pore diameters were estimated to be around 9 nm and 4 nm, respectively. The refractive index selectivity and sensitivity were found to depend on the relative proportion (layer thickness ratio) of the two mesoporous oxides in the Bragg stack (Figure 11.1). This behavior was related to different hydrophilicity properties of the oxides.

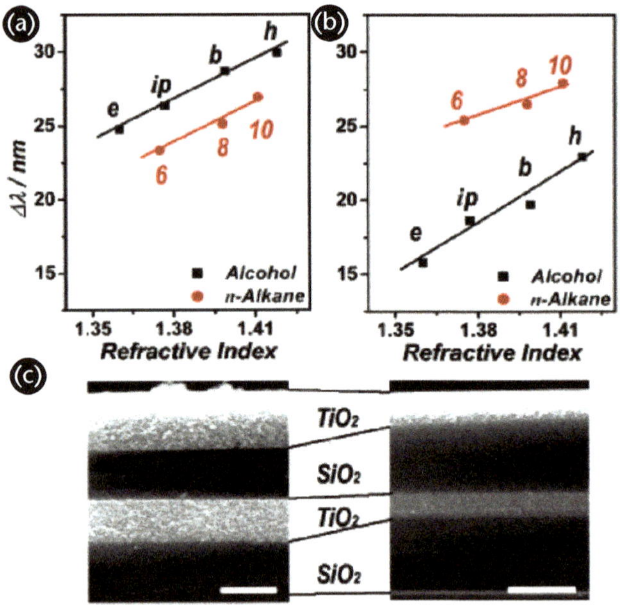

Figure 11.1 Bragg wavelength shift (a,b) $\Delta\lambda$ from two different four-layer mesoporous TiO_2/SiO_2 stacks in a series of alcohols (e: ethanol, ip: isopropanol, b: 1-butanol, h: 1-hexanol) and a series of n-alkanes (6: hexane, 8: octane, 10: decane) as a function of the analyte's refractive index. Cross-section SEM images of both stacks (c). Scale bars: 200 nm (c). (Reproduced from [3], with permission from the American Chemical Society.)

In another work, Ozin's group added chemical functionality to Bragg stacks by using smectite clay Laponite for one material of the bilayer and dense, or mesoporous, TiO_2 for the other material [4]. The clay layers were subject to swelling and ion exchange reactions in such a way that clay-analyte interactions translated into a Bragg wavelength shift (i.e., color change). Mesoporous SiO_2 was used as well, in combination with smectite clay Laponite and the layer thicknesses of both clay and metal oxide were tuned across a wide range [5].

The use of colloidal suspensions of mesoporous silica or titania nanoparticles is another approach to synthesizing Bragg stacks via spincoating [6]. The method allows a hierarchical porosity to be created in the layers by combining mesoporosity (intrinsic to nanoparticles) and textural porosity (caused by the packing of the individual nanoparticles).

The control of the sorption of molecules by functionalization of the pores with organic groups, as demonstrated by Fuertes and coworkers, allows the optical properties of mesoporous Bragg stacks to be tailored and selectivity toward small-sized molecules to be achieved [7].

The Hercules beetle (*Dynastes hercules*), the hygrochromic property [8] of which was discussed in Chapter 5, inspired Kim and coworkers with the fabrication of a humidity sensor [9]. A colloidal templating method was used to mimic the three-dimensional porous structure of the beetle's cuticle. The polymer structure obtained was an inverse opal film made of poly (ethylene glycol)-diacrylate (PEG-DA), which changed color from blue-green to red, as relative humidity increased from 25% to

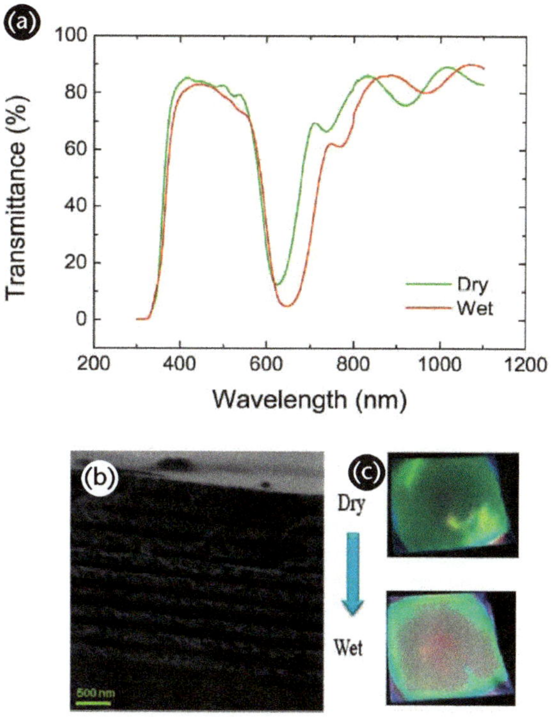

Figure 11.2 Transmittance spectra (a) of mesoporous TiO_2/SiO_2 Bragg stack in dry and wet states. Cross-sectional SEM images (b) of the periodic multilayer stack (10 bilayers). Photographs (c) of the sample in the dry state and the wet state. (Reproduced from [10], with permission from the Royal Society of Chemistry.)

98%, due to the filling of the pores and the swelling of the polymer. Other devices sensible to hygroscopy will be presented in the following chapter (Section 12.2.1 related to vapor-induced color changes in bioinspired devices).

Ghazzal and coworkers synthesized periodic multilayer stacks by means of a sol-gel preparation method, in which surfactants were used to produce porosity and a spin-coating deposition method. The materials used were mesoporous dielectric oxides (i.e., TiO_2 and SiO_2) on a glass substrate. They were also able to demonstrate the hygrochromic response of the coatings (Figure 11.2) [10]. Accessibility of the pores across the whole multilayer stack was checked by infiltration of Rhodamine 6G fluorescent dye from the top. Homogeneous distribution of the dye across all the layers was observed in cross section by confocal microscopy. By examining the Bragg peak change during water soaking in samples with either different or similar pore sizes in TiO_2 and SiO_2 layers, it was inferred that water gradually filled all the pores of the former sample, whereas it could not infiltrate the pores of the latter sample (Figure 11.2). From this experiment, the researchers concluded that capillary attraction induced by different pore sizes between adjacent layers plays a key role in liquid infiltration throughout the entire mesoporous dielectric structure.

By using mixed oxides in the high-refractive index layers ($x TiO_2(1-x)Al_2O_3$ or $x TiO_2(1-x)SiO_2$) and by adjusting the porosities, Ghazzal and coworkers managed to match effective refractive indices of mixed oxide layers and SiO_2 layers. This

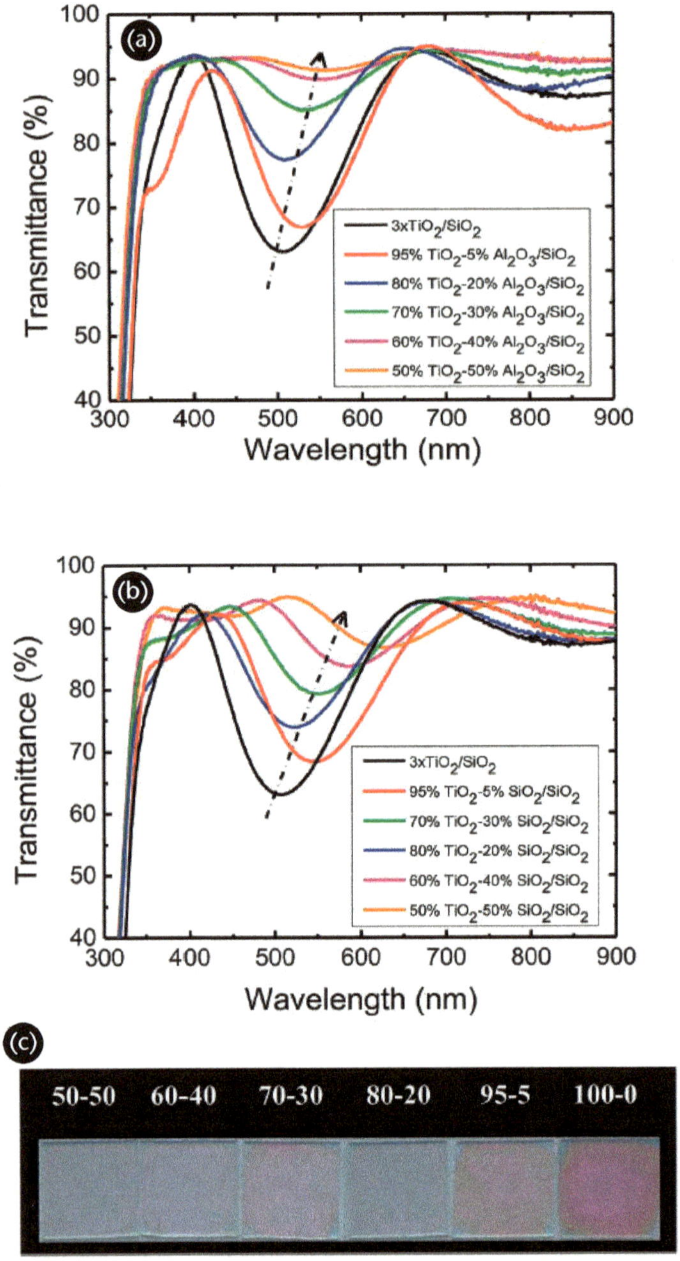

Figure 11.3 Transmittance spectra (a,b) at normal incidence of mesoporous multilayer coatings in which various ratios of alumina (a) and silica (b) were added to the high-refractive-index titania. Photographs of samples with various ratios of alumina mixed with titania in dry state (c). The increase in alumina content (from right to left) reduces the contrast of effective refractive index between layers. (Reproduced from [11], with permission from the Royal Society of Chemistry.)

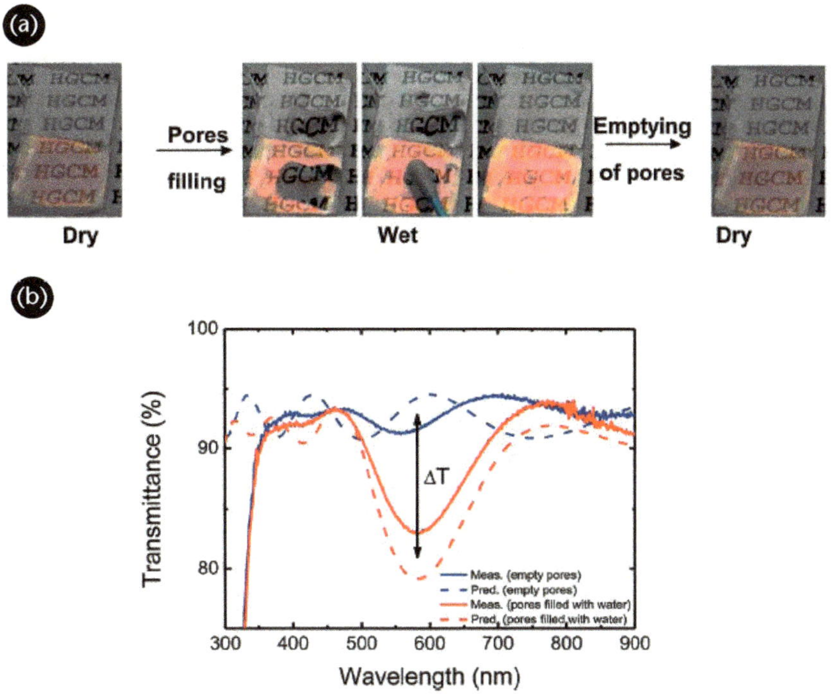

Figure 11.4 Hygrochromic coating reversely switches from transparent to colored states upon contact with water (a). Transmittance spectra upon contact with water (pores filled with water) and in the dry state (empty pores) (b). Solid curves: measurements, dotted curves: theoretical predictions. The coating consists of three bilayers of $0.5TiO_2 0.5Al_2O_3$ and SiO_2 on a glass substrate. (Reproduced from [11], with permission from the Royal Society of Chemistry.)

adjustment process allowed them to reach transparency [11]. High transmission in the visible spectrum close to that of the bare glass substrate was achieved for $x = 0.5$ in both mixed oxide systems (Figure 11.3).

In accordance with theoretical predictions [2], these multilayer coatings were able to switch, reversibly, from transparency to structural color upon water absorption (Figure 11.4).

References

[1] Vigneron, J.-P., et al., "Switchable Reflector in the Panamanian Tortoise Beetle *Charidotella egregia* (Chrysomelidae: Cassidinae)," *Physical Review E*, Vol. 76, No. 3, 2007, 031907.

[2] Deparis, O., et al., "Theoretical Condition for Transparency in Mesoporous Layered Optical Media: Application to Switching of Hygrochromic Coatings," *Applied Physics Letters*, Vol. 104, No. 2, 2014, 023704.

[3] Choi, S. Y., et al., "Mesoporous Bragg Stack Color Tunable Sensors," *Nano Letters*, Vol. 6, No. 11, 2006, pp. 2456–2461.

[4] Lotsch, B. V., and G. A. Ozin, "Clay Bragg Stack Optical Sensors," *Advanced Materials*, Vol. 20, No. 21, 2008, pp. 4079–4084.

[5] Lotsch, B. V., and G. A. Ozin, "Photonic Clays: A New Family of Functional 1D Photonic Crystals," *ACS Nano*, Vol. 2, No. 10, 2008, pp. 2065–2074.

[6] Kobler, J., et al., "Vapor-Sensitive Bragg Mirrors and Optical Isotherms from Mesoporous Nanoparticle Suspensions," *ACS Nano*, Vol. 3, No. 7, 2009, pp. 1669–1676.

[7] Fuertes M. C., et al., "Photonic Crystals from Ordered Mesoporous Thin-Film Functional Building Blocks," *Advanced Functional Materials*, Vol. 17, No. 8, 2007, pp. 1247–1254.

[8] Rassart, M., et al., "Diffractive Hygrochromic Effect in the Cuticle of the Hercules Beetle *Dynastes hercules*," *New Journal of Physics*, Vol. 10, 2008, 033014.

[9] Kim, J. H., et al., "Biologically Inspired Humidity Sensor Based on Three-Dimensional Photonic Crystals," *Applied Physics Letters*, Vol. 97, No. 10, 2010, 103701.

[10] Ghazzal, M. N., et al., "Porosity Control and Surface Sensitivity of Titania/Silica Mesoporous Multilayer Coatings: Applications to Optical Bragg Resonance Tuning and Molecular Sensing," *Journal of Materials Chemistry*, Vol. 22, No. 48, 2012, pp. 25302–25310.

[11] Ghazzal, M. N., et al., "Tailored Refractive Index of Inorganic Mesoporous Mixed-Oxide Bragg Stacks with Bio-Inspired Hygrochromic Optical Properties," *Journal of Materials Chemistry C*, Vol. 1, No. 39, 2013, pp. 6202–6209.

CHAPTER 12
Bioinspired Applications in Photonics

More than ever in this book, the word "bioinspiration" makes sense while making a selection among various reported applications of natural photonics. Indeed, the degree of proximity between a natural photonic structure and a bioinspired device may vary considerably from one application to another. Antireflective coatings, which are first discussed in this chapter, belong to the class of basic photonic devices. Devices in this class benefit the most from direct transfer (i.e., templating) of the structure to real-world industrial applications, for instance, as cover layers to reduce front glass reflection in solar cells. The case of nipple arrays, as found on moth eyes, is well-known. Moth eye surface structures and similar ones (such as cicada wings) have been reproduced by different technological routes, such as nanoimprinting and self-assembly, both of which will be discussed hereafter. Surface textures are also important for sensing. For instance, optical gas and vapor sensors have been inspired by the gas- and vapor-induced color changes observed in several natural organisms. Similarly, the optical detection of trace chemicals at extremely low concentrations by surface-enhanced Raman scattering relies on noble-metal-coated textured substrates, which are complex and expensive to produce. Gold-coated butterfly wing surfaces could be an interesting alternative source of such substrates, as explained below. Harvesting solar energy is a grand challenge today and necessary for the development of sustainable, highly efficient energy conversion processes, such as those involved in photovoltaic cells, thermal solar collectors, photocatalysis, or artificial photosynthesis. As examples, we will discuss the applications of bioinspired photonic structures as solar concentrators, light-trapping devices, and photocatalysts. Harvesting as much light as possible into an absorber is a challenge; letting it go out from an emitter is another, although with similar principles: one cannot avoid using a textured interface to couple light in or out. We will show how the textured lantern of a firefly has inspired a novel, efficient light extraction device that helps to increase radiance of commercial LEDs. The bioinspired device simply comprises a patterned cover layer that is glued on top of the LEDs. This solution inspired from nature, hence, contributes to saving on an LED's electrical consumption and, in a kind of virtuous circle, saving natural resources. Since its introduction in 1970, the Universal Product Code, which adopted the barcode symbology, has been used ubiquitously in industry and commerce. Optics, as in many other areas, has contributed to increasing the economical benefits from this technology further, thanks to optical technologies enabling the automatic reading of barcodes, including, recently, the popular QR codes. Together with holograms on banknotes or passports, such physical codes perform security functions at different levels and for different purposes. This chapter will end with a brief presentation of anticounterfeiting systems that have been inspired by the polarization properties of certain butterfly scales.

In general, one of the main advantages of natural photonic structures is their robust optical response that is known for not being greatly affected by defects [1]. This robustness against defects is a clear advantage from bioinspired, technological, and industrial points of view. However, many other devices are currently under development through different approaches. Bioinspired solutions are, hence, in competition with them. For instance, optical gas sensors are being developed through various technologies, including localized plasmon resonances, Bragg diffraction, or Fabry-Pérot interference [2–4]. Depending on the applications, the main challenges in development are reaching a higher selectivity, a higher sensitivity, a better resolution, a shorter response time, a larger measurement dynamics, a longer stability, a smaller size, a lower energy consumption, an easier maintenance, a higher scalability, a lower cost of production and use, or a higher commercial viability. For each application, the suggested solution needs to offer a better trade-off between these requirements than the competing solutions.

12.1 Antireflective Coatings

Moth eye nanostructures [5, 6] (see Chapter 4) were among the natural devices that were mimicked early in order to produce antireflection coatings using various fabrication methods [7]. These non-close packed hexagonal arrays of surface protuberances (known as nipples) are found similarly on, for example, cicada and moth wings, where they take the more elongated form of truncated cones [8–18]. Such gradual refractive index depth profiles have inspired the design and fabrication of antireflective coatings with superior performances in terms of broad spectral bandwidth and angular independence, in comparison with planar multilayer designs. Often, the antireflection properties of these structures are combined with (super)hydrophobic and self-cleaning properties [19, 20], which is of great interest for many industrial applications.

This section describes a few examples of bioinspired antireflective coatings among many examples of synthesized devices. Applications of antireflective devices in optics are numerous. For instance, antireflective layers are ubiquitous elements of solar cells. In this field, simple fabrication from reusable biological templates could be an interesting alternative for low-cost mass production. Antireflective layers are also important for increasing performance of thermochromic smart windows that are used to mitigate heat loss from building envelopes. It has been predicted numerically that a single layer of TiO_2 nano-cones on top of a thermochromic ViO_2 layer covering a glass substrate can increase the luminous transmittance by 39% and the solar modulation between hot (metal) and cold (insulator) states of vanadium dioxide phase transition by 72%, compared with a ViO_2 window without an antireflective layer [21].

12.1.1 Self-Assembled Coatings Inspired by Moth Eyes

Self-assembly of spherical particles results in a hexagonal compact arrangement that approximately resembles the surface patterns of a moth's eye. Linn and coauthors cleverly exploited this property to synthesize bioinspired coatings from self-assembled silica particles [22].

Conventional self-assembly, however, produces close-packed (i.e., compact) hexagonal arrays, whereas mimicking moth eye structures implies being able to produce non-close packed arrays. In the work cited above [22], this was achieved by shear aligning concentrated colloidal suspensions of silica particles (with a diameter of 360 nm) that had been deposited by spin-coating. The biomimetic silica moth-eye template was used to cast a polymer (namely PDMS) mold which, once peeled off, was then pressed on a UV-curable monomer (i.e., ethoxylated trimethylolpropane triacrylate (ETPTA)) film that had been deposited on glass. The depth of the replicated nipples was controlled by adjusting the height of the protruded portions of the silica spheres through reactive ion etching (RIE) before PDMS casting. A replicated nipple array with 180-nm high hemispherical caps (produced with 45-s RIE duration) showed excellent antireflection performance with a reflectance lower than 0.5% over the visible range (Figure 12.1). Simulations showed that, at wavelengths longer than the distance between neighboring nipples, a lower reflectance is achieved with non-close-packed structures than with close-packed ones (Figure 12.1). This confirms the fitness of the natural moth-eye design for synthesizing highly efficient antireflective devices.

Soon thereafter, these bioinspired moth-eye templates were used as masks for etching silicon wafers [23]. Silicon wafer reflectance (approximately 40% in visible-NIR range) was reduced to a level lower than 5% thanks to a silicon nipple array that had been formed by RIE with SF_6 (Figure 12.1); the array acted as a thin antireflective layer. Because this bottom-up approach for producing subwavelength nipple arrays directly on silicon is compatible with standard microfabrication, it could be applied to low-cost fabrication of antireflective layers on silicon solar cells. The same templating technique has been used to produce wafer-scale antireflective coatings on gallium antimonide (GaSb) susbtrates for thermophotovoltaic cells [24]. Drastic reduction of reflectance was obtained that was similar to those on silicon wafers.

Using such a soft-lithography-like process, Jiang's group also explored the combination of antireflection and superhydrophobicity by transferring the PDMS mold to a UV-curable perfluoroacrylate polymer with low surface energy [19]. The water contact angle was increased from approximately 106° (on a flat surface) to approximately 128° (on a hemispherical nipple array templated from 360-nm silica spheres), thereby enhancing the hydrophobicity of the fluoropolymer surface while simultaneously reducing its reflection.

In the process of creating the moth-eye template, an oxygen plasma step is used, after the spin-coating step, in order to remove partially the polymer matrix in which silica particles are embedded [23]. Moreover, a thin (100 nm) polymer wetting layer is present between the shear-aligned silica colloidal monolayer and the substrate, which acts as a support for the silica particles [19]. By increasing the oxygen plasma etching time, disorder can be introduced in the colloidal template while keeping the same density of particles (Figure 12.1). In keeping with the effective medium theory, which predicts that reflection should not be changed as long as the particle (i.e., nipple) density is unchanged, the introduced disorder was shown not to affect antireflection performance in a significant manner (Figure 12.1).

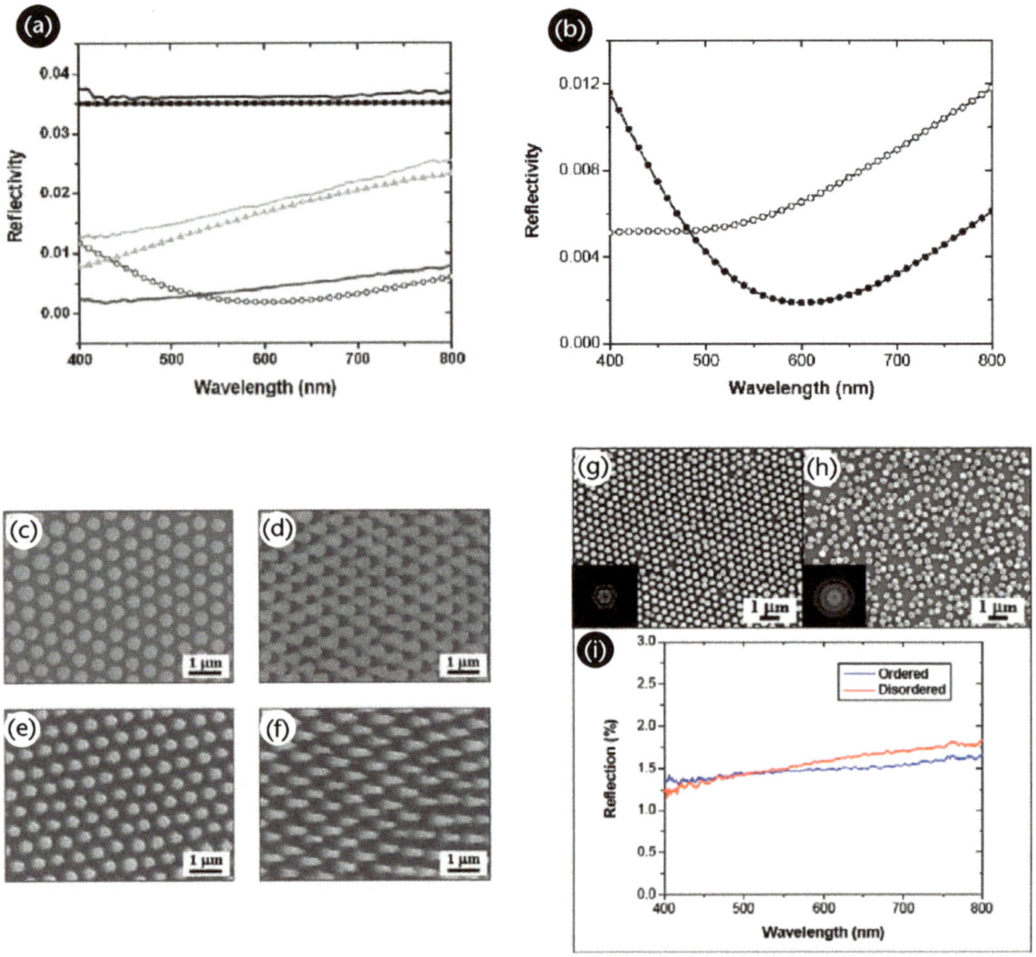

Figure 12.1 Reflectance spectra (a) at normal incidence (measured: solid lines, calculated: dotted lines) of flat PETPTA substrate (black curves) and replicated nipple arrays with 110-nm spherical cap or 180-nm hemispherical cap (light gray and dark gray curves, respectively). Reflectance spectra (b) calculated for non-close packed (filled circles) and close-packed (open circles) hemispherical nipple arrays (with a diameter of 360 nm). SEM images of spin-coated non-close packed monolayer of 360-nm silica spheres (c,d) and of silicon nipple arrays after 6-min (e) and 9-min (f) SF_6 etch using the monolayer as a mask. SEM images of ordered (g) and disordered (h) self-assembled, colloidal SiO_2 monolayer templates. Measured reflectance spectra (i) of corresponding nipple arrays molded in UV-curable perfluoropolyether tetraacrylate. (Reproduced from [22] (a,b), from [23] (c–f) and from [19] (g–i), with permission from AIP Publishing.)

Another reason for the interest in replicating moth-eye patterns on polymers with good elastomeric properties is the potential this offers to create efficient antireflective coatings on curved surfaces such as lenses.

12.1.2 Nanoimprinted Coatings from Cicada Wing Templates

Beside moths' eyes, cicada wings, with their arrays of nanometer-scale high-aspect-ratio pillars that cover the wing surface on both sides, are another interesting template for the fabrication of bioinspired antireflective devices. Zhang and coworkers have used hot embossing to create inverse replicas of the wings of *Cryptympana atrata* [25,26]. Contrary to the works discussed previously on mimicking moth eye patterns with self-assembled silica particles, the approach here is to use the biological structures directly as the template. The microscopic structures of the wings consist of ordered hexagonal arrays of 400-nm-high pillars, which are spaced about 190 nm apart. The wing, which acts as a rigid stamp courtesy of the good thermomechanical properties of its chitinous material, is heated and pressed on poly(methyl metacrylate) (PMMA) [25].

This technique has all the advantages of nanoimprint lithography—low cost, high throughput, and high spatial resolution—while using a readily available biological stamp. The negative structure of the stamp consists of an array of wells in PMMA with very similar spacing and depth to the original pillars. This inverted replica can then be used for various purposes. In a proof-of-principle experiment, Zhang and coworkers used it as a mask for etching silicon [25]. After removal of PMMA, the nanostructured silicon surface exhibited antireflective ability, as confirmed by the darker appearance of the surface, as observed by optical microscopy [25]. A noninverted replica of the cicada wing in PMMA (Figure 12.2) was obtained later by the same team using a slightly adapted process [26]. To achieve this, they used an intermediate inverted replica, obtained by thermally evaporating gold onto the wing. This negative gold replica was then used to cast PMMA, which was finally peeled off from the gold mold (Figure 12.2(a)). Mechanical stability of the PMMA replica was found to be affected by the heating step that was used to peel off the film. When it was heated only below 60°C, severe structural deformation was observed in the replicate (Figure 12.2(c)). The reflectance of the cicada wing replicated PMMA film was reduced from circa 6% to circa 2% over the visible-NIR range because of the nipple array (Figure 12.2(d)). The use of PMMA offers the advantage of producing flexible soft films, which can be put on curved optics, to serve as antireflective layers.

12.2 Sensing

In addition to the bioinspired hygrochromic devices that are sensitive to water and liquids in general (described in Chapter 11), several optical sensors have been inspired by natural organisms [27]. They are typically sensitive to chemical and biochemical entities. Such sensors should be able to distinguish between various substances and to give an indication of their concentrations within a limited time of response. Often, the quantification of these concentrations can be performed via spectrometry, colorimetry, and image processing. Many optical sensors are based on human vision, which is one of humankind's most accurate senses, in order to highlight changes that cannot be perceived by other human senses (e.g., smell). Detection with the naked eye is straightforward for humans. Besides the visual signal, remote sensing capabilities may be added to perform measurements for analysis. Such sensing does not require electronics or a power supply to operate in situ, which is advantageous in some environments, such as explosive or flammable atmospheres.

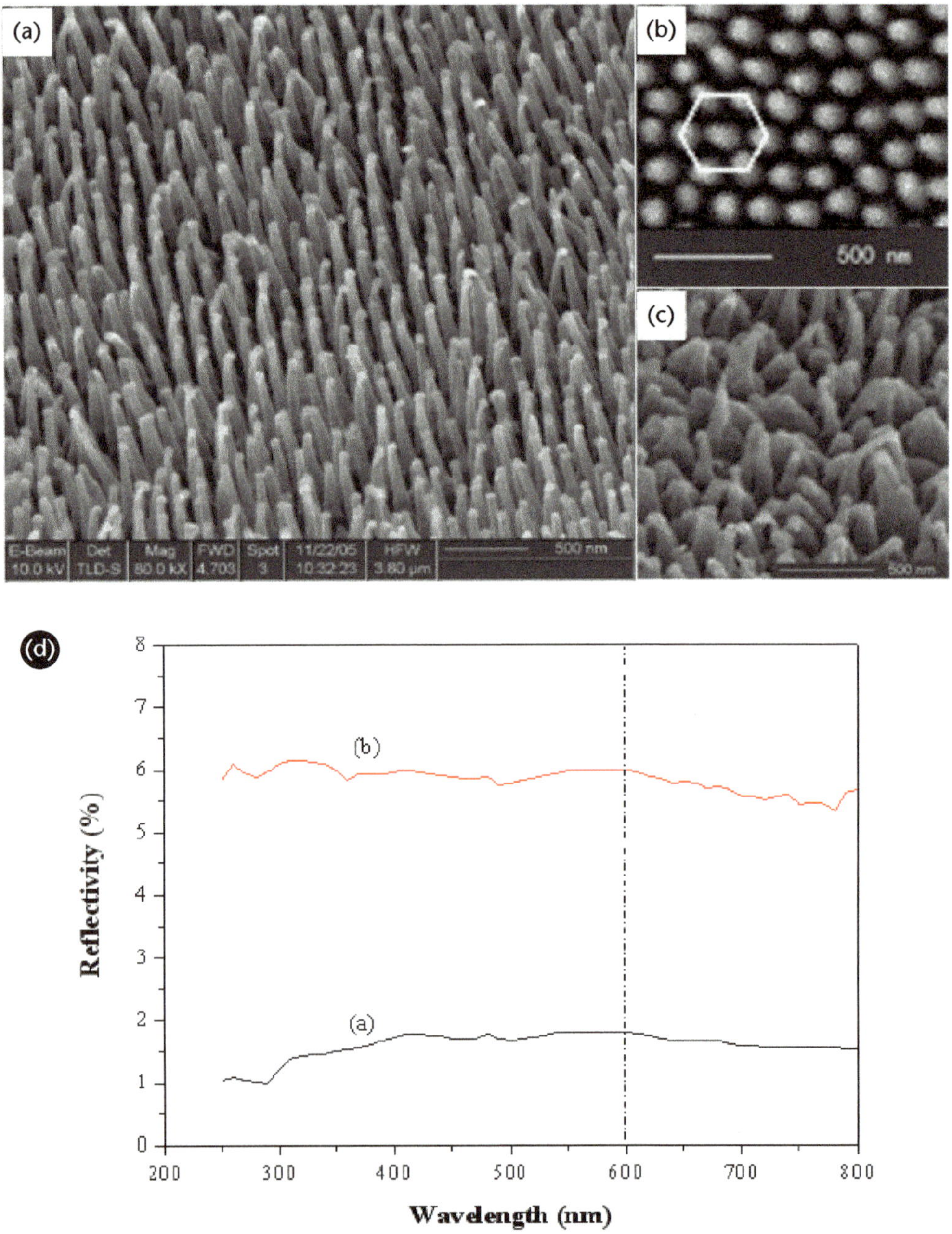

Figure 12.2 SEM images (a–c) of replicated PMMA film with nipple arrays; images obtained by casting the polymer into a gold mold with the inverted pattern: large-scale view (a) and higher magnification top view showing a hexagonal pattern (b). Images (a,b) were taken after the PMMA film was heated at 90°C for 30 min, whereas image (c) was taken after it was heated at 60°C for 30 min. Reflectance spectra (d) of replicated PMMA with nipple arrays (a) and nonpatterned, flat PMMA film. (Reproduced from [26], with permission from IOP Publishing.)

12.2.1 Gas and Vapor Sensing

Due to their response to gases and vapors [28, 29], the photonic Christmas tree structures occurring in the wings of *Morpho* butterflies have inspired designs of gas and vapor sensors. A very accurate analoguous device was fabricated by Potyrailo and coworkers [30]. It was produced by electron beam lithography through the patterning of a periodic stack made of 100-nm thick layers of PMMA and a copolymer of methyl methacrylate (MMA) and methacrylic acid (MAA). As described in Chapter 6, a polarity gradient was found along the height of *Morpho* butterflies' structures, leading to their selective response to vapors. This gradient was mimicked with chemical functionalization of the structure, relying on the vapor-phase deposition of silanes. The latter were consecutively exposed to an electron beam, which gradually removed the silane coating from the structure top. The device was shown to detect selectively and quantify vapors of benzene, methyl ethyl ketone, acetonitrile, methanol, and water in mixtures. Vapor sensors inspired by *Morpho* butterfly scales were also synthesized by Atomic Layer Deposition (ALD) of Bragg mirrors made of metal oxide layers (i.e., Al_2O_3, TiO_2, and HfO_2) that were sculpted by dry and wet etching (Figure 10.7, Chapter 10) [31]. They displayed responses to vapors of ethanol and isopropanol.

In addition to the gas and vapor sensors arising from the modified insect wings mentioned in Chapter 6 and humidity sensors inspired by the structures occurring in the elytra of the beetle *Dynastes hercules* [32] described in Chapter 11, the paradise whiptail fish *Pentapodus paradiseus* served as a base for the development of a sensor comprising periodic organic/inorganic hybrid multilayers [34]. Its iridescent bands change, reversibly, from blue to red, following hormone release [33] (Figure 6.9, Chapter 6). The bioinspired sensor was made of TiO_2 layers and poly(2-hydroxyethyl methacrylate-co-glycidyl methacrylate) (PHEMA-co-PGMA) layers. While the color change of the fish's bands is not humidity-induced, the polymer layer of the bioinspired device does swell upon contact with water vapor, giving rise to a change in the displayed color.

Vapor-sensitive, colloidal photonic crystals [35, 36] were described as being inspired by the porous multilayer located in the scales of the longhorn beetle *Tmesisternus isabellae* [37] (Figure 5.8, Chapter 8). These mesoporous structures were fabricated in various ways, such as, using inkjet printing of SiO_2 nanoparticles [35], which assembled as the ink dried, or using a microchannel assisted dip-coating and a rotational drying method [36]. The interstices allow water vapor to condensate, leading to changes in the refractive index contrast and the optical response. Moreover, bioinspired silk-fibroin inverse opals have been produced [38]. In this case, condensed water changes the refractive index contrast and swells the structure. Both phenomena lead to color changes.

12.2.2 Butterfly Wings as SERS Substrates

SERS is known to be a very efficient process for trace chemical detection at nanomolar concentration levels, even enabling single-molecule spectroscopy. Localized surface plasmon resonances associated with rough metallic surface or substrates covered with metal nanoparticles are responsible for huge electromagnetic field enhancement, which translates into a million-fold increase in the Raman signal of molecules grafted on the surface. Searching for the most efficient substrates has been the objective of many studies in this field [39].

Ordered papillary structures covering cicada wings not only provide excellent antireflection devices as discussed in Chapter 4 but their roughness is also comparable with those of typical SERS substrates. Stoddart and coworkers observed that cicada wings of *Cicadetta celis* covered with silver and gold exhibited up to 10^6 SERS enhancement factors for thiophenol and rhodamine 3G molecules, respectively [40]. Butterfly wings have, therefore, been considered as a low-cost alternative to conventional SERS substrates. Later, Garrett and coworkers used both naturally derived (butterfly wing) and biomimetic SERS substrates for biochemical analysis [41]. The selected butterfly species (*Graphium weiskei*) had nano-cone arrays on its wings, whereas the biomimetic substrate was fabricated in PMMA from self-assembly of polystyrene spheres, which served as a mask for reactive ion etching. Avidin/biotin model immunoassays were performed both with directly metal (gold and silver) coated butterfly wings and with biomimetic substrates. Comparable SERS enhancement factors were found in both cases (results for the Au-coated butterfly wing are shown in Figure 12.3).

A recent study highlighted the complexity of the interaction between the photonic structure of a butterfly wing used as the SERS substrate and the plasmonic response of metal nanoparticles deposited on such a substrate [42]. First, the intricate surface relief (ridges, cross ribs, etc.) of the butterfly wing was shown to promote aggregation of nanoparticles, which is a key ingredient for localized electromagnetic field enhancement (plasmonic hot spots). Second, coherent optical wave scattering occurring within the wing's photonic structure, which is responsible for species-specific structural color, was hypothesized to provide wavelength-selective SERS enhancement due to constructive interference between the incident and Raman scattered fields. The increase in the SERS signal that was observed when the excitation wavelength matched the reflectance peak wavelength (wing's structural color) was large and nonlinear. It closely tended to support this hypothesis of coupling between the photonic structure and the plasmonic response at the origin of SERS.

12.3 Light Harvesting

Harvesting solar light energy is currently of high importance for our modern societies. Nature, however, has evolved many devices that provide vital functions to organisms, ranging from plants to insects and birds. Observing nature has taught us some lessons in this matter (Chapter 7). Several solar concentrators and light-trapping devices have, thus, been designed and fabricated through a bioinspired approach.

12.3.1 Solar Concentrators

Basking refers to the V-shaped rest posture adopted by butterflies in cloudy conditions, which allows them to warm their flight muscles quickly. White butterflies (*Pieris genus*) are quicker than others to take flight after basking because of the high reflectance of their wings, which gives them excellent ability to concentrate solar

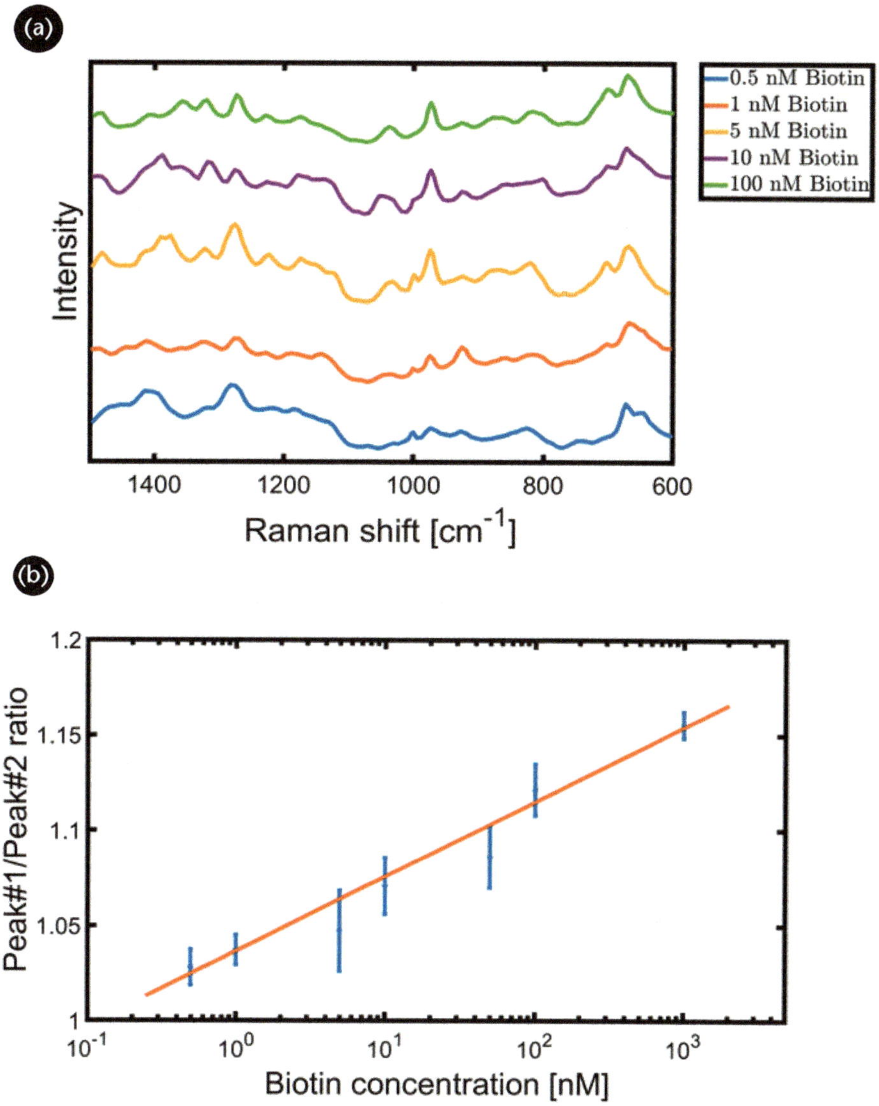

Figure 12.3 SERS spectra (a) at various biotin concentrations (peak 1 at 975 cm^{-1} and peak 2 at 1000 cm^{-1}), indicating avidin-biotin complex formation and calibration curve (b) of Au-coated wing biosensor, showing Raman peak ratio as a function of biotin concentration in buffer solution. (Data from [41].)

light through reflections at the apex of the V-shape where the butterfly's thorax is located. This warming mechanism has been studied by Shanks and coworkers who aimed at evaluating the potential of white butterfly wings as low-weight coatings for solar photovoltaic cells [43]. They first set up an experiment to evaluate the heating efficiency by placing detached wings of the large white butterfly (*Pieris brassicae*) in a V-shape configuration. Then they measured the temperature at the apex with an infrared camera at variable wing opening angles (Figure 12.4). An optimal angle of 17° (from the normal direction, i.e., closed wings) was found to increase the temperature by 7.3°C, thereby confirming the effect of basking. Next, they selected the large white butterfly for its high wing reflectance (R = 78.6%) and used its

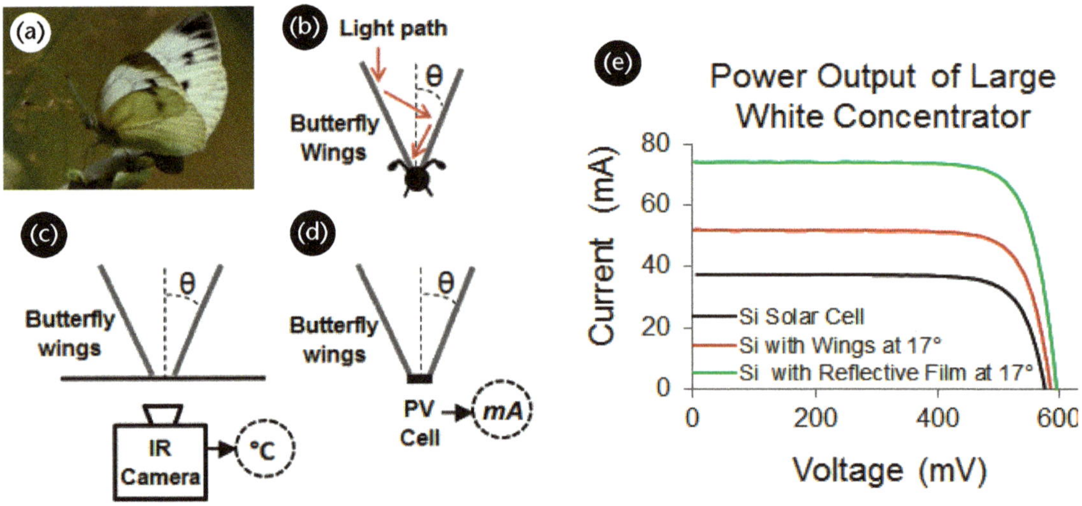

Figure 12.4 Large white (*P. brassicae*) butterfly (a) in V-shaped, basking posture, schematic of solar light concentration on butterfly's thorax (b) due to reflection by the opened wings; sketches of experimental setups for measuring temperature (c) and solar cell output power (d) as a result of light concentration in V-shaped mirror configuration; current-voltage curves (e) of a monocrystalline Si solar cell either alone, or with large white wings, versus reflective film held at optimal angle of 17°. (Reproduced from [43], License CC-BY-4.0.)

forewings in a V-shape configuration to concentrate light onto a 10×10 mm^2 monocrystalline silicon solar cell at the apex. An increase of 42.3% in output power was obtained, which represents an effective concentration ratio of 1.3×. This performance might seem modest when compared with the factor of 2× obtained in the same experiment with common reflective film coatings. However, in terms of power-to-weight ratio, the wing coating was 17× more efficient than the reflective films, thereby highlighting its potential for use in low-weight solar concentrations. Even more interestingly, a simple monolayer of scale cells removed from the wing on adhesive tape was found to maintain a high reflectance ($R = 62.0\%$). This raises the possibility of mimicking the specific nanostructure of pierid scale cells (quasi-random pattern of pterin beads) in order to produce bioinspired highly reflective thin-film coatings for low-weight mirrors in solar concentrators.

12.3.2 Solar Light-Harvesting Structures

In petals or leaves of some plant species, the epidermal cells exhibit microstructures that not only reduce light reflection (thereby sustaining their metabolism in low-light intensity environment or strengthening their coloration) but also potentially act as light-trapping devices [44–46]. Hünig and coworkers replicated the hierarchical structures found on epidermal cells of *Rosa* "El Toro" into a polymer film and applied it directly onto the front glass of thin-film organic solar cells [44]. The replication process involved transfer of the epidermal surface structures into an elastomer (namely, PDMS) mould, imprinting of the negative structures into an UV-photosensitive resist and UV curing of the resist to form the positive replica. With the cover layer on the front glass, they found an increase in the solar cell's short-circuit current of 13% at normal incidence: an enhancement that exceeds the mere antireflection effect of the corrugated surface (4.6% for typical subwavelength moth-eye

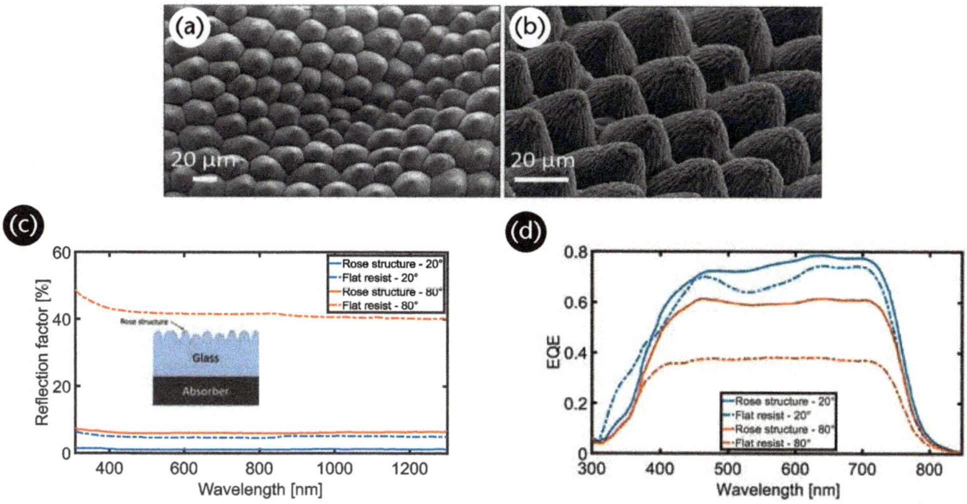

Figure 12.5 SEM images (a, b) of replicated rose petal cover layer to be put on solar cells. Reflectance spectrum (c) of a glass substrate coated with a black absorber on the backside and either the replicated rose structure or a planar resist layer (reference) on the front side, at incidence angles equal to 0°and 80°. External quantum efficiency (d) as a function of wavelength of a thin-film organic solar cell with either the replicated rose structure or a planar resist layer (reference) on the front side, at incidence angles equal to 20°and 80°. (Reproduced from [47] (a,b), License CC-BY-4.0. Data (c,d) from [44].)

antireflective coating). This indicated an additional contribution of a light-trapping effect. The petal's surface consists of a disordered array of domes, called micropapillae, the aspect ratio of which is about 0.6. Their top- and side-walls are decorated by nanometer-scale folding. Both micro- and nanostructures were well replicated into the resist (Figure 12.5). The antireflection property was assessed by applying the replicated structure onto a glass substrate with a highly absorbing back surface; an arrangement that is representative of glass encapsulated thin-film organic solar cells. The biomimetic coating causes substantial reduction in the reflectance over the whole spectral range at normal incidence and this reduction is even larger at grazing incidence (Figure 12.5). Not only is the rose structure an efficient antireflective coating, but it also plays the role of a light-trapping device. Indeed, replicated micropapillae act as imperfect microlenses, which deviate incoming rays at large angles and, thereby, increase the optical path length. This angular broadening of light propagation was evidenced by confocal laser scanning microscopy imaging of the sample's cross section, as well as by far-field measurement of the transmitted intensity under white light illumination.

Such a coating is particularly useful for thin-film organic solar cells, which suffer from insufficient solar light absorption and detrimental dips in their spectrum due to Fabry-Pérot interferences. In order to demonstrate its usefulness, the rose structure replica was applied directly onto the front side of thin-film solar cells with the following stacking: glass/ITO/ZnO/OAL/MoO_x/Ag (where OAL stands for organic absorber layer). The solar cells' external quantum efficiencies are increased over the whole spectrum in comparison with a flat resist coating (Figure 12.5). The increase in the short-circuit current J_{SC}, 13% at normal incidence reaches 44% at grazing incidence. This angular enhancement is particularly interesting for fixed solar panel

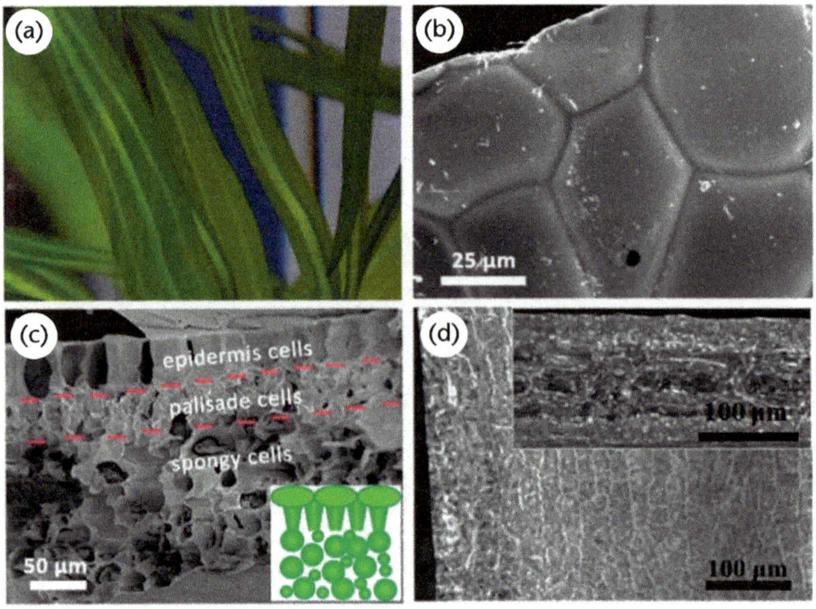

Figure 12.6 Photograph of the aquatic grass plant *Vallisneria* (a), SEM image of upper convex shaped epidermal cells (b), SEM image of a leaf's cross section (c) showing epidermis, palisade, and spongy layers (spongy mesophyll); a schematic illustration of the leaf's structure is inset, 3D-reconstructed image of leaf's structure (d) obtained by X-Ray microcomputed tomography (MCT). (Reproduced from [48], with permission from the Royal Society of Chemistry.)

systems, the efficiency of which drops as the sun moves to more inclined positions over the day. Nanofolding atop micropapillae is also believed to bring an additional contribution to light trapping.

Epidermal cell microstructure is just one template for bioinspired solar light harvesting devices. Actually, the structure of a plant leaf is even more intricate than that described above. Leaves have evolved exquisite hierarchical architecture, allowing them to perform the most efficient solar energy conversion process known (i.e., photosynthesis). Let us take, as an example, the thin and soft leaves of *Vallisneria*, an aquatic grass plant [48]. Their anatomical structure (Figure 12.6) comprises, starting from the surface and going deeper, (1) convex, lens-like epidermal cells, which focus light inside, (2) palisade parenchyma, a layer of elongated columnar cells that act as waveguides, and (3) a spongy mesophyll layer made of irregularly shaped, loosely packed mesophyll cells with lots of large interconnecting veins that scatter light and enhance its effective light path. Let us now focus our attention on leaf veins. The hierarchical interconnected branching network they form serves to supply water, biomechanical support, and nutrient delivery—all functions that are essential to the photosynthesis reaction (i.e., the production of sugars and oxygen from water and carbon dioxide).

Enhanced light scattering, as well as the large surface area of *Vallisneria*'s leaves' spongy structure, make it a template of choice for photocatalysis. Liu and coworkers fabricated a rather faithful mixed metal oxide replica of a *Vallisneria* leaf using a sol-gel method [48]. Silicon oxide was used in combination with titanium oxide—the latter being known for its photocatalytic activity—in order to bring thermal stability and, hence, to maintain the leaf's macroporosity (vein network). In order to compensate for the loss of intricate (nanometer-scale) structure after sample calcination, a porogenic agent (namely, P123 surfactant) was introduced in Ti-Si sols to create mesoporosity. Photocatalytic performance was evaluated in terms of rate constant of the degradation reaction of an aqueous solution of methylene blue under UV exposure in the presence of catalysts [48]. The leaf-template Ti-Si catalyst was found to be better (three times higher rate constant) than commercial P25 Degussa nanopowder TiO_2 catalyst. The artificial leaf concept was also explored by Zhou and coworkers, who synthesized perovskite titanates with hierarchical porosity from cherry tree leaves and demonstrated the photoreduction of CO_2 into hydrocarbon fuels (CO and CH_4) [49]. Such an artificial photosynthesis system, using sunlight as the energy source, atmospheric carbon dioxide as the carbon source, and water as the electron donor, is promising for achieving a global carbon-neutral cycle.

12.4 Light Extraction

For any light-emitting devices, photon escape to the outside is an issue. Indeed, in the case of a simple planar interface between the photon-emitting material and the surrounding environment (usually air), the light extraction efficiency is not only limited by Fresnel optical loss at the interface, but also by total internal reflection (TIR), since the optical material is unavoidably more refringent than air. In nature, light extraction puts selection pressure on natural organisms that emit light through bioluminescence. In the case of the *Photuris* sp. firefly we examined in Chapter 8, the bioluminescent organ (i.e., the firefly lantern) has evolved a unique structure on its cuticle surface in order to improve light extraction efficiency. To some extent, bioluminescent marine animals are luckier while facing this problem since, at equivalent refractive index, the critical angle for total internal reflection is reduced in water compared with air.

Artificial light emitters, such as LEDs, do not escape this physical limitation imposed by TIR. Different surface corrugation strategies (roughness, nanostructures, microlens patterns) have, therefore, been used to improve their light extraction efficiency. In view of the existing technological solutions, the *Photuris* sp. firefly's surface corrugation (Figure 12.7) is unique in the sense that it employs a grating with asymmetric profile (tiles-on-a-roof structure) and a period much longer than the bioluminescence emission wavelength [50–53]. The enhancement of light extraction efficiency with respect to a planar interface (i.e., extraction gain), originates essentially from geometrical effects, as explained in Chapter 8.

When such a solution, which has been optimized by nature, is intended to be transferred to LEDs, it is anticipated to be more challenging, since the refractive index of typical LED materials is significantly higher than the refractive index of their natural counterparts (i.e., the firefly's lantern) [52–54]. For instance, gallium nitride (GaN) has a refractive index of 2.53 at 425 nm (corresponding to the GaN LED emission wavelength), whereas the chitinous lantern material has a refractive index of 1.56 at 560 nm (i.e., the wavelength of the firefly's yellow flashes). As a result, the TIR critical angle, which defines the escape cone for rays from a planar

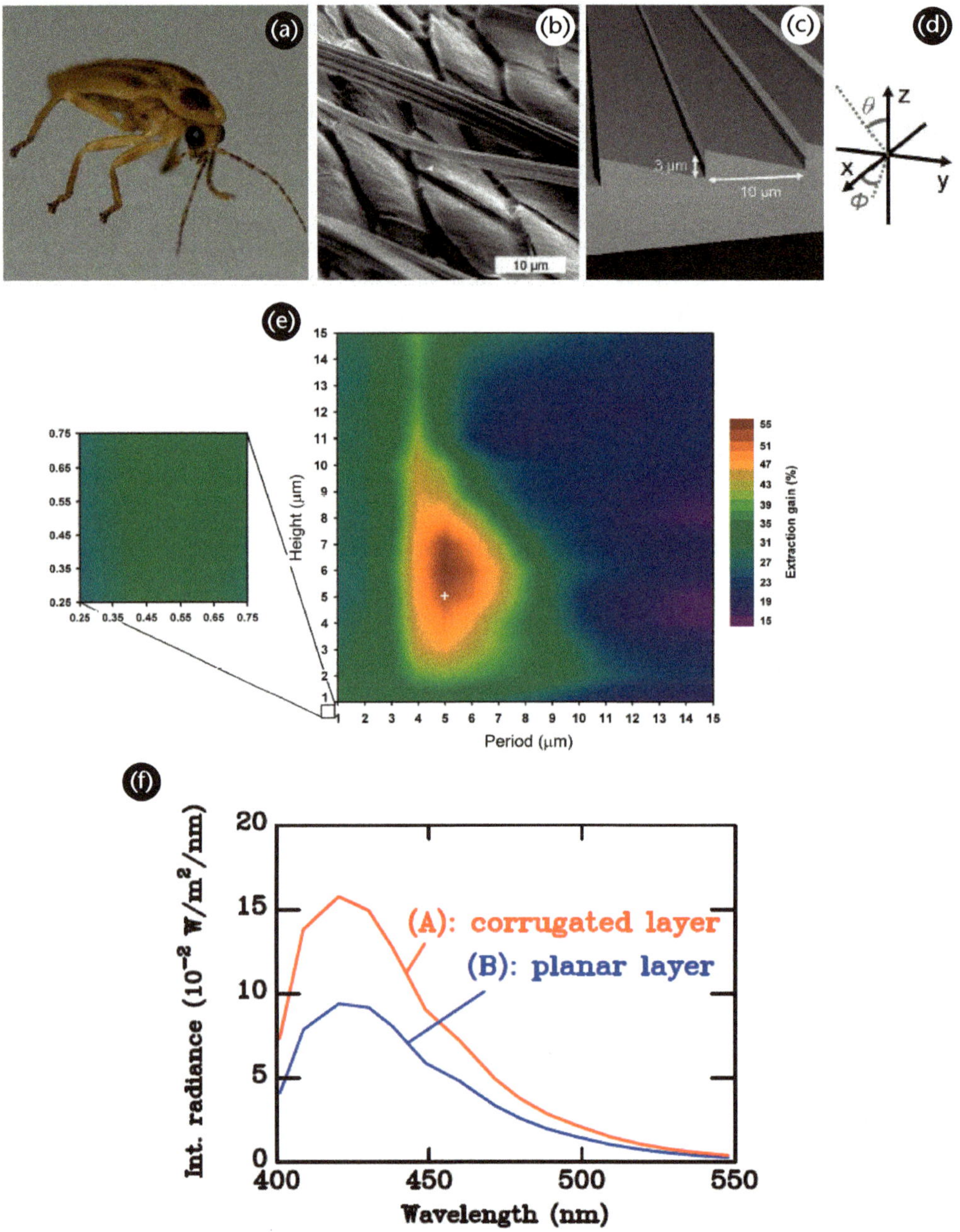

Figure 12.7 Photograph of a *Photuris* sp. firefly (a), SEM image of the lantern's cuticle (b), tiles-on-a-roof model of the cuticle surface used in simulations (c) and coordinate system (d). Map of theoretical light extraction gain with respect to a flat surface as a function of period and height of the surface corrugations (e). The white cross indicates parameter values used for experimental demonstration of surface patterning (the inset on the left shows a parameter subspace corresponding to corrugations at the scale of the emission wavelength). Measured radiance spectra (f) integrated over the full hemispheric solid angle for two identical LEDs covered by a patterned (A) or flat (B) photoresist layer. (Reproduced from [53], with permission from the Optical Society.)

interface, is almost twice smaller (23°) for GaN LED than for firefly lantern (40°). This narrower cone increases optical transmission losses (i.e., it impedes light extraction efficiency, as confirmed by numerical simulations) [55]. Instead of processing the LED surface during its fabrication, Bay and coworkers applied a photoresist overlayer and patterned it in order to improve the light extraction from existing GaN light-emitting diodes [55]. Before doing that, however, while taking the refractive index of GaN into account, they optimized the geometrical parameters (period p and height h) of the firefly's surface pattern and predicted a theoretical *extraction gain* of 55% for $p = 5\ \mu$m and $h = 6\ \mu$m (Figure 12.7). With respect to the original firefly template, the optimal structure for GaN LED is less asymmetrical and more spiky. By comparison, similar patterns at the wavelength scale ($p, h \in [1–2]\ \mu$m) were predicted to provide only a 33% extraction gain. Using direct-writing laser lithography (with a wavelength $\lambda = 365$ nm), patterned overlayers were fabricated in photoresist deposited on AlGaN/GaN LED heterostructures. Thereafter, absolute radiance $R(\theta, \varphi, \lambda)$ spectra were measured for two LEDs of identical sizes, both coated with a photoresist over-layer, of which one was patterned and the other not. The LED output power spectrum was determined by integrating radiance measurements over the full hemisphere; that is, $P = \int_0^{2\pi} d\varphi \int_0^{\theta_{\max}} \sin\theta d\theta R(\theta, \varphi, \lambda)$. The direct application of the bioinspired patterned photoresist overlayer turned out to increase the LED power by 68% with respect to the reference (i.e., a flat LED surface) (Figure 12.7).

12.5 Anticounterfeiting Patterns Inspired by Butterfly Scales

Berthier and coworkers proposed anticounterfeiting devices that were inspired by the iridescent, concave, hemispherical, multilayered structures found in the scales of *Papilio blumei* and *Cicindela campestris*, as well as by convex cover scales of *Suneve coronata* [56], the latter being notable as they form a V-groove-like multilayered grating. These proposed artificial devices rely on the polarization properties arising from the double reflection of light at the Brewster angle (Figure 12.8). The resultant anticounterfeiting system consists of a conformal multilayered structure with alternated triangular grooves and flat sections at the macroscopic level. The multilayer provides the surface with iridescent colors. On direct observation (with the naked eye), the reflected light is green and results from mixing of both yellow and blue reflected light (Figure 12.8, on the left). Yellow and blue reflections are assumed to correspond to the multilayer's iridescence at normal incidence and Brewster angle incidence, respectively. Therefore, yellow reflected light is nonpolarized (just like the incident light), whereas blue, doubly reflected light is entirely TE-polarized. Yet this is a simplified view, since complete suppression of TM reflection at the Brewster angle is only possible for a single interface, not a multilayer, and the sketch further assumes that the Brewster angle is equal to 45°. Now, if the pattern is observed through a linear polarizer, the axis of which is perpendicular to the groove axis, only the yellow reflected light will be perceived, the other (blue, polarized) component being blocked by the polarizer. Because this color change is invisible to the naked eye, it can be regarded as a hidden embedded tag that is useful for counterfeiting detection. It is noteworthy that this Brewster angle polarizing effect does not need

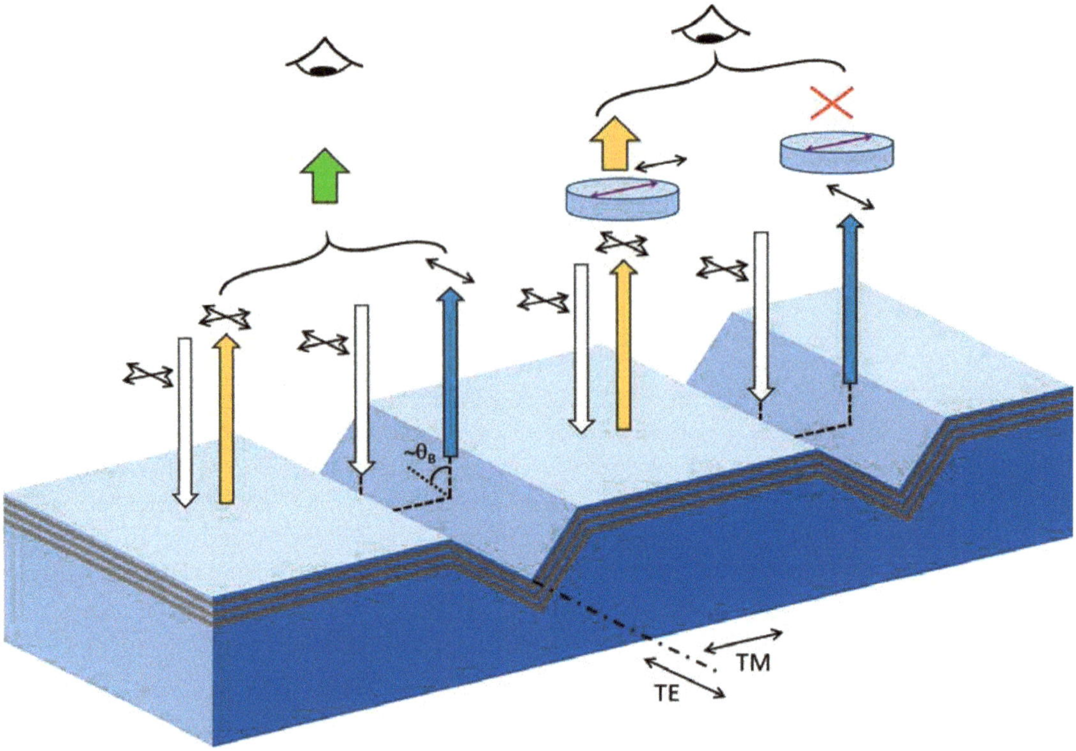

Figure 12.8 Anticounterfeiting patterns. Thin gray stripes represent a multilayered dielectric structure that runs along alternated triangular grooves and flat sections. Transverse-electric (TE) and transverse-magnetic (TM) refer to light polarization states, the directions of which are, respectively, parallel and perpendicular to the groove axis. The discs represent a linear polarizer, the axis of which is perpendicular to the grooves' axes. In the grooves, only blue TE light reflects twice with an angle equal to the Brewster angle θ_B. Without a polarizer, the observed green color results from the addition of yellow and blue reflected light. When a linear polarizer is used with its axis perpendicular to the groove axis, only the yellow reflected light is perceived.

a Bragg mirror (multilayer) to take place: a single dioptre may suffice. However, iridescence brings an additional feature for encoding information: the color of the anticounterfeiting pattern changes as reflected light is observed, either by the naked eye or through a polarizer.

Berthier noticed that a hemispherical cavity cannot produce a macroscopic polarization effect because of its symmetry [56]. Indeed, if two opposite sides are TE-polarized, two other sides, the directions of which are perpendicular to those first mentioned, are TM-polarized; that is, the polarizing effect associated with Brewster angle cancels out overall. This explains why the hemispherical reflectivity measured from *Suneve coronata* wings exhibits very little polarization dependence [56].

Symmetry breaking is the route Poncelet and coworkers followed to synthesize triangularly ridged structures such as those depicted in Figure 12.8, which they called V-grooves [57, 58]. They also fabricated cylindrically ridged structures (C-grooves). Isotropic and anisotropic etching techniques were used to carve, respectively, C-grooves and V-grooves into a silicon substrate through a patterned mask. After that, ALD was used to deposit TiO_2/Al_2O_3 multilayers (10 layers) in order to form the Bragg reflector. After reflection on the grooves, the state of

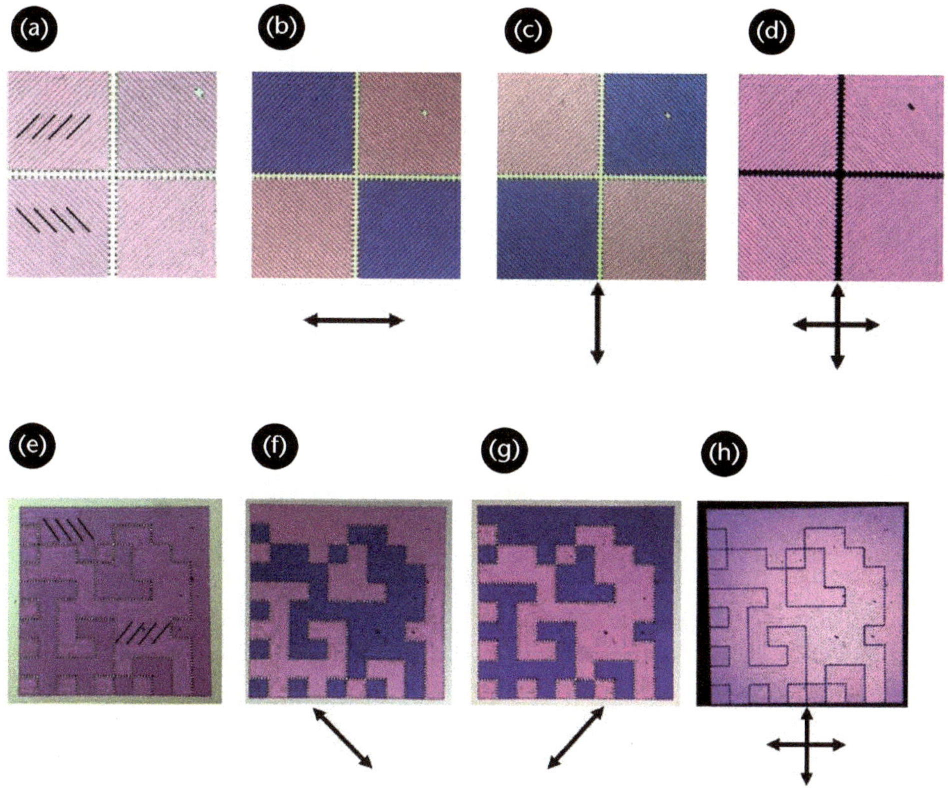

Figure 12.9 Optical microscope images of anticounterfeiting patterns. Patterns consisted of four quadrants with arrays of V-grooves (axis orientation indicated by parallel black lines) on a silicon substrate that was coated with Bragg mirrors. Images were taken without a polarizer (a,e), with a single polarizer aligned in orthogonal directions ((b,f) and (c,g)), and between crossed polarizers (d,h). (Reproduced from [58], with permission from O. Poncelet.)

polarization of the light, with respect to a reference axis system, is determined by the orientation of the groove axis in that system (Figure 12.8). Therefore, by fabricating a pattern with different groove orientations in distinct areas, different polarization states are produced upon reflection. Specifically, perpendicular groove arrays give rise to orthogonal polarization states. Without a polarizer (i.e., with the naked eye), areas with perpendicular arrays are undistinguishable; regardless of the groove axis orientation, the same color is observed (Figure 12.9(a)). With a polarizer, one or another area shows up with a distinct color according to the polarizer orientation (Figure 12.9(b,c)). When placed between crossed polarizers, the pattern is again undistinguishable (Figure 12.9(d)). In the latter case, the uniform color results from the rotation of polarization—a pure geometrical effect due to double reflection [57,58]. The light transmitted through the analyzer (hence the color brightness) is maximized when the polarizers directions and the grooves axes form a 45° angle (Figure 12.9(d)). The same observations can be done for more elaborate patterns, such as QR code type figures (Figure 12.9(e–h)).

Small variations from perfect photonic structures (e.g., surface- or interface-irregularities, faults in the optical properties of the materials, positional disorder, size or shape disorder) arise in natural organisms [1] from cellular noise during the tissue growth and development (morphogenesis) [59, 60]. These imperfections make each specimen unique. Recently, they were exploited to develop optical anticounterfeiting devices [61]. The scales covering the ventral side of the butterfly *Issoria lathonia*'s wings were laser-cut and bleached in order to imprint cryptographic information and to fabricate an optical memory tag that was described as inimitable.

12.6 Conclusion

Although we have touched on only a few bioinspired optical devices in this book, a great many applications for bioinspired products are already evident. As mentioned above, these range from solar panel and LED coatings, to anticounterfeiting devices and carbon capture technologies. Beyond this, a plethora of other uses for bioinspired approaches are either already being explored, or indicated as clear possibilities for future work from just the techniques we have discussed here and the biological photonic devices already studied in nature. Given the broad scope of applications of this work to numerous, diverse fields, scientists choosing to draw upon bioinspiration are somewhat spoiled for choice of research subjects. The enormous diversity of forms and functions found within creatures of the natural world provides a vast and rich resource for inspiration. In concluding this overview of the field, we are therefore very excited about the potential of bioinspired devices in the years ahead. It is rare to find an approach that offers such self-evident applications to help with so many of humankind's technological and engineering challenges.

References

[1] Mouchet, S. R., et al., "Optical Costs and Benefits of Disorder in Biological Photonic Crystals," *Faraday Discussions*, Vol. 223, 2020, pp. 9–48.

[2] Potyrailo, R. A., et al. "Materials and Transducers Toward Selective Wireless Gas Sensing," *Chemical Reviews*, Vol. 111, No. 11, 2011, pp. 7315–7354.

[3] Liu, X., et al., "A Survey on Gas Sensing Technology," *Sensors*, Vol. 12, No. 7, 2012, pp. 9635–9665.

[4] Potyrailo, R. A., "Multivariable Sensors for Ubiquitous Monitoring of Gases in the Era of Internet of Things and Industrial Internet," *Chemical Reviews*, Vol. 116, No. 19, 2016, pp. 11877–11923.

[5] Bernhard, C. G., et al., "The Insect Corneal Nipple Array. A Biologica, Broad-Band Impedance Transformer that Acts as an Antireflection Coating," *Acta Physiologica Scandinavica*, Vol. 63 (Suppl. 243), No. 1, 1965, pp. 1–79.

[6] Stavenga, D. G., et al., "Light on the Moth-Eye Corneal Nipple Array of Butterflies," *Proceedings of the Royal Society B*, Vol. 273, No. 1587, 2006, pp. 661–667.

[7] Han, et al., "Antireflective Surface Inspired from Biology: A Review," *Biosurface and Biotribology*, Vol. 2, No. 4, 2016, pp. 137–150.

[8] Yoshida, A., et al., "Nanoprotuberance Array in the Transparent Wing of a Hawkmoth, *Cephanodes hylas*," *Zoological Science*, Vol. 13, No. 4, 1996, pp. 525–552.

[9] Yoshida, A., et al., "Antireflective Nanoprotuberance Array in the Transparent Wing of a Hawkmoth, *Cephonodes hylas*," *Zoological Science*, Vol. 14, No. 5, 1997, pp. 737–741.

[10] Yoshida, A., et al., "Antireflection of the Butterfly and Moth Wings through Microstructure," *Forma*, Vol. 17, No. 2, 2002, pp. 75–89.

[11] Stoddart, P. R., et al., "Optical Properties of Chitin: Surface-Enhanced Raman Scattering Substrates Based on Antireflection Structures on Cicada Wings," *Nanotechnology*, Vol. 17, No. 3, 2006, pp. 680–686.

[12] Deparis, O., et al., "Assessment of the Antireflection Property of Moth Wings by Three-Dimensional Transfer-Matrix Optical Simulations," *Physical Review E*, Vol. 79, No. 4, 041910.

[13] Sun, M., et al., "Wetting Properties on Nanostructured Surfaces of Cicada Wings," *Journal of Experimental Biology*, Vol. 212, No. 19, 2009, pp. 3148–3155.

[14] Sun, M., et al., "A Study of the Anti-Reflection Efficiency of Natural Nano-Arrays of Varying Sizes," *Bioinspiration & Biomimetics*, Vol. 6, No. 2, 2011, 026003.

[15] Dellieu, L., et al., "A Two-in-One Superhydrophobic and Anti-Reflective Nanodevice in the Grey Cicada *Cicada orni* (Hemiptera)," *Journal of Applied Physics*, Vol. 116, No. 2, 2014, 024701.

[16] Deparis, O., et al., "Nanostructured Surfaces: Bioinspiration for Transparency, Coloration and Wettability," *Materials Today: Proceedings*, Vol. 1, No. S, 2014, pp. 122–129.

[17] Stavenga, D. G., "Thin Film and Multilayer Optics Cause Structural Colors of Many Insects and Birds," *Materials Today: Proceedings*, Vol. 1, No. S, 2014, pp. 109–121.

[18] Siddique, R. H., et al., "The Role of Random Nanostructures for the Omnidirectional Anti-Reflection Properties of the Glasswing Butterfly," *Nature Communications*, Vol.6, No. 1, 2015, 6909.

[19] Sun, C.-H., et al., "Templated Biomimetic Multifunctional Coatings," *Applied Physics Letters*, Vol. 92, No. 5, 2008, 051107.

[20] Vüllers, F., et al., "Bioinspired Superhydrophobic Highly Transmissive Films for Optical Applications," *Small*, Vol. 12, No. 44, 2016, pp. 6144–6152.

[21] Liu, S., et al., "Bio-Inspired TiO_2 Nano-Cone Antireflection Layer for the Optical Performance Improvement of VO_2 Thermochromic Smart Windows," *Scientific Reports*, Vol. 10, 2013, 11376.

[22] Linn, N. C., et al., "Self-Assembled Biomimetic Antireflection Coatings," *Applied Physics Letters*, Vol. 91, No. 10, 2007, 101108.

[23] Sun, C.-H., et al., "Broadband Moth-Eye Antireflection Coatings on Silicon," *Applied Physics Letters*, Vol. 92, No. 6, 2008, 061112.

[24] Min, W.-L., et al., "Bioinspired Broadband Antireflection Coatings on GaSb," *Applied Physics Letters*, Vol. 92, No. 14, 2008, 141109.

[25] Zhang, G., et al., "Cicada Wings: A Stamp from Nature for Nanoimprint Lithography," *Small*, Vol. 2, No. 12, 2006, pp. 1440–1443.

[26] Xie, G., et al., "The Fabrication of Subwavelength Anti-reflective Nanostructures Using a Bio-template," *Nanotechnology*, Vol. 19, No. 9, 2008, 095605.

[27] Potyrailo, R. A., and R. R. Naik, "Bionanomaterials and Bioinspired Nanostructures for Selective Vapor Sensing," *Annual Review of Materials Research*, Vol. 43, No. 1, 2013, pp. 307–334.

[28] Potyrailo, R. A., et al., "*Morpho* Butterfly Wing Scales Demonstrate Highly Selective Vapour Response," *Nature Photonics*, Vol. 1, No. 2, 2007, pp. 123–128.

[29] Potyrailo, R. A., et al., "Discovery of the Surface Polarity Gradient on Iridescent *Morpho* Butterfly Scales Reveals a Mechanism of Their Selective Vapor Response," *Proceedings of the National Academy of Sciences of the United States of America*, Vol. 110, No. 39, 2013, pp. 15567–15572.

[30] Potyrailo, R. A., et al., "Towards Outperforming Conventional Sensor Arrays with Fabricated Individual Photonic Vapour Sensors Inspired by *Morpho* Butterflies," *Nature Communications*, Vol. 6, No. 1, 2015, 7959.

[31] Poncelet, O., et al., "Vapour Sensitivity of an ALD Hierarchical Photonic Structure Inspired by *Morpho*," *Bioinspiration & Biomimetics*, Vol. 11, No. 3, 2016, 036011.

[32] Kim, J. H., et al., "Biologically Inspired Humidity Sensor Based on Three-Dimensional Photonic Crystals," *Applied Physics Letters*, Vol. 97, No. 10, 2010, 103701.

[33] Mäthger, M. F., et al., "Rapid Colour Changes in Multilayer Reflecting Stripes in the Paradise Whiptail, *Pentapodus paradiseus*," *Journal of Experimental Biology*, Vol. 206, No. 20, 2003, pp. 3607–3613.

[34] Wang, Z., et al., "Bioinspired Water-Vapor-Responsive Organic/Inorganic Hybrid One-Dimensional Photonic Crystals with Tunable Full-Color Stop Band," *Advanced Functional Materials*, Vol. 20, No. 21, 2010, pp. 3784–3790.

[35] Bai, L., et al., "Bio-Inspired Vapor-Responsive Colloidal Photonic Crystal Patterns by Inkjet Printing," *ACS Nano*, Vol. 8, No. 11, 2014, pp. 11094–11100.

[36] Seo, H.-b., and S.-Y. Lee, "Bio-Inspired Colorimetric Film Based on Hygroscopic Coloration of Longhorn Beetles (*Tmesisternus isabellae*)," *Scientific Reports*, Vol. 7, 2017, 44927.

[37] Liu, F., et al., "Structural Color Change in Longhorn Beetles *Tmesisternus isabellae*," *Optics Express*, Vol. 17, No. 18, 2009, pp. 16183–16191.

[38] Diao, Y. Y., et al., "Multiple Structural Coloring of Silk-Fibroin Photonic Crystals and Humidity-Responsive Color Sensing," *Advanced Functional Materials*, Vol. 23, No. 43, 2013, pp. 5373–5380.

[39] Langer, J., et al., "Present and Future of Surface-Enhanced Raman Scattering," *ACS Nano*, Vol. 14, No. 1, 2020, pp. 28–117.

[40] Stoddart, P. R., et al., "Optical Properties of Chitin: Surface-Enhanced Raman Scattering Substrates Based on Antireflection Structures on Cicada Wings," *Nanotechnology*, Vol. 17, 2006, pp. 680–686.

[41] Garrett, N. L., et al., "Spectroscopy on the Wing: Naturally Inspired SERS Substrates for Biochemical Analysis," *Journal of Biophotonics*, Vol. 2, No. 3, 2009, pp. 157–166.

[42] Palmer, L. D., et al., "Probing the Coupling of Butterfly Wing Photonic Crystals to Plasmon Resonances with Surface-Enhanced Raman Spectroscopy," *Journal of Materials Chemistry C*, Vol. 7, 2019, pp. 13887–13895.

[43] Shanks, K., et al., "White Butterflies as Solar Photovoltaic Concentrators," *Scientific Reports*, Vol. 5, 2015, 12267.

[44] Hünig, R., et al., "Flower Power: Exploiting Plants' Epidermal Structures for Enhanced Light Harvesting in Thin-Film Solar Cells," *Advanced Optical Materials*, Vol. 4, No. 10, 2016, pp. 1487–1493.

[45] Schmager, R., et al., "Texture of the Viola Flower for Light Harvesting in Photovoltaics," Vol. 4, No. 11, 2017, pp. 2687–2692.

[46] Fritz, B. et al., "Upscaling the Fabrication Routine of Bioreplicated Rose Petal Light Harvesting Layers for Photovoltaic Modules," *Solar Energy*, Vol. 201, 2020, pp. 666–673.

[47] Fritz, B. et al., "Bioreplicated Coatings for Photovoltaic Solar Panels Nearly Eliminate Light Pollution that Harms Polarotactic Insects," *PLoS ONE*, Vol. 15, No. 12, 2020, e0243296.

[48] Liu, J., Yang, Q., Yang, W., Li, M., Song, Y., "Aquatic Plant Inspired Hierarchical Artificial Leaves for Highly Efficient Photocatalysis," *Journal of Materials Chemistry A*, Vol. 1, No. 26, 2013, pp. 7760–7766.

[49] Zhou, H., et al., "Leaf-Architectural 3D Hierarchical Artificial Photosynthetic System of Perovskite Titanates Towards CO_2 Photoreduction into Hydrocarbon Fuels," *Scientific Reports*, Vol. 3, 2013, 1667.

[50] Bay, A., and J.-P. Vigneron, "Light Extraction from the Bioluminescent Organs of Fireflies," *Proceedings of SPIE*, Vol. 7401, 2009, 740108.

[51] Bay, A., M. Sarrazin, and J.-P. Vigneron, "Light Extraction: What We Can Learn from Fireflies," *Proceedings of SPIE*, Vol. 8480, 2012, 84800G.

[52] Kim, J.-J., et al., "Biologically Inspired LED Lens from Cuticular Nanostructures of Firefly Lantern," *Proceedings of the National Academy of Sciences of the United States of America*, Vol. 109, No. 46, 2012, pp. 18674–18678.

[53] Bay, A., et al., "Improved Light Extraction in the Bioluminescent Lantern of a *Photuris* Firefly (Lampyridae)," *Optics Express*, Vol. 21, No. 1, 2013, pp. 764–780.

[54] Chen, L., et al., "Bioinspired Photonic Structures by the Reflector Layer of Firefly Lantern for Highly Efficient Chemiluminescence," *Scientific Reports*, Vol. 5, 2015, 12965.

[55] Bay, A., et al., "Optimal Overlayer Inspired by *Photuris* Firefly Improves Light-Extraction Efficiency of Existing Light-Emitting Diodes," *Optics Express*, Vol. 21, No. S1, 2013, pp. A179–A189.

[56] Berthier, S., et al., "Multiscaled Polarization Effects in *Suneve coronata* (Lepidoptera) and Other Insects: Application to Anti-Counterfeiting of Banknotes," *Applied Physics A*, Vol. 86, No. 1, 2007, pp. 123–130.

[57] Poncelet, O., et al., "Synthesis of Bio-Inspired Multilayer Polarizers and Their Application to Anti-Counterfeiting," *Bioinspiration & Biomimetics*, Vol. 10, No. 2, 2015, 026004.

[58] Poncelet, O., "Synthèse, Caractérisation et Modélisation de Nanostructures Photoniques Bio-inspirées," PhD Thesis, UCLouvain, Belgium, 2015.

[59] Johnston, I. G., et al., "Mitochondrial Variability as a Source of Extrinsic Cellular Noise," *PLoS Computational Biology*, Vol. 8, No. 3, 2012, e1002416.

[60] Tsimring, L. S., "Noise in Biology," *Reports on Progress in Physics*, Vol. 77, No. 2, 2014, 026601.

[61] Pavlović, D., "Naturally Safe: Cellular Noise for Document Security," *Journal of Biophotonics*, Vol. 12, No. 12, 2019, e201900218.

About the Authors

Sébastien R. Mouchet received his BSc, MSc, and PhD degrees in physics from the Univcrsity of Namur, Belgium. Since his MSc and PhD degrees, he has studied the physical origins of various visual effects evolved by natural organisms: from additive colors in photonic polycrystals and fluid-induced color changes to fluorescence emission in natural photonic structures and UV scattering in integuments. After the completion of his PhD degree in 2015, he carried out his research as a postdoctoral researcher at the University of Exeter, UK, for about 5 years. He is now a postdoctoral researcher at the University of Namur, where he was appointed honorary lecturer and later senior honorary lecturer. He has given the lectures for the "Natural Photonics and Biomimetism" module of the master's programs in physics since 2018 and for the "Analytical Methods for Solving Problems in Physics" module of the bachelor's program in physics since 2019. Since the beginning of his PhD degree, his research has mainly been supported by the Belgian National Fund for Scientific Research (FRS-FNRS) through fellowships he secured, as well as a Postdoctoral Fellowship for Excellence program WBI.WORLD, a grant from the Maturation Fund of the Walloon Region, and two grants from the Agathon De Potter Fund of the Royal Academy of Science, Letters and Fine Arts of Belgium. Additionally, Dr. Mouchet is an editorial board member of *Data in Brief* (Elsevier). He co-organized the Belgian editions of *Photonics Online Meetup* (bePOM) in 2020 and 2021, as well as the 2021 edition of the *Living Light Conference* series dedicated to early career researchers.

Olivier Deparis received his electrical engineer's degree in 1991 from the Faculté Polytechnique de Mons (FPMs), Belgium, with majors in fiber optic communications. His PhD thesis (1997) was dedicated to the study of radiation effects on optical fibres and was carried out in collaboration with the Belgian Nuclear Research Centre (CEN•SCK). From 1997 to 2001, he was a postdoctoral researcher in the Electromagnetism & Telecommunications Department of FPMs, where he studied fundamental aspects of mode-locking in erbium-doped fiber lasers, properties of fiber Bragg or long-period grating, radiation effects on fiber Bragg grating sensors as well as instabilities of semiconductor lasers subject to optical feedback. In January 2002, he joined the Optoelectronics Research Centre (ORC) at the University of Southampton, UK, where he conducted experimental research on thermal electric-poling of glass for applications in nonlinear optics and fiber laser sources. During his stay at

ORC, he initiated original research on metal-glass nanocomposites and pursued in this direction after he came back to FPMs. In October 2005, he moved to the Physics Department of the Facultés Universitaires Notre-Dame de la Paix (now University of Namur, Belgium), where he was hired as a postdoctoral researcher for the FP6 BioPhot European project led by Prof. Jean-Pol Vigneron. This project (entitled "Complexity and Evolution of Photonic Nanostructures in Bio-Organisms: Templates for Material Sciences") was dedicated to the study of photonic crystals found in biological organisms and to the development of bioinspired photonic devices. In a following project, he contributed to the design and synthesis of bioinspired hygrochromic materials. In 2009, he became assistant professor and, one year later, full professor at the University of Namur (Department of Physics). Fifteen years of experience in the field of photonics in natural organisms led him to coorganize the 2014 edition of the *Living Light Conference* series and now to coauthor this book. Over the last ten years, he pursued his research in optics, with an emphasis on optical materials, nanostructures, solar energy, photocatalysis, and, more recently, cultural heritage science. Since 2019, he is the director of the solid-state physics laboratory (UR-LPS) of the Department of Physics, which has a well-established expertise in nanostructured materials, surface science, nanophysics, and photonics. He has been a member of the Optical Society of America since 1999. As an OSA member, he has been a reviewer for journals such as *Optics Express*, *Optics Letters*, *Applied Optics*, and *JOSA B*. From 2013 to 2019, he served as an associated editor for *Optics Express*. He has authored more than 170 articles in peer-reviewed journals, a large majority being related to photonics. He holds two patents related to optical materials.

Index

Artech House
Applied Photonics Library

Brian Culshaw and Marc Wuilpart, Series Editors

The ABCs of Fiber Optic Communication, Sudhir Warier

Acousto-Optic Signal Processing: Fundamentals and Applications, Pankaj K. Das and Casimer De Cusatis

Advanced Optical Communication Systems and Networks, Milorad Cvijetic and Ivan B. Djordjevic

Applications of Modern RF Photonics, Preetpaul Singh Devgan

Bistabilities and Nonlinearities in Laser Diodes, Hitoshi Kawaguchi

Chemical and Biochemical Sensing with Optical Fibers and Waveguides, Gilbert Boisdé and Alan Harmer

Coherent and Nonlinear Lightwave Communications, Milorad Cvijetic

Coherent Doppler Wind Lidars in a Turbulent Atmosphere, Viktor Banakh and Igor Smalikho

Coherent Lightwave Communication Systems, Shiro Ryu

Digital Optical Measurement Techniques and Applications, Pramond Rastogi, Editor

DWDM Fundamentals, Components, and Applications, Jean-Pierre Laude

Electro-Optical Systems Performance Modeling, Gary Waldman

Elliptical Fiber Waveguides, R. B. Dyott

Engineering Optical Networks, Sudhir Warier

Fiber Bragg Gratings: Fundamentals and Applications in Telecommunications and Sensing, Andrea Othonos and Kyriacos Kalli

Fiber Optics Communications, Henry F. Taylor

The Fiber-Optic Gyroscope, Second Edition, Hervé C. Lefèvre

Fiber-Optic Sensors for Biomedical Applications, Daniele Tosi and Guido Perrone

Field Theory of Acousto-Optic Signal Processing Devices, Craig R. Scott

Finite Element Modeling Methods for Photonics, B. M. Azizur Rahman and Arti Agrawal

Frequency Stabilization of Semiconductor Laser Diodes, Tetsuhiko Ikegami, Shoichi Sudo, and Yoshihisa Sakai

Fundamentals of Multiaccess Optical Fiber Networks, Dennis Mestdagh

Germanate Glasses: Structure Spectroscopy and Properties, Alfred Margaryan and Michael Piliavin

Handbook of Distributed Feedback Laser Diodes, Second Edition, Geert Morthier and Patrick Vankwikelberge

Highly Coherent Semiconductor Lasers, Motoichi Ohtsu

High-Power Optically Activated Solid-State Switches, Arye Rosen, Fred Zutavern

Integrated Optics: Design and Modeling, Reinhard Marz

Introduction to Electro-Optical Imaging and Tracking Systems, Shaheen A. Hovanessian, Khalil Seyrafi

Introduction to Glass Integrated Optics, S. Iraj Najafi

Introduction to Infrared and Electro-Optical Systems, Second Edition, Ronald G. Driggers, Melvin H. Friedman, and Jonathan Nichols

Introduction to Radiometry and Photometry, Second Edition, William Ross McCluney

Introduction to Semiconductor Integrated Optics, Hans P. Zappe

Laser Communications in Space, Stephen Lambert

Liquid Crystal Devices: Physics and Applications, Vladimir G. Chigrinov

Lithium Niobate Photonics, James E. Toney

Military Laser Technology and Systems, David H. Titterton

Natural Photonics and Bioinspiration, Sébastien R. Mouchet and Olivier Deparis

New Photonics Technologies for the Information Age: The Dream of Ubiquitous Services, Shoichi Sudo and Katsunari Okamoto, editors

Optical Document Security, Third Edition, Rudolf L. van Renesse

Optical FDM Network Technologies, Kiyoshi Nosu

Optical Fiber Amplifiers: Materials, Devices, and Applications, Shoichi Sudo, editor

Optical Fiber Communication Systems, Leonid Kazovsky, Sergio Benedetto and Alan Willner

Optical Fiber Sensors, Volume One: Principles and Components, Brian Culshaw and John P. Dakin

Optical Fiber Sensors, Volume Two: Systems and Applications, Brian Culshaw and John P. Dakin

Optical Fiber Sensors, Volume Three: Components and Subsystems, John Dakin and Brian Culshaw, editors

Optical Fiber Sensors, Volume Four: Applications, Analysis, and Future Trends, John Dakin and Brian Culshaw, editors

Optical Interconnection: Foundations and Applications, John Caulfied and Christopher S. Tocci

Optical Measurement Techniques and Applications, Pramod Rastogi

Optical Network Theory, Yitzhak Weissman

Optical Transmission for the Subscriber Loop, Norio Kashima

Optical Transmission Systems Engineering, Milorad Cvijetic

Optoelectronics for Low-Intensity Conflicts and Homeland Security, Anil K. Maini

Optoelectronic Techniques for Microwave and Millimeter-Wave Engineering, William M. Robertson

Passive Optical Components for Optical Fiber Transmission, Norio Kashima

Polarization in Optical Fibers, Alan Rogers

Photonic Applications for Radio Systems and Networks, Fabio Cavaliere and Antonio D'Errico

Principles of Modern Optical Systems, Volume 1, Deepak Uttamchandani and Ivan Andonovic, editors

Principles of Modern Optical Systems, Volume 2, Deepak Uttamchandani and Ivan Andonovic, editors

Reliability and Degradation of III-V Optical Devices, Osamu Ueda

Securing Information and Communications Systems, Stephen Furnell, Sokratis Katsikas, Javier Lopez, and Ahmed Patel

Semiconductor Raman Lasers, Ken Suto and Jun-Ichi Nishizawa

Semiconductors for Solar Cells, Hans Joachim Moller

Semiconductor Quantum Wells and Superlattices for Long-Wavelength Infrared Detectors, M. Omar Manasreh

Signal Processing and Performance Analysis for Imaging Systems, S. Susan Young, Ronald G. Driggers, and Eddie L. Jacobs

Single-Mode Optical Fiber Measurement: Characterization and Sensing, Giovanni Cancellieri and Franco Chiaraluce

Smart Structures and Materials, Brian Culshaw

Substrate Surface Preparation Handbook, Max Robertson

Surveillance and Reconnaissance Imaging Systems: Modeling and Performance Prediction, Jon C. Leachtenauer and Ronald G. Driggers

Tunable Laser Diodes, Markus-Christian Amann and Jens Buus

Ultrafast Diode Lasers: Fundamentals and Applications, Peter Vasilev

Understanding Optical Fiber Communications, Alan Rogers

Wavelength Division Multiple Access Optical Networks, Andrea Borella, Giovanni Cancellieri, and Franco Chiaraluce

*Wired and Wireless Seamless Access Systems for Public Infrastructur*e, Tetsuya Kawanishi

For further information on these and other Artech House titles, including previously considered out-of-print books now available through our In-Print-Forever® (IPF®) program, contact:

Artech House
685 Canton Street
Norwood, MA 02062
Phone: 781-769-9750
Fax: 781-769-6334
e-mail: artech@artechhouse.com

Artech House
16 Sussex Street
London SW1V 4RW UK
Phone: +44 (0)20 7596-8750
Fax: +44 (0)20 7630-0166
e-mail: artech-uk@artechhouse.com

Find us on the World Wide Web at: www.artechhouse.com